Predator upon a Flower

Predator upon a Flower

LIFE HISTORY AND FITNESS
IN A CRAB SPIDER

DOUGLASS H. MORSE

HARVARD UNIVERSITY PRESS

Cambridge, Massachusetts

London, England   2007

Library of Congress Cataloging-in-Publication Data

Morse, Douglass H., 1938–
    Predator upon a flower : life history and fitness in a crab spider /
Douglass H. Morse.
        p.   cm.
    Includes bibliographical references and index.
    ISBN-13: 978-0-674-02480-9 (cloth : alk. paper)
    ISBN-10: 0-674-02480-X (cloth : alk. paper)
    1. Crab spiders.   I. Title.
QL458.4.M678   2007
595.4'4—dc22   2007060724

Dedicated to my students:

*For your keen insights, support, and enthusiasm—
and for helping to keep a brain from aging*

Contents

Acknowledgments

$\mathcal{D}$URING THE GENESIS OF THIS book I have profited from the collegiality, insights, and assistance of many people. First, I must thank Professor Mark Elgar and his colleagues in the Department of Zoology at the University of Melbourne for providing such a stimulating atmosphere for me during a sabbatical year spent there. The opportunity for long, interruption-free days is a priceless commodity for someone accustomed to a full schedule of teaching and research overlain by an often-crushing load of administrative responsibilities. Notwithstanding, the Ecology and Evolutionary Biology department at Brown University has been a wonderful venue in which to work for much of my career. I have learned a great amount from my colleagues here, which, combined with the enormously supportive and egalitarian atmosphere they have provided, makes this site an academic person's dream come true. And, the Brown students have been a constant source of stimulation and inspiration.

I have conducted most of this research at and in the immediate vicinity of the Darling Marine Center in Walpole, Maine, the marine station of the University of Maine. In spite of its name and primary function, the Center contains protected fields that I have been free to manipulate as needed for the experiments and long-term observations described in this work. I am particularly indebted to its director, Kevin Eckelbarger, and to key staff members Tim Miller and Linda Healy, for providing such a welcoming environment for my students and me.

I am particularly indebted to Reuven Dukas, Mark Elgar, Linden Higgins, Johanna Schmitt, and Jonathan Waage, who have generously commented on major parts of the manuscript at various stages of its development. My research on *Misumena* has been supported by the National Science Foundation and that of several of my students by the Howard Hughes Foundation. Elizabeth Farnsworth, herself once a field assistant on this project, performed the artwork. Elizabeth Morse produced the graphics. G. B. Edwards, Robert Edwards, and Daniel Jennings identified spiders. Ann Downer-Hazell, Alissa Anderson, and Kate Brick of Harvard University Press, and John Donohue, Rebecca Homiski, and Betty Pessagno of Westchester Book Services, guided the project through to completion.

I thank the following publishers for use of copyrighted materials in their care: American Arachnological Society *(Journal of Arachnology)*, Blackwell Publishing *(Biological Journal of the Linnean Society)*, Ecological Society of America *(Ecology)*, Elsevier Publishers *(Animal Behaviour)*, and Springer-Verlag Berlin Heidelberg *(Oecologia)*.

Predator upon a Flower

1

Introduction

$\mathcal{I}$N THIS BOOK I FOCUS on the life cycles of animals and
on how they are sculpted by the combination of environmental and interactive factors they encounter. The central theme of this book, lifetime fitness, is treated here as the combined factors that determine an
individual's contribution to the next generation's breeding pool. I concentrate on a sit-and-wait predator, the crab spider *Misumena vatia*
(henceforth *Misumena*), which hunts its prey on flowers. An ideal
species for such studies, *Misumena* allows me to address some of the
most important, and often vexing, variables related to fitness and the
consequences of variation in decision making for some of the most important events of an individual's life. The ability to quantify key stages
in terms of their lifetime fitness consequences makes *Misumena* a valuable model for studies of populations.

In the best of all worlds, an animal would most enhance its represen-
tation in the next generation by maximizing both its survival and fecundity. Obviously, that level of accomplishment is unrealistic: if carried to its extreme, it would produce eternal animals that reproduced at
an exponential rate (Seger and Stubblefield 1996), the so-called Darwinian demon (Partridge and Harvey 1988). The fittest individual in
any population achieves far less than perfection, but consideration of
that unattainable goal may help to visualize the difficulties facing such
an animal. A colleague once pictured the problem extremely well to me

by noting his amazement, not that animals were not remarkably successful, but that they managed to survive at all, given the odds stacked
against them! Of greatest interest to us is the similarity of the members
of a population to the most fit member of that population. Though divided into survival and fecundity, the standard l_x and m_x of life tables, a
myriad of variables will often dictate success (see Roff 1992; Stearns
1992), and the basis for difference in success may often be complex.

Over its lifetime, an animal must respond to such potentially conflicting variables as spatial differences in food density or differences in
predator pressure in a variety of circumstances, as well as in certain
combinations. Be they environmental or population driven, these factors mold populations over time, with the likely outcome differing
from one instance to the next. Such comparisons permit study of the
opportunities open to members of a population, the constraints on
their lifestyles, and the resulting limits. The limits experienced in any
such population are modest in comparison to the set of conditions
compatible with life of some kind. Of course, one would not expect a
mouse to adapt to conditions guaranteeing the success of thermophilic
bacteria, but, more importantly, constraints are typically far more
subtle and less extreme than those of the almost trivially broad bacterium–mouse comparison. Constraints acting on these extremes of an
animal's tolerances may arise most predictably from the physical conditions that it experiences and that its ancestors experienced—the temperature and its range, the moisture regime, and the like.

Still, most organisms occupy ranges far narrower than those predicted if limited solely by such physical factors as climatic extremes. Although the frequency with which physical and other factors ultimately
limit a species is a source of utmost importance, this frequency differs
with the conditions in question as well as the resources at hand. Com-
pounding the difficulty of interpreting any given situation, limiting
factors may assert themselves only at certain times, sometimes only
once over several generations. Moreover, individuals will be differentially sensitive to potentially limiting factors at different stages of their
life cycles.

In addition to physical environmental factors, an individual's presence and success will depend on its relationships with conspecifics, potentially competing species, predators, and parasites—and, potentially,
with symbionts and mutualists. Interactions with competitors, either

conspecifics or other species, may rein in the size of populations, with a progressively stronger effect on individuals as resources decline. Predators and parasites impinge on prey in a way that potentially drives their numbers below the point at which resources are limiting, so that competition declines or ceases. Prey species may also respond to predators in such a way as to decrease their vulnerability to them, but at the cost of compromising their efficiency as foragers and competitors. These conflicting variables have recently received considerable attention (e.g., Dukas 1998a; Lima 2002).

Lifetime fitness places all of the activities of an animal's lifetime into a single convenient currency that permits us to relate those activities to the perpetuation and propagation of its genes. However, common life history patterns in animals, such as multiple breeding seasons and complex foraging routines, render such analyses extremely complex or impractical for many species. Operationally, lifetime fitness can be studied and measured most easily in short-lived animals that reproduce but once, thereby minimizing such complications. However, the same principles hold for less easily studied species. *Misumena* fits the criteria for study particularly well.

Why This Subject Is Important and Interesting

Life histories (Roff 1992; Stearns 1992) incorporate some of the most important variables that determine whether an individual (or type of individual) can integrate into a given population and how it can accomplish this act. Life histories provide insight into how individuals adapt to a given set of circumstances, often including many of the variables that drive selection on a trait, and they help to determine in natural situations whether a combination of traits will permit change in some particular direction. They include some of the central forces by which selection acts, forces that will determine stasis or change. Selection itself will not generate change without the necessary genetic propensities, however. Phenotypic variation as well as heritability, the extent to which phenotypes are determined by their parents' genes (Falconer 1989), are also important. All of these factors contribute to an individual's success in recruiting productive offspring to the next breeding generation, its lifetime fitness as I will operationally define it for the purposes of this study. However, in spite of their importance for understanding the evolution of populations, data sets for populations in the wild are few and far between (Endler 1986). The basis for this deficiency is quite simple: several of the sorts of data required for such evaluations are difficult or impossible to obtain for most kinds of animals in the field. In this book I will emphasize the lifetime fitness consequences of certain behavioral decisions made by foragers and mate-choosers in the field—for instance, choice of a foraging site or the acceptance or rejection of a potential mate.

The Target Species, *Misumena vatia*

Misumena vatia is a member of the family Thomisidae. Crab spiders are small to medium-sized, squat spiders with two anterior pairs of large raptorial forelimbs and a tendency to scuttle sideways if startled or attacked. Thus, the term *crab spider* describes them well. Rather than using webs, *Misumena* select sites that attract many insects and capture prey that often approximate or even exceed them in size.

Many of *Misumena*'s life history traits are unusually amenable to study. In particular, the lifetime fitness consequences of differences in some fundamentally important behavioral variables, such as choice of hunting site or patch, can readily be studied experimentally in the field. Patch choice is intimately related to foraging success and singularly important in determining the reproductive success of adult female *Misumena*. The ability to distinguish among sites takes on major importance for the earlier instars as well, helping them to avoid starvation or to avoid subsequent dispersal events that could banish them to completely unfavorable sites. In addition to these survivorship factors, traits such as a strongly female-biased secondary sex ratio and extreme sexual dimorphism are highly amenable to analysis, making the species a natural for studies that attempt to integrate male and female perspectives into the calculation of lifetime fitness.

Since *Misumena* is so well suited to experimental manipulation under natural or realistic laboratory conditions, it has considerable potential as a model species on which to test basic questions and principles associated with lifetime fitness, foraging theory, and sexual selection. I will present insights from work on *Misumena* in the context of current theory and then suggest appropriate steps for further study.

How I Came to This Set of Problems

My current attention to lifetime fitness problems derives from my ear-lier exploration of somewhat different questions that focused on pre-dictions about how animals should forage. I shifted my research from bumble bees *Bombus* spp. to *Misumena* after discovering that this spider permitted me to address certain basic questions relating to foraging theory more easily than by using the bees themselves. The irony of this story is that *Misumena* preys heavily on the bumble bees upon which I had previously worked. Bumble bees did not turn out to be as easy to propagate under field conditions as I had originally anticipated. How-ever, success in addressing the foraging theory questions with *Mis-umena* revealed that I could explore subsequent stages of its life cycle from the same perspective, thereby obtaining progressively more di-rect estimates of the fitness consequences resulting from adult females initially choosing different hunting sites (Morse and Stephens 1996). The ability of some adult females to choose sites heavily visited by bumble bees enhanced their egg production severalfold. Since each fe-male lays a single clutch and guards it for relatively long periods, it is feasible to compare the success of the resultant different-sized clutches up to emergence of the young from their nests (Fritz and Morse 1985; Morse 1988a). This work in turn led to studies of predation on the young as they left the nest (Morse 1989), patterns of dispersal from nests by young of different-sized clutches, the role of the nest site in determining how offspring dispersed (Morse 1993a), and nest-site choice in relation to the mothers' size (Morse 1993a).

All of these results reinforced the importance of the females' choice of hunting sites, focusing attention on the paradox of how variance in choice behavior remained in the population despite the strong fitness disadvantage to poor choosers. The recurrent variance in patch choice, notwithstanding the resultant cost in fitness to the poor choosers, raised the possibility that other stages of the life cycle might provide insight into why some adults repeatedly made apparently poor choices of hunting sites when better ones were readily available (Morse 2000a). Perhaps as young they enjoyed counterweighing advantages from be-havior that made them poor patch-choosers as adults. So, although they might not perform well as adults, juveniles might prosper from making such choices, in this way countering adult success. I do not have the answer to this question yet, but it is an intriguing hypothesis that clearly warrants testing.

Male *Misumena* are by no means as easy to study in the field as the females. Not surprisingly, then, most of the initial work focused on the adult females as the ideal "foraging automata." However, the striking differences in size between the sexes, and subsequently the recognition that adult sex ratios were strongly female biased, led slowly to the recognition that much was to be gained by extending the work to the males. This realization was greatly helped along by my Honors students, whom I encourage to select novel topics of their own choosing for study. For some reason, perhaps related to the opportunity to strike out on a trajectory distinct from my own, several of these students focused on the males, generating the background of natural history that made the males far more amenable to subsequent work than previously. Since we could already evaluate the factors most important in determining female fitness, this enhanced the attractiveness of working on the males, for we could then relate any information of interest about the males to the facts known about the females. Eventually it became clear that *Misumena* would prove an ideal system in which to explore the potentially competing interests of the males and females. When we can tie these factors directly to the basic fitness variables of the participants in question, the value of these studies, together with the insights they may generate, are magnified many times over. Since we are only embarking on the study of these issues, we have more than enough projects to keep us occupied for the foreseeable future.

Before proceeding further, it is important to comment on the use of "I" and "we" in this book. I wrote this book by myself, though benefiting immeasurably from the invaluable critiques of several generous colleagues acknowledged here. Hence, I use "I" in relation to this process of writing. However, I am indebted to a long list of collaborators for sharing in the research and contributing to its progress. First of all, during early collaborations, Robert Fritz, my former graduate student, convinced me of the major advantages of pursuing life history work with *Misumena*. Most of my subsequent collaborators have been undergraduates at Brown University. Even when I have written single-authored papers, they have been based on work to which these people contributed. I fear that some of what I consider to be my best ideas were placed in my memory bank during discussions with these capable and enthusiastic people. Thus, in relating the research itself, I have found it appropriate to use "we." In the meantime, I hope that few editorial "we's" have crept into these discussions.

One revelation in these studies followed another, such that the research ended up taking a path markedly different from that initially anticipated. However, among researchers with some freedom to chart their own directions, this agenda of following an attractive, but ever-changing, goal may constitute the rule rather than the exception. It may also account for why seemingly undirected research produces so many novel discoveries.

Layout of This Book

This book introduces topics related to resource exploitation and fitness, male-female relationships, population and community implications, how *Misumena*'s attributes compare with those of related species, and, finally, potentially fruitful areas for future research. Chapter 2 describes key traits of *Misumena* and relates them to other arthropods, particularly spiders. Chapter 3 establishes the importance of patch-choice decisions for the lives of *Misumena* and addresses several other aspects of its foraging. Chapter 4 explores lifetime fitness consequences of patch-choice decisions, emphasizing the importance of lifetime, as opposed to partial, fitness estimates. Chapter 5 investigates factors (constraints) that prevent individuals from achieving perfection in these exercises. Chapter 6 considers the role played by experience and learning in effecting foraging repertoires. Chapter 7 evaluates some of the cues that spiders likely use in selecting important sites over their lifetimes. Chapter 8 addresses the extreme sexual dimorphism of *Misumena* and considers possible explanations for it. Chapter 9 explores the interactions and relationships between males and females in sexual encounters. Chapter 10 investigates the interactions and possible impact of *Misumena* on other members of its community and, conversely, the effects of these members on *Misumena*. Chapter 11 compares traits of *Misumena* with those of another crab spider, *Xysticus emertoni*, that occupies the same old fields as *Misumena*. Lastly, Chapter 12 provides some profitable future directions based on the findings of this work.

I began this book during a sabbatical year spent at the University of Melbourne. Time on my hands allowed me to sort out thoughts about the ongoing spider project, which had commenced well before my first published paper on the subject in 1979. While there I was able to rough out an initial draft of this book. Although I hope that the legibility and logic of the prose have proceeded immeasurably since that time, I managed then to put in place most of the thoughts expressed in this book.

ᔥ 2

Some Basic Biology

$\mathcal{I}$ HAVE ARGUED THAT THE crab spider *Misumena* presents an unsurpassed opportunity to evaluate factors such as predators, competitors, and environmental variables that favor or counter success in resource exploitation endeavors. Variation in the spider's responses to these factors permits us in turn to estimate the lifetime fitness consequences of fundamentally important behaviors, in particular those that govern critical resource exploitation, be the resources food or mates. In the following chapters I will focus on how various behavioral traits, primarily those of *Misumena*, contribute to overall lifetime fitness. I will pay particular attention to foraging variables, both in the usual context of finding conventional resources, predominately food and in the sense of finding females. These quests lead from individually based actions to those that involve interactions among the young, between males, between females, and between males and females. Initially, however, to provide an adequate context for this inquiry, I will provide an introduction to the biology of *Misumena*, the principal character of this story. This narrative is developed in further detail in subsequent chapters, supplemented with information on other species of crab spiders wherever appropriate.

General Arthropod and Arachnid Characteristics

First, like all arthropods, *Misumena* is a molting species that periodically must cast off its exoskeleton to grow. Since it must first generate a new exoskeleton inside the old, this exoskeleton must be flexible when formed and can be allowed to harden only after its inhabitant has escaped its former exoskeleton. Therefore, it cannot function normally during a certain period, making it especially vulnerable both to predators and to desiccation. This constraint simultaneously curtails foraging, with a consequent temporary cessation in growth rate. Including the lead-up period to the actual molt event and subsequent recovery, *Misumena* exhibits reduced activity, or no activity at all, up to one-third of the time (Sullivan and Morse 2004). Although terrestrial species with exoskeletons experience severe restrictions on how large they may ultimately become (see Dudley 2000), *Misumena*, as a medium-sized spider, is unlikely to be constrained by these limitations. Very few adult *Misumena* even manage to approach the species' maximum recorded size (Fritz and Morse 1985).

As a spider, *Misumena* employs external digestion, injecting powerful enzymes into its prey (Mommsen 1978). In this way, it breaks down the tissues to a soupy consistency, which the spider subsequently consumes somewhat as a person would suck on a straw. Since *Misumena* does not tear apart its prey, this mechanism permits it to process large insects that might otherwise be difficult to handle. Many spiders (but not *Misumena* or other members of its family) tear apart the exoskeletons of their prey before abandoning them, thereby exposing remaining food that cannot be suctioned initially from the outside (Gertsch 1979). However, all spiders must imbibe their food in liquid form.

As an arthropod, *Misumena* proceeds through several discrete stages, or instars, punctuated by molts. The young grow significant amounts at each molt, such that successive instars (periods between molts) may encounter different experiences: different prey, different predators, and different habitats. It is important to study key variables, such as foraging capabilities or vulnerability to predators, over these different stages to obtain an accurate sense of how they impinge on *Misumena's* life history strategies. In being independent of their parents over most or all of the life cycle, *Misumena* resemble a wide variety of taxa, including many insect groups (particularly forms with no distinct larval stage—hemimetabolous). However, they differ from such intensively studied groups as birds and many insects that undergo a complete metamorphosis (bees, wasps, etc.), most of which do not forage for themselves until they have reached adult size. For that reason it is essential that studies of important life history variables not be confined to adults, as in model species such as insectivorous birds or bees. *Misumena* presents a graphic illustration of why it is necessary to consider the entire life cycle when exploring fitness issues.

Familial Characteristics

Crab spiders (family Thomisidae) are commonly encountered but seldom recognized as such by the layperson. Those that frequent flowers, such as *Misumena*, often remain inconspicuous because their color matches that of their background. Often they are only discovered when flowers are brought into the house for bouquets. When they begin to move about on lines between the flowers or down onto the table on which the bouquet has been placed, they become conspicuous, sometimes to the discomposure of the interior decorator!

Crab spiders of the family Thomisidae, the true crab spiders, are characterized by their two large anterior pairs of limbs, squat body form, and tendency to move sideways in a way somewhat reminiscent of crabs. The raptorial forelegs, combined with potent venom, allow these sit-and-wait predators to capture prey that are unusually large in relation to their body size. Crab spiders vary widely in color, ranging from species like *Misumena* that can change between white and yellow depending on their background, complemented by permanent red striping, to the lime green of some *Diaea* species and the browns that characterize members of litter and bark-dwelling genera, such as *Xysticus* and *Coriarachne*. Crab spiders are entelegynes; that is, their females possess complex genital organs with paired fertilization ducts ending in sperm-storing organs, the spermathecae or seminal receptacles, within a plate-like epigynum (Foelix 1996a). Of particular interest to us, entelegyne spiders often exhibit first-male sperm priority, a factor that fundamentally affects male reproductive behavior. Although exceptions to first-male sperm priority are now known within this group (Masumoto 1993; Elgar 1998), *Misumena* almost certainly exhibit first-male priority.

The family Thomisidae numbers over 2000 described species (Platnick 2004) and has a worldwide distribution, even being well represented in such geographically isolated areas as Australia (Main 1976)
and New Zealand (Forster and Forster 1973). Some species have a
wide range; the focal species of this book, *Misumena vatia*, one of 55
currently recognized species in the genus (Platnick 2004), enjoys a Holarctic distribution, though it is one of only two *Misumena* species
found in North America north of Mexico. *Misumena vatia* itself has
been studied in both the Palearctic and Nearctic parts of its range.
Most *Misumena* are of Old World distribution, and the genus most
likely originated there. Several misumenine genera have broad ranges;
Misumena's is nearly worldwide. It will be interesting, however, to see
whether future taxonomic work will enable *Misumena* to retain this distinction, as new revisions of other groups (e.g., Australian lycosids by
Framenau and Vink [2002]) have revealed that these species were not
closely related to the Palearctic or Holarctic genera to which they were
traditionally assigned. Careful analysis may be in order to determine
whether *Misumena* itself is a recent introduction to the New World
that arrived along with many of the ruderal plants of the Palearctic that
now carpet large areas of the New World, and which it frequents. Most
characteristic in the Nearctic is the numerical dominance of members
of the similar genera *Misumenoides* and *Misumenops*, generally somewhat to the south of *Misumena*'s areas of greatest abundance.

In addition to these so-called misumenine thomisid crab spiders, a
second group, the Philodromidae, has frequently been placed within the
family Thomisidae. They largely consist of wandering predators that
comb the vegetation for arthropod prey, and they may often occupy foliage and bark as well as the litter (Haynes and Sisojevic 1966). In contrast to the thomisids, all of their legs are of similar length. As such, they
are considerably more cursorial than misumenines and occupy a hunting
style quite distinct from those of the misumenines (e.g., Abraham 1983).
I will not consider them further in discussions of crab spiders.

Thomisid crab spiders *sensu stricto* are highly sexually size-dimorphic
as a group; among the large families they are second only to the orb-
weavers (Araneidae and Tetragnathidae) in this regard (Vollrath 1998;
Hormiga, Scharff, and Coddington 2000). One of *Misumena*'s most
striking characteristics is its especially high degree of sexual dimorphism, with large gravid females sometimes attaining a mass two or-

Figure 2.1. Adult female *Misumena vatia* feeding on worker bumble bee *Bombus terricola*. Illustration by Elizabeth Farnsworth.

ders of magnitude greater than that of adult males (400 vs. 4 mg: Figures 2.1 and 2.2). This difference places them at the extreme limit of their family, along with the presumably closely related genera, *Misumenoides* and *Misumenops* (LeGrand and Morse 2000). Although their raptorial limbs facilitate the capture of large prey, these spiders depend on finding high-quality hunting sites that provide them with many opportunities to attack the large prey that they capture with only low success rates, a trait shared with most other predators that hunt large prey (Morse 1980).

Other members of the family—for instance, members of the speciose genus *Xysticus* (351 sp.: Platnick 2004)—are sexually dimorphic as well, although their degree of difference does not approach that between male and female *Misumena*, *Misumenoides*, and *Misumenops*. The strikingly small size of males in the three latter genera is probably in large part a consequence of the extreme decrease in numbers of instars, in contrast to *Xysticus*. Typically, male *Misumena* experience two fewer molts than the females, while male *Xysticus* species molt only one less time than their females. Accordingly, females of some of the larger

Figure 2.2. Adult male *Misumena*. Illustration by Elizabeth Farnsworth.

species of *Xysticus* approximate the size of female *Misumena*, but their males approximate the size of penultimate female *Misumena* (LeGrand and Morse 2000). Over 30 years of working with *Misumena*, I have found two males that were so strikingly much larger than any of the thousands of other males I have handled (50 percent and 100 percent greater mass) that I suspect they had experienced an extra molt.

Thomisid crab spiders are solitary sit-and-wait or ambush predators, and *Misumena* hunt habitually on flowers that attract great numbers of extremely large prey, the adult females concentrating on large bees and moths. This widespread family probably qualifies as the premier group of flower-hunting spiders—certainly that is so in our study areas. Others that hunt regularly on flowers in our study areas include jumping spiders (Salticidae) and sac spiders (Clubionidae).

The crab spiders' relatively sedentary sit-and-wait lifestyle makes them a minority lifestyle among spiders, most of which either spin webs or actively hunt for their prey. However, similarities with the orb-weavers, which often rebuild webs in a single site for considerable periods of time, are of particular interest, especially since crab spiders share with orb-weavers the distinction of including the most extremely sexually dimorphic species of spiders, and, indeed, of free-living terrestrial animal groups as well. *Misumena*'s extended periods of inhabiting particularly favorable flower-hunting sites (Morse and Fritz 1982) are often comparable to the residence times of solitary orb-weaving spiders at favored hunting sites (Enders 1974).

The crab spiders' sit-and-wait foraging habits make them extremely convenient subjects for study and experimentation. Adult females are not highly mobile, and when they do shift sites, they move relatively slowly and over relatively short distances. Because they capture such large prey, the numbers taken are relatively low and possible to monitor, often in their entirety. Thus, one can gather considerable amounts of information on known individuals moving about freely in the field, which is essential for calculating food budgets.

Other members of the family may hunt on tree trunks and leaves. Although their foraging strategies have not been studied in as much detail as the habits of sit-and-wait predators, they, too, probably experience strong selective pressure to choose sites that attract relatively large numbers of prey. Many crab spiders hunt in the litter, and their abundance and diversity at a site are often very high. *Xysticus*, a genus of dull-colored, usually brown, spiders, can attain high numbers in some habitats, at times dominating the spider catches from pitfall traps in our study areas. Though most often found in the litter, members of this genus sometimes hunt on flowers; however, they do not spend as large a proportion of their time on flowers as do *Misumena*. Furthermore, they are not as successful in capturing prey at these sites as *Misumena*. Members of the largest species in our study areas, *X. emertoni* (Chapter 11), did not spend as long on flower-hunting sites as did *Misumena*, and they experienced much less success than *Misumena*. Other species of *Xysticus* in our old fields only occasionally visited flowers, or were only found in pitfall traps and, occasionally at their nests.

Characteristics of Sites with *Misumena* Populations

Misumena exhibit an extremely patchy distribution over the range of our study sites along the central Maine coast, which fall within an area of roughly 15×35 km (over 500 km^2), the vast majority of which consists of forests or water and hence is completely unsatisfactory for the spiders. The satisfactory areas, which constitute no more than a few km^2 in total, range from the edge of the sea to sites several kilometers inland. They consist of old fields, open fields, roadsides, other edges, and bogs to pocket-sized sites either in the midst of a forest or along an otherwise wooded roadside. The common element appears to be a variety of flower species that collectively bloom over much or all of the spring, summer, and autumn. Although this progression of flowers is probably continuous or nearly continuous, we do not yet know the minimal characteristics required to make a habitat satisfactory for *Misumena*. *Misumena*, primarily newly emerged spiderlings, are capable of significant dispersal via ballooning on silken threads. Although the probability of success for any given ballooning individual may be extremely low, their numbers are great enough that "hits" in isolated areas should not be infrequent. As juvenile dispersers, however, it will be necessary that both males and females colonize and survive until they become adults. Mated females move no more than a few meters from their hunting sites (Morse 1993a).

Probably these potential colonists will only respond to the resources present when they recruit—flowers appropriate for second-instars (likely goldenrod), along with a large insect population. However, these sites may not be satisfactory for other stages of the life cycle, and if flowers critical for later developmental stages do not grow at the site, early success of the spiderlings will be for naught. The presence of goldenrod itself may provide adequate hunting sites for the adults, but these flowers only bloom late in the season. Even at that time they are not as profitable for the adults as some of the other flowers (Morse 1981). Consequently, even if such spiders managed to produce an egg mass, it would most likely be both small and extremely late. Both factors would operate against the success of adults and make it highly unlikely that *Misumena* will maintain a stable population at such a site. Such sites as these would merely be sinks—that is, sites that cannot maintain populations without regular immigration.

We do not yet understand how variables such as these affect *Misumena* populations. Some of the sites with flowers blooming throughout the growing season harbor a population of *Misumena* and others do not, and it often is not clear why they differ. Some sites may not have been colonized, others may have received recruits but are deficient in flower species, and yet others may periodically support populations, but populations so small that the probability of random extinction is high. Thus, these populations effectively function as metapopulations, as envisioned by Hanski and Simberloff (1997)—that is, discrete local breeding populations connected by migration. The sites differ from each other in both size and distance, as well as in inherent quality. The appearance of new populations, together with the disappearance of others, suggests the importance of both migration and extinction. Although small populations have a maximum risk of random extinction, larger populations may be subject to limitations that do not occur in the smaller ones solely because the small sites do not provide satisfactory toeholds for other potentially key species—competitors, predators, and parasitoids.

Densities of adult *Misumena* usually appear much higher in small, linear sites, such as those along roadsides, than in large fields, which appear to contain far more resources than the marginal sites. We have not ascertained the basis for these differences or, for that matter, whether they are only apparent. The linear roadside habitats with their frequently low flower densities may merely concentrate these individuals at certain sites, where we discover them more readily than in large habitats. Detailed examination of some large habitats reveals that they contain far more individuals than is initially apparent, although their density per satisfactory hunting site may be much lower. Densities at larger sites may be more reliably controlled by the predictable presence of predators and parasitoids. Dispersal may also be particularly successful along roadsides, often the most densely settled sites, via localized wind sources, especially the vortices of passing vehicles. These forces should tend to disperse the spiderlings along this linear environment, and with fewer losses than the less efficient, and more dangerous, alternative of free ballooning. Roadsides are primarily one-dimensional areas, and air movement here is often predominantly one-directional.

Size

If one surveyed the diversity of spiders, adult female *Misumena* would probably qualify as medium-sized and adult males as small, although their sizes may vary in accordance with their success in prey capture. I hypothesized earlier that two extremely large outlier males found during our work with *Misumena* had proceeded through an extra instar. Similarly, one large outlier female, 20 percent heavier than any other adult female, may have done the same. Unfortunately, we have no way of verifying those suppositions, but they do present interesting potential exceptions to the largely consistent number of instars, to which most individuals conform. In other spiders, differences in size are often attributed to differences in numbers of instars (e.g., Higgins 1992; Vollrath 1998), which are rather flexible in some species.

The usual constraint in adult size may thus reflect the number of instars and individual experiences—normally six instars in males and eight in females (Gabritschevsky 1927). The number of instars, in combination with the growth per instar, may dictate the final size of that spider. Many spiders do not molt unless they have made a substantial gain in mass (e.g., twofold in well-fed house spiders *Tegenaria agrestis:* Agelenidae [Homann 1949]). The time spent in an instar may thus be heavily influenced by the availability of food. In middle-instar *Misumena*, these intermolt periods may regularly differ threefold among individuals (Sullivan and Morse 2004). They experience a significant period of molt-related inactivity, during which they do little feeding, and their mass declines somewhat over this period, partly reversing the gains necessary to reach maturity. These periods substantially slow down the overall growth rate, and it is of interest to ask why the spiders do not simply decrease their number of instars and increase the gains possible within an instar. This tactic apparently is not easily accomplished. Although we have observed considerable differences in mass among individuals of a given instar, this differential apparently results from modest incremental gains over a number of instars, rather than a single gain at some point.

Exceeding some level of minimal gain within an instar may be a difficult and dangerous strategy. Higgins and Rankin (2001) have shown that juvenile *Nephila clavipes* that gained unusually large amounts of mass over an instar exhibited significantly higher mortality at molt than did individuals with conventional gains. *Nephila clavipes* are strictly limited in how much they can grow before molting into the next instar (Higgins 1992, 2000). This restriction is probably related to the inconvenient strategy of producing a new exoskeleton inside the old one and should select for conservative feeding habits. We have yet to discover such an effect in *Misumena*, but so far have only searched for it in newly emerged second instars, which, occasionally take relatively gigantic prey (4-mg hover flies) that have resulted, at an extreme, in an instant increase in mass of more than three fold! These early gains have strong implications for subsequent size gains, including eventual adult size. We would expect to see gains larger than the roughly 50 percent gains of *Misumena* during the second instar (Morse 2000b) if they did not result in some disadvantage, even if the spiderlings only fed on much smaller individual food items, which are abundant under some circumstances.

Life Cycle

Most *Misumena* appear to have a life cycle of two years in the areas where we work. Their single egg mass, which usually makes up nearly two-thirds of a female's gravid mass, is most often laid in midsummer. Young emerge about three and one-half weeks later in their second instar (one molt occurs while still in the egg sac). Survival is then a question of whether the young can quickly find favorable hunting sites, such as the ubiquitous composites, goldenrods (*Solidago* spp.), that bloom at this time. If they recruit quickly to such sites, they are likely to remain there for considerable periods; if not, they typically disperse by ballooning, a rather drastic procedure that often wafts them far out of the range of sites that will sustain them. Although spiders are renowned for their ability to withstand long periods of starvation (Anderson 1974), this trait applies to adults and later instars of some species and does not apply to these newly emerged young (see Vogelei and Greissl 1989). Young *Misumena* still contain considerable resources in their yolk sacs when they emerge from their natal nests. This resource does not, however, sustain them for more than a few days, so they soon enter a vulnerable state if they fail to find a satisfactory foraging site (Morse 1993b). Those individuals that do find favorable conditions grow rapidly and pass through one or two molts before the cold sets in and they retreat into the litter. Not surprisingly, the larger individuals enjoy considerably higher overwintering survival than those that have emerged only shortly before the weather forces an end to their season. This difference is revealed by the decreased ratio of early instars at the beginning of the following spring. Species in other spider families also exhibit this mortality (Edgar 1971; Schaefer 1977).

In the following summer, these individuals pass through several molts and usually overwinter a second year in the penultimate or antepenultimate stage (Figure 2.3). They molt into the adult stage early the next summer, in synchrony with the large prey approaching maximum

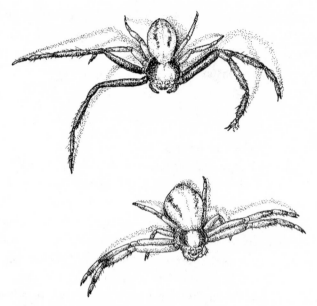

Figure 2.3. Penultimate male (bottom) and adult male (top)
Misumena of similar mass. Note difference in length of
forelimb and size of abdomen. Illustration by Elizabeth
Farnsworth.

abundance, thereby enhancing the spiders' chances for success. They
apparently cannot overwinter as adults, at least not in our study areas.

The life cycle of the males roughly follows that of the females, although they undergo fewer molts and reach the adult stage somewhat
before the females. Though experiencing two or more molts less than
the females (Gabritschevsky 1927), the males mature only shortly before the females. Whereas the adult females concentrate their activities
on foraging and often register prodigious gains in mass at this time,
males do not increase in mass after maturing. In fact, they do not significantly increase in overall mass between the penultimate and adult
stages. They do, however, take enough prey as adults to maintain their
body mass, in contrast to males of many other spiders whose males
often do not feed. In common with males of other species, their main
activity as adults is searching for females. Given that virgin adult females are seldom abundant (LeGrand and Morse 2000), this search
presents a considerable challenge, which is compounded by a dearth of
location cues from the females (Anderson and Morse 2001).

The two-year life cycles may be a consequence of the short season at the northern site where I work. In the laboratory, females will mature over a single winter, and some males reach adulthood considerably before the females, though, as noted, in the field this difference in maturation time is not as great. By feeding young males on a richer diet than they would likely obtain in the field, we have managed to rear them to their penultimate stage by the end of their natal season. The short season also appears to be responsible for the single clutch laid by the females. One can readily manipulate them to produce a second successful clutch by initial supplemental feeding, removing them from their initial clutch as soon as they have laid it, and following with further supplemental feeding (Morse 1994). Although we have no explicit evidence for a three-year cycle, some individuals grow so slowly that they seem unlikely to complete a cycle in two years. It is possible to find individuals of most instars throughout the summer, though their relative abundance undergoes marked shifts over the season.

Color and Its Relevance

Probably the majority of those familiar with *Misumena* know of its ability to change color; it freely shifts from white to yellow and back, depending on the substrate it occupies. The yellow phase is the basis for its common name of "goldenrod spider" in North America, where it frequently occupies those flowers in late summer. At such times individuals adopt a rich yellow coloration that closely matches the color of their substrate. The middle-instar young emerging from goldenrod bouquets usually are yellow. Color change is the subject of several papers on *Misumena* in the older literature, which describes the phenomenon before and around 1900 (e.g., Packard 1905), as well as a major study by Weigel (1941). More recently, color change in crab spiders has again become a subject of intense interest (Chittka 2001; Théry and Casas 2002; Heiling, Herberstein, and Chittka 2003).

Color change is most readily observed in adult females. Indeed, the older literature states, incorrectly, that the ability to change color is confined to adult females (Holl 1987). Packard (1905) demonstrated experimentally that adult females can change color, but was skeptical that juveniles could change color as well. Perhaps echoing his period, Packard did not believe that color change could be under the control of natural selection, instead preferring a Lamarckian interpretation. Gabritschevsky (1927) also reported that only adult females have the ability to change color, a statement that has been dutifully repeated by several subsequent authors, including Gertsch (1939). This stricture does not hold for the spiders we study, and I assume, following Schmalhofer (2000), that Gabritschevsky's conclusion follows from not working with them under conditions in which color changes would be predicted. However, there is no doubt that, at least to human eyes, these color differences and changes are most conspicuous in large adult females, a simple consequence of their size.

Color changes result from the concentrations of pigments in the hypodermis, a thin epithelial layer immediately underlying the exoskeleton. Ommochromes in the hypodermis produce the bright yellow coloration typical of female *Misumena* on substrates such as goldenrod. White coloration results from the refraction of guanine crystals in this layer, which is revealed following the displacement of the ommochromes (Weigel 1941; Holl 1987). Color change does not take place almost instantaneously as in a chameleon or cuttlefish (Hanlon and Messenger 1996); rather, it occurs gradually, requiring anywhere from two to seven or more days. The basis for the variance in time to change is not clearly understood, but probably the characteristics of their substrate drive it to some extent. The color of *Misumena* (yellow or white) on a particular substrate is highly predictable. Exceptions to these matches can be readily assigned to their recent immigration from flowers of different species; the nonmatches thereby often convey useful ecological and behavioral information. Most stages probably change color, though it is less obvious in the tiny second instars (to human eyes, at least), and to date we have no information to suggest that adult, penultimate, and antepenultimate males share this ability (Weigel 1941). These males exhibit a different color pattern, best described as dirty white, with prominent deep red to mahogany stripes on the dorsal and lateral parts of the abdomen and candy-striped legs of the same colors. Since we have not sexed earlier juvenile stages in the field, we do not know whether younger males can change color.

Juvenile *Misumena* are sometimes said to have a green phase as well (Gabritschevsky 1927). This color pattern probably is simply a modification of the white phase, in which some of the skeletal elements can be seen, distorted via complex refraction patterns, through the integument. However, Schmalhofer (2000) found that the abdomens of the closely related *Misumenops asperatus* turned brilliant green when they fed on certain green bugs (Hemiptera), only to fade back to the previous coloration within two days.

Schmalhofer (2000) also noted that upon being fed red-eyed *Drosophila melanogaster* the abdomens of second- to fourth-instar *Misumena*, *M. asperatus*, and *M. formosipes* changed to a pink color, which faded back to the previous white after four to six days. The intensity of color and rate of fading differed with the number of flies eaten. In similar experiments, older instars did not exhibit this color change. Schmalhofer also noted that high proportions of *M. asperatus* exhibited pink, orange, or brown abdominal color changes in the field, a pattern we have observed in *Misumena* in the field as well.

In the simplest of situations for the human eye, the color of the spider appears to match the background flower substrate closely, as in goldenrod and milkweed. In other instances, however, the match is not apparent, such as when adult female *Misumena* occupy flowers of pasture rose *Rosa carolina*, a species with large pink petals and a golden central disc composed in large part of stamens containing large amounts of yellow pollen. *Misumena* habitually rest on the petals, or sometimes at the interface of petals and stamens, and invariably are white to the human eye, or soon change to white when they are on this substrate and so are strikingly conspicuous to human eyes. However, the visual system of arthropods is skewed toward wavelengths lower than ours, with receptors in the ultraviolet, blue, and green, and in many instances without red receptors. The result is a strikingly different receptive pattern from ours. We have shown that both the spiders and rose petals reflect ultraviolet, so that the spiders appear dark against a dark background to typical arthropods, both their most likely prey and predators. Since many of their insect prey probably cannot see in the red range, *Misumena*'s conspicuous dorsolateral red abdominal stripes are probably invisible to them (Hinton 1976). Although recent information demonstrates that certain insects can see into the red range (Tovée 1995; Briscoe and Chittka 2001) and certain birds can see into the ultraviolet range (Cuthill et al. 2000; Eaton and Lanyon 2003), these results seem unlikely to alter the relationships reported here.

Ironically, this information scotches the frequent supposition that under such circumstances these spiders must be highly vulnerable to predators. If vertebrates were a dominant predatory risk, one would not expect these spiders to adopt backgrounds that made them highly conspicuous to these visual predators. We have yet to record a case of bird predation in 30 years of intensive field research on these spiders. Although predatory pressure from these vertebrates might nevertheless be an evolutionary factor affecting color matching (see Lima 2002), the difficulty experienced by vertebrates in hunting on these herbaceous substrates, most of which support little weight, suggests that they are unlikely to be major players dictating color patterns of *Misumena* (Chapter 5). This conclusion differs from that of Théry et al. (2004), who assumed an important direct effect of bird predation on the color of the spiders.

Chittka (2001) treated background matching, typically adult females on flowers, in finer detail than had been previously attempted, considering the quantitative spectral reflective differences of the spider and the background, as well as the visual capabilities of the prey themselves. Chittka noted that even a modest difference between the color of the spider and its background, as perceived by its prey, could provide the prey with the information necessary to avoid the dangerous site. To complicate the issue, he noted that the insect prey may not use all of their sensory capabilities to determine the suitability of a potential foraging site. Honey bees, for instance, may only use their green detectors to make decisions about flower visitation (Giurfa and Lehrer 2001; Spaethe, Tautz, and Chittka 2001). This information suggests how variable these discriminatory systems may be. They also point to the importance of including other sensory systems in the mix—we have always found that flower visitors were more sensitive to movement of the spiders than to color patterns, though we have not performed the necessary experiments to back up that impression. Youngs and Stephens' (2004) conclusion that *Misumena* did not use color as a cue in selecting hunting sites, though not backed by spectral studies, is consistent with our impression.

Female *Misumena's* conspicuous red abdominal stripes differ widely in their completeness among individuals, ranging from total absence to two pairs of complete stripes, one lateral and one nearly mesial. We have routinely recorded the striping patterns of *Misumena* females when collecting from several populations in the study areas. In addition to the "standard" pattern of a single, complete outer pair of stripes, these spiders exhibit a wide range of variation on this basic theme. It includes individuals with broken stripes; some have a shortened stripe on the anterior part of the usual site, plus a posterior spot; others have a single anterior spot; and still others, roughly 0.5 percent of the individuals we have studied, have no red markings at all, although unmarked individuals are apparently more common in some populations than in any of ours (Gertsch 1939). A small minority of individuals possesses part of a second pair of stripes that runs mesial to the lateral markings. In the rarest instances (only two out of the thousands we have observed over the past 30 years), these inner stripes form a complete "V," converging at their anterior end on the midline of the spider.

These patterns may have a relatively simple genetic basis, although they have not been explored from that viewpoint. The wide variation in the patterns, which seems nearly continuous among individuals for both a common outer pair of stripes and a less frequent inner pair, suggests that their expression is under the influence of modifiers that affect the degree to which the color is expressed. It is possible, though not tested, that the two sets of stripes are each under the main control of a single locus, though with modifiers. However, since we have yet to find an individual with a partial or complete inner set of stripes and no outer stripes, it seems more likely that the control of this relationship is somewhat more complex. Furthermore, individuals that possess any part of the inner set of stripes have robust, complete outer stripes as well.

Gabritschevsky reported that females only gain their bright red abdominal stripes as adults, which is true for nearly all individuals we have studied. However, we have occasionally found much smaller females with bright red stripes, in one instance as small as 9 mg (vs. a normal minimum among adult females of about 35 mg), and hence probably a sixth instar. This individual molted once in the laboratory and still had not reached adult condition when we released it at the end of the summer field season. We have also recorded several penultimate females with bright stripes that later molted into adults while in the laboratory. For example, in 2005, two of the 64 females (3.1 percent) initially identified as putative adults because of their bright stripes turned out to be penultimate females. In addition, the red stripe can be seen through the translucent exoskeleton of penultimates about to molt, though we perceive these stripes to be dull reddish-brown and would not mistake them for the bright red of adults that had just molted. As the abdomen of an adult female swells with increase in mass, the pigment color shifts to pink (to the human eye), probably a consequence of a finite amount of pigment covering an increasingly large area.

One morph of the co-occurring candy-stripe spider, *Enoplognatha ovata* (Theridiidae), possesses a very similar pair of dorsolateral red stripes on its abdomen. It is thus of interest to compare this species with *Misumena*. Striped individuals ("redimita") make up only one of three color morphs of this species at some sites. In addition, *E. ovata* also possesses a pure yellowish-white morph ("lineata") and a morph with a complete red shield on the dorsal part of its abdomen ("ovata"). Extensive work in both Britain (Oxford 1976, 1983, 1985) and Finland (Hippa and Oksala 1979, 1981) has shown that these color patterns result from three alleles at a single autosomal locus, and that the North American populations, apparently introduced from Europe (Oxford and Reillo 1994), are similar (Reillo and Wise 1998a). Although tending toward a frequency of lineata > redimita > ovata, this pattern differs among populations (Oxford 1985; Reillo and Wise 1988b), suggesting that selection acts on them. However, the only adequate tests of this proposition (Oxford and Shaw 1986; Reillo and Wise 1988b) failed to detect selection-related variation, perhaps a consequence of swamping by widely dispersing offspring. Oxford (2005) has recently suggested that these populations are usually subject to drift and that only when gene frequencies exceed a certain level does selection act. To date we have found no suggestion that the striping patterns of *Misumena* are closely correlated with any set of environmental conditions, which is consistent with the small, frequently disappearing metapopulation-like nature of *Misumena* in our region. Under these circumstances, founder effects may dominate the patterns seen in characters such as these.

Hinton (1976) suggested that these stripes enhance *Misumena*'s crypticity, even though the colors are striking when taken out of the appropriate background. However, they may serve to break up the otherwise relatively large, continuous mass gravid females would otherwise project. Under these circumstances, the most plausible explanations are that they provide cover either from their predators or from their prey. Relevant observations are few, and it is apparently not known whether predators or prey are more likely to respond to striped or unstriped spiders. The rarity of individuals completely lacking stripes (fewer than 1 percent) makes a purely naturalistic set of experiments infeasible, though it is possible, but unlikely, that their very rarity is a consequence of differential predation. Oxford's (2005) drift/selection model for *E. ovata* could well explain the striping patterns seen in *Misumena*, though it is not known whether the three *Misumena* color morphs (one pair, two pairs, or no red stripes) even involve three alternative alleles at a single locus.

These spiders are highly sensitive to rapid movements, which is readily demonstrated by the brisk movement of a hand; response to motion probably plays a major role in predator avoidance. Movement, of course, potentially provides nonvisual information as well, including vibration and air movement, senses that spiders are well equipped to intercept (Barth 2002). Many of *Misumena*'s likely arthropod predators or prey are more apt to respond strongly to one or more of these potential cues (see Barth 2002) than to a stationary object, notwithstanding its color. The relative importance of body color in the context of trophic relationships therefore remains largely open at this point.

Characteristics of Nests

Female *Misumena* typically construct their nests with a single living ovate leaf, whose distal tip is initially drawn under the rest of the leaf with a thick silken line that connects the underside of the leaf's midvein with the tip. Subsequently, the female lays her eggs, suspended in flocculent silk, within this shelter; then she secures the sides tightly with silk and guards the nest, spending most of her time on its underside (Figure 2.4). Gravid females are not particularly mobile and build their nests close to where they last foraged (Morse 1993a), which sometimes limits the types of leaves available for nest construction. Since the quality of leaves used as nest sites varies markedly (Chapter 4), this lack of mobility may significantly affect the success of a nesting female and a local population as well.

The spiders prefer milkweed *Asclepias syriaca* leaves to those of other species and typically use these leaves when available (Morse 1989). When confined experimentally to other types of leaves, *Misumena*'s nests vary widely in form, ranging from nests with several small leaflets pulled together, as on pasture rose *Rosa carolina*, to nests in which

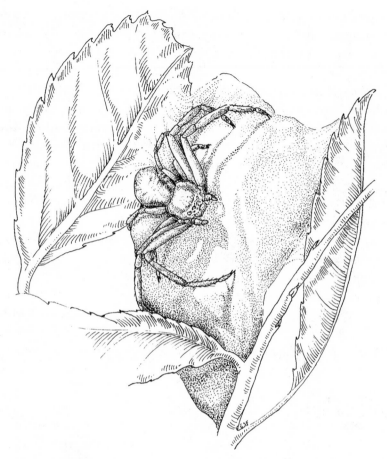

Figure 2.4. Adult female *Misumena* guarding her nest. Illustration by
Elizabeth Farnsworth.

pinnae of sensitive ferns *Onoclea sensibilis* are drawn together (Morse
1989). These nests are not as tightly constructed, and considerable
areas are covered only by silk, which likely increases their vulnerability
to excessive desiccation and predation (Morse 1987, 1989).

Of particular interest are *Misumena* nests that are built without access even to the leaves present in their normal environment. Individuals confined to small, cylindrical 7-dram vials (5 cm long, 3 cm diameter) in the laboratory will often produce a distinct funnel similar to
those of funnel-web spiders (Agelenidae) at the vial entrance and sus-
pend their egg mass in flocculent silk behind the funnel and guard
there. Since the evolutionary relationship between thomisids and age-

lenids is unclear (Coddington and Levi 1991), it may be best to assume provisionally that this similarity is an ancestral character to be found in other species constrained from producing their characteristic derived nests.

Although several other spiders in the study area build similar nests on ovate leaves (e.g., *Xyticus emertoni* [Thomisidae], *Pelegrina insignis* [Salticidae], *Enoplognatha ovata* [Theridiidae]: Morse 1989), only *Misumena* guard on the outside of their nests. The others retreat into the nest itself without leaving a passageway to the exterior. Although one might expect this nest type to have phylogenetic value in linking these species with relatives, the species using this type of nest form a heterogenous group. This suggests that they are merely responding to the opportunities available in the herbaceous canopy that they occupy. Our observation that members of the crab spider genus *Xysticus* build nests of this sort in the canopy, and a different type of nest on the ground or litter layer, strongly supports the latter point.

Sex Ratios and Related Issues

Adult sex ratios of *Misumena* populations are highly female-skewed—between 2.6 and 5.1 adult females per male where we have exhaustively sampled natural populations (Holdsworth and Morse 2000). This inequality is apparent whenever one collects, marks, or censuses these spiders, although a major concern about casual counts is that males and females are not equally conspicuous. In the ratios we just noted, we either systematically marked all individuals until we ceased to find additional unmarked individuals in the population, or we systematically removed all individuals until no more were to be found.

Sex ratios of young *Misumena* at birth are also female-skewed at a population level, although the ratio is less extreme than that of the adults in the field, averaging 1.5:1 (LeGrand and Morse 2000). To make this measure, we used the criterion of rearing individuals from birth until they could be externally sexed (when males reach their antepenultimate stage, their legs become striped). However, the ratios within broods differed widely, ranging from a few entirely female broods to two with strongly male-biased ratios. The mode hovered around the mean, however (LeGrand and Morse 2000). (Since estimates of single broods were based on samples of 20 eggs randomly removed from the egg masses, it probably is safer to say that these broods exhibited a very strong sex bias than to claim that they consisted exclusively of one sex or the other.)

The difference between the observed sex ratio in the field and the estimated primary sex ratio implies that males and females must experience differential mortality between emergence and adulthood. The difference is unlikely to occur over the early juvenile stages, since males and females are then of similar size and early instars of the two sexes do not differ in activity. Although Edgar Leighton's initial measurements suggested that second-instar males were more active than second-instar females, subsequent additions to his sample evened out this relationship, though allowing the possibility that this trait differs among broods (pers. comm.).

Much of the literature suggests that the mortality rates of adult males significantly exceed those of females (e.g., Morse 1980; Andersson 1994), a pattern routinely described in other spiders as well. In spiders, this difference is taken to result from males spending most of their time and effort hunting for females, which is perceived to be an extremely risky business (e.g., Vollrath 1998). The extremely high activity of the males is presumed to expose them to predators or cause their physiological condition to deteriorate prematurely. Fighting with other males may result in injury or mortality, and failure to feed adequately if at all may compromise their condition. Males of some species suffer greatly at mating time in one or more of these ways, though most of this information is to be found for large mammals with polygynous mating systems (e.g., Clutton-Brock, Guinness, and Albon 1982). How frequently these patterns actually hold in small invertebrates like spiders is much less clear, though the literature similarly argues that high adult mortality rates of males occur here as well (Vollrath and Parker 1992).

Whatever the rates of mortality among male spiders in general, we have no evidence that male *Misumena* suffer increased mortality rates over most of their adult life. Our census data argue strongly against accelerated mortality in *Misumena* (LeGrand and Morse 2000). This discrepancy with other species may be a consequence of male *Misumena* being so mobile that they are extremely difficult to capture. Furthermore, they may fail to provide a particularly rewarding target to most potential enemies because of their small size. Their rates of movement significantly exceed those of other age or sex categories of *Misumena*

(Holdsworth and Morse 2000; Sullivan and Morse 2004), and only when they have reached apparent senescence does their mortality rate increase rapidly (Morse and Hu 2004). Adult male *Misumena* capture prey regularly, in contrast to the stated situation with males of other spiders (Vollrath and Parker 1992), which should enhance their bodily condition over this period of high demand.

Differential overwintering mortality is the most likely basis for the change in *Misumena*'s sex ratio over its lifetime. Since *Misumena* likely overwinter twice, these losses could occur during either year. However, most of the disparity may occur between the end of their second fall and the following spring, at which point most are penultimates or antepenultimates and differ clearly from the females. From a physiological standpoint, this would be a logical time for differential mortality, since the factors that have occasioned morphological changes should reflect associated physiological change as well. Since the behavior and activity patterns of penultimate and adult males differ markedly, however, with those of the penultimates resembling that of juvenile females (Holdsworth and Morse 2000; Sullivan and Morse 2004), the basis for this difference is by no means resolved. It is of interest to note in passing that males are the heterogametic sex in *Misumena* (spiders in general have an $X_n{:}O$ system). In many instances the heterogametic sex suffers higher inherent mortality rates than the homogametic one (Haldane's "rule": White 1973).

These explanations do not account for why a female-biased sex ratio occurs at the emergence of the second instars. Mortality could occur at the embryonic level, or within the nest itself during the nest-confined part of the second instar (the first instar is a semi-embryonic form that does not feed). Although we have recorded examples of cannibalism in a small percentage of the broods at this stage (see Chapter 4), their low frequency cannot account for the differences we have recorded. For the most part mortality of eggs is virtually nil in these broods, and it is exceptional to find more than a handful of unhatched eggs or dead early instars. Gunnarsson (1987) also noted that male sheetweb weavers *Pityohyphantes phrygianus* (Linyphiidae) experienced higher overwintering mortalities than females.

Of late, it has become popular to account for female-biased sex ratios of arthropods as a consequence of infection by *Wolbachia pipientis*, a bacterium that may have a feminizing or male-killing effect on offspring. Effects of *Wolbachia* are widespread among insects and have also been found in other arthropods, including isopods, mites, and a few spiders (Oh et al. 2000; Cordaux, Michel-Salzat, and Bouchon 2001; Rowley, Raven, and McGraw 2004). Thus, the question arises as to whether the skewed sex ratio in *Misumena* could result from *Wolbachia* infection. Sampling 10 broods from several populations in our study area, we find that the answer appears to be "no," for we have not yet detected *Wolbachia* (M. Palmer, personal communication). The sample, though admittedly too small to permit a definitive statement, suggests that skewed sex ratios are not a consequence of *Wolbachia* infection.

Unusual Behavior of Crab Spiders

A few crab spiders from the genus *Diaea* exhibit some interestingly variant behaviors. Certain species of Australian *Diaea* have attained a level of sociality comparable to that of other social spiders (Main 1988; Evans 1998). They are unusual in being the only social spiders thus far reported that are not web builders. Lacking access to an extended capture surface like the orb-weavers, sociality presents these spiders with some unusual problems. Their nests of bound-together leaves bear some resemblance to certain natal nests of *Misumena*, being reminiscent of occasional nests in which the females have access to substantial numbers of adjacent leaves. They bind these leaves about the nest, thereby providing additional possible protection for the nest and its contents.

Nest sites such as those of some *Misumena* could easily provide physical conditions that facilitated the development of social behavior. Since these nests probably provide more moderate ambient temperatures and humidities than those in the open, one would expect them to attain considerable utility in particularly hot and dry conditions, such as those in which some of the Australian *Diaea* nest. *Misumena* young remain within their nests for a few days after molting into their second instar. After emerging, however, they exhibit no apparent tendencies to remain together and are solitary through the rest of their lives. Nevertheless, they provide certain of the conditions necessary for the development of sociality. Probably the interposition of autumn and winter prevents further change of this sort.

Jackson et al. (1995) have reported that an as-yet undescribed species of *Diaea* from New Zealand builds webs, an otherwise unreported behavior for the family. These webs are small and apparently function largely by "tripping" and slowing or restraining prey long enough for the nearby spider to attack. This behavior brings to mind the tendency of guarding female *Misumena* to make lines about their nests, which is probably a simple consequence of habitually leaving draglines behind them wherever they go. The *Misumena* sites bear a rough resemblance to some of the most rudimentary theridiid webs, although our spiders show no sign of using them to capture prey. We have never seen insects attacked as a result of being tripped by these lines, but the spiders do capture insects that venture onto their nests, which usually are placed in sites that do not attract many prey.

These *Diaea* webs also bring to mind the lines that *Misumena* almost invariably make about their hunting sites. Although these lines are usually far less numerous than those likely to be constructed about a nest (probably a consequence of the different amounts of time they occupy these sites), they are located at sites that attract many insects. Although we have never seen a female *Misumena* capture a prey item tripped by a line made about its hunting site, it is easy to imagine that it might happen on occasion. It could also represent a possible route to web construction such as that seen in the New Zealand *Diaea* sp.

These two behaviors, sociality and webmaking, occur in isolated areas with highly distinctive, and, in one instance, depauperate faunas. The social *Diaea* produce colonies that differ greatly from those of any other Australian social spiders (Buskirk 1981). New Zealand has a taxonomically depauperate group of web-building spiders (Forster and Forster 1973), which might facilitate evolution of the web-building trait. The parallels to behavior of *Misumena* in both instances suggest that development of such novel behavior may not be difficult under supportive conditions. I assume that in both instances the *Diaea* behavior is a derived trait rather than a trait retained from an ancestral form. These *Diaea* species provide some interesting possibilities for study and comparison with species that exhibit more traditional behavioral patterns (Evans and Goodisman 2002).

Similarities to Other Groups

The sit-and-wait strategies of the thomisids resemble those of other invertebrate predators, especially various true bugs (Hemiptera) that also hunt large insect prey from flowers and other sources of concentrated large prey (Cohen 1995). Ambush bugs (Phymatidae) occupy the same kinds of flowers as those upon which *Misumena* hunts and sometimes can be seen hunting on the same sites as *Misumena*. The ambush bug *Phymata americana* shows many foraging similarities to *Misumena*, though exhibiting some differences in choosiness that seem associated with differences in their mobility (Greco and Kevan 1994; Greco, Weeks, and Kevan 1995), differences that reverse over their lifetimes (Kevan and Greco 2001).

Synthesis

Misumena is a relatively common predator over a broad geographical range. It exhibits widely varying traits that make it an extremely attractive subject for field studies and experimental work. As a sit-and-wait predator it can be followed easily, and because it takes extremely large prey relative to its size, which can be easily observed, one can obtain accurate information on prey intake under natural conditions. Being among the most strikingly sexually dimorphic of free-living terrestrial species also makes *Misumena* an excellent subject for investigation of this factor. Furthermore, having relevant information from the entire life cycle of both sexes makes possible the study of interactions between male and female reproductive strategies.

ᛨ 3

Foraging Strategies

$\mathcal{H}$OW ANIMALS ACHIEVE DIFFERENT LEVELS of fitness
depends on several variables. Foraging, the search for and procurement of resources, provides fundamental requirements of life for most
animals. Arguably the most widespread and basic of such needs is sustenance and promotion of bodily condition. Only upon achieving a satisfactory maintenance level does reproduction become possible, and to
reproduce requires further success, allowing an animal to build up a
cushion of resource reserves. This resource is food, often supplemented by water, when food itself lacks adequate moisture. Naturally,
foraging performance may be constrained in varying ways, which I will
consider in subsequent chapters. Here I focus on obtaining food (diet)
and, particularly, where and how animals obtain it (patch and hunting
site, tactics). In some instances, the search for mates may take on many
aspects of foraging, but I defer that issue to Chapter 9.

Introduction to Foraging Theory

Foraging is currently thought about in the context of optimal foraging
theory, which states that animals forage in a way that maximizes their
fitness, the ultimate criterion of optimization. Since its formal introduction into the ecological literature (MacArthur and Pianka 1966;
Emlen 1966), optimal foraging theory, or, as it is now more frequently

called, foraging theory, has made a major impact on how ecologists look at animal economies. Subsequently, optimization methodology has gained broad use in ecology, behavior, and evolutionary biology (e.g., Stephens and Krebs 1986; Houston and McNamara 1999). Optimality principles, however, are not new, having long been treated analytically by economists (see Schoener 1987) and, in principle, even by biologists. In fact, Darwin himself employed these general principles in his reasoning (Seger and Stubblefield 1996). Thus, though not new, the formal application of optimality theory to biology has had an important impact on the field over the past 35 years.

Foraging theory has had a somewhat checkered and controversial history in ecology, behavior, and evolutionary biology as a result of its uses, misuses, and excesses. On the positive side, it has introduced rigorous hypothesis generation to areas that often lacked an explicitly deductive focus. By its very nature it has driven the formal erection of hypotheses and the identification of specific predictions (Seger and Stubblefield 1996). On the other hand, as often happens when a new area of interest (or, more cynically, a "bandwagon") develops, some practitioners have appeared determined to fit their results into the predictions resulting from the standard foraging models available at the time. At worst, differences were explained away, rather than being addressed as entities worthy of exploration in their own right. In the process, alternative hypothesis testing often did not receive the attention it deserved, to the detriment of both the subject and the theory, even engendering the charge that the theory was circular or untestable (Oaten 1977; Lewontin 1979; Ollason 1980; Gray 1987, etc.). Although the charge may be appropriate for some of the studies they criticized, it does not fit an objective analysis of the topic. Maynard Smith (1978) pointed out that what was being tested by optimality principles was not whether or not animals were optimal, but how well they performed in relation to others and to the world about them. What optimality theory did do was to provide a rigorous new way of doing things.

Foraging theory is but part of a much broader set of procedures that uses phenotypic (i.e., nongenetic) models to ask what alternative state or strategy is optimal or most fit under a particular set of circumstances (see Seger and Stubblefield 1996). Fitness maximization in this particular context does not imply that these animals are perfect foragers. Many constraints, ranging from the laws of physics to evolutionary legacies and complex tradeoffs among individuals, will prevent the attainment of absolute perfection. The optimal individual would live for-ever and reproduce at an exponential rate (Seger and Stubblefield 1996), which makes the point that we are not specifically testing for whether animals are optimal. It is thus important to think of fitness in its intended context—that is, the success of an individual relative to other members of its population (relative fitness). At the same time, the animal must perform successfully enough for it to maintain an energy and resource balance sufficient to replace itself in the population. In this sense, fitness refers specifically to lifetime fitness, here defined for convenience as the success with which an individual places its offspring into the next breeding generation. Unless noted otherwise, I will use lifetime fitness as my operational definition of fitness.

Measures of Fitness

In most of the foraging literature, fitness, if explicitly discussed, is usu-ally taken to be some extremely indirect measure such as energy gain/time. Energy gain/time provides a relatively easy and convenient estimate of fitness, but a far from direct one. For most foraging deci-sions, lifetime fitness can only be measured appropriately on the basis of what happens over an entire life cycle, and, to understand it inti-mately, more than a single life cycle. The production and subsequent success of offspring thus provide a much more sensitive criterion of an individual's ultimate success (lifetime fitness) than its energy gain from one or a set of foraging actions. The accomplishments of these young can only be completely evaluated at the end of their lifetimes.

Energy gain/time has probably received widespread attention as a standard for foraging theory predictions because it is the currency built into the basic optimal foraging models. The justification for use of this currency is that it can be calculated for most of the animals studied in this context. In fact, it is usually the information specifically sought in such studies. It may also come nearly as close as can be reasonably ex-pected to direct measures of lifetime fitness for many of the most assid-uously studied subjects of foraging conducted to date, small songbirds analyzed over short segments of their lifetimes. However, an investi-gator's inability to make more direct estimates of lifetime fitness does not justify the casual use of such indirect measures as sufficient proxies for lifetime fitness, notwithstanding the arguments by proponents of work on birds and other difficult-to-study species (e.g., Pyke, Pulliam, and Charnov 1977; Krebs, Stephens, and Sutherland 1983). Probably the most serious consequence of using these proxies has been that they have led some workers to assume that the proxies are an end in themselves. In contrast, the ideal research program might make initial use of these models and then follow up with studies testing the adequacy of the energy gain/time proxy for lifetime fitness. We have used that tactic in our work with *Misumena* (Morse and Fritz 1987; Morse and Stephens 1996), and in this book I wish to evaluate the importance of key foraging-related behavior (e.g., patch choice, prey choice) in determining lifetime fitness.

Key Variables

The initial foraging models were extremely simple, including an absolute minimum of variables (prey handling time, net energy gain, prey encounter rate, and probability of attacking prey for the diet or prey model). They experienced a mixed predictive success, which occasioned charges that they were so oversimplified as to be virtually useless (see Pyke 1984; Schoener 1987). However, these charges miss their mark if the purpose of simple deterministic models is to capture the essence of a relationship with the absolute minimum of variables. Viewing the models in this context, we find that they have been useful: when they appeared to work accurately, one could argue that they captured the critical behavior within their few variables (which should, of course, then be subjected to further critical testing). If they did not approximate the situation being modeled, however, much was to be gained by relating the nonconformities to the biology of the situation being tested, for they might well imply the presence of some interesting biological relationship. Alternatively, such an approach could lead to the substitution or addition of variables to the model, yielding more accurate results, and perhaps focusing on the important variables driving the relationship. A caveat: excessive model development resulting from the addition of terms always decreases the generality of the model and its ability to predict relationships under new combinations of contingencies faced by an animal. Therefore, the justification of building from the simple to the more complex seems amply justified, but only to the limited degree needed.

Our own approach (Morse and Stephens 1996) has been to test a system initially with the simplest models. These models make predictions for the circumstances our system presents, but predictions that are not completely accomplished by any of the individuals in the population. We have viewed the failure of these individuals to reach criterion as the basis for asking a whole set of new and important questions of our system—the results from testing the initial model suggested the direction for this modified line of investigation. The optimality paradigm forced us to set up explicit hypotheses and to test them with the quantitative performances of the subjects. In that way it has clarified our thinking and provided a series of protocols that have been most useful in progressing along lines of inquiry that have allowed us to uncover key fitness factors.

Our efforts along this line have focused on *Misumena* because we can make far more direct estimates of its lifetime fitness than those characterizing the vast majority of animals found in the literature. This task has not been an easy one, requiring a span of several years, but we have had the advantage of working with a species in which these measures can be made with a realistic effort. Once accomplished, they add to the confidence of interpreting other work that has not provided as direct measures as these.

Theorists have identified several different factors likely to be important in determining how animals choose resources. Probably the most frequently considered are prey choice and patch choice, and I will focus primarily on them. Other variables, such as optimal giving-up time, are clearly related to patch or diet considerations, and such topics as central-place theory depend on a forager being confined to a focal point for reasons such as having to provide food to young confined at a nest, which is not a concern with these spiders. Stephens and Krebs (1986) have provided a useful primer to foraging theory that serves as a convenient entry into this subject. Schmitz (1997) and Houston and McNamara (1999) provide new developments in the topic.

Diet (Prey) Choice

Diet choice, often synonymous with prey choice, has been one of the most frequently explored aspects of foraging theory. Much of the original interest in the theory involved predicting which food items a forager should accept and which ones it should eschew. Since formal equations for prey choice have been derived and discussed in detail elsewhere (see Stephens and Krebs 1986), I will confine my comments here to certain salient features of the basic model, especially to features relevant to *Misumena*. In the simplest situation (no constraints), prey of high rank (energy/time) should always be taken when encountered, and items of lesser rank should also be taken, in descending order of profitability, until their addition to the diet decreases the animal's net rate of food intake to below the mean rate to be obtained from all the prey species, taken at large. What is to be taken and what is to be eschewed likely differ with the individual in question, depending on its ability to handle and process these items. This model does not allow for partial preferences; that is, a particular type of food should always be taken or never be taken under a given set of circumstances. The simple diet model assumes the complete absence of constraints and the forager's complete knowledge of the nature of the available food supply, so that the forager can make accurate decisions about which prey to take and which to pass by. Whether or not an item is taken is independent of its frequency. Thus, search time for a particular type of item, often a vital factor in the foraging of specialists, is not included in this model. Only when an encounter takes place will selectivity be exhibited, and only then to the value of the prospective prey independent of any earlier search costs. Thus, if encountered, any prey item would be evaluated solely on the basis of the resources it provides at that spot; no earlier travel or search costs are added to the assessment of acceptability. Since this model does not recognize or process constraint, issues of familiarity or novelty do not arise.

In this instance, the smallest number of variables that could dictate whether to accept or eschew a prey item should be energy (e) gained and time involved in handling (h) a particular food object (e/h). Energy gained is a simple concept, although its actual measurement may not be straightforward. Nor may the value obtained for a particular item be similar for different species (or even different individuals of the same species). The measure involves simple gains/simple costs.

In contrast, total time expended by searchers and pursuers incorporates other facets of foraging, including hunting or search time, as well as handling or processing time. This nonprocessing aspect is not included in the basic model discussed earlier. The costs of searching and handling will differ markedly with the resources involved. Work on diet choice has concentrated on predators *(sensu stricto)* that combine relatively long searching time with relatively short processing time, since they are eating partially preprocessed food (meat). Herbivores, however, may spend little time in searching, and processing time will often be the dominant aspect of feeding (as in ungulates or koalas), in which potential input rates far exceed those allowed by gut processing times. Although the basic model may thus be so simple that it does not fully characterize the phenomenon being tested, its rationale, often misunderstood, is to capture the essence of the fundamental, broadly underlying features, employing the fewest variables possible. And even if we would be hard pressed to name an organism that completely follows the ideal foraging program dictated by this model (certainly *Misumena* does not), it is reasonable to envision such an organism as one that exhibited the simplest searching protocol possible.

In the real world, constraints, and partial preferences as well, may modify performances, the latter perhaps a consequence of these constraints. Not surprisingly, these predictions often do not fit the data particularly well. This recognition can lead to more detailed models, which may be more accurate, but not accommodate as many organisms. The most important point, however, is to note the fit of the original, general model before attempting to change it. If the model addresses realistic variables, identifying the nonconformity to prediction may go a considerable way toward establishing the acting constraints. Unfortunately, in early studies some workers seemed more concerned with fitting their results to the models than in exploring the results relevant to the biology of their subjects. Thus, advocates sometimes championed interpretations that would have profited from a more balanced treatment (see critique by Lewontin [1979]).

Patch Choice

Patch choice is a second fundamental aspect of foraging theory. It is concerned with the sites at which animals forage rather than explicitly with what items they take, though the availability of such items will dictate in part or whole the quality of the patch. Probably the two most basic concerns of a patch-choosing animal relate to which patches it should exploit and when it should leave a site in search of another (giving-up time). One may consider the first (which patches to exploit) as a special case of the diet model. In common with the basic prey-choice models, the basic patch-choice models assume that the forager has complete knowledge of the resources about it.

Patch-choice models contain other similarities to prey-choice models. Patches should be exploited as long as they yield an energy balance exceeding the mean for the habitat. Over the short term, prey abundance may be taken as a constant in patches that the forager does not deplete of prey. *Misumena* do not deplete their patches, though the patches have an important temporal aspect—flowers senesce and cease to attract nectar and pollen-seeking insects. In the study areas, *Misumena* capture fewer than 1 percent of these available prey (Morse 1986a). Since the patches exploited by the spiders usually consist of a single flowering species, they typically attract these potential prey for no more than several days to a few weeks (Morse and Fritz 1982).

Much of the literature on patch choice is based on depleting patches, however. Although depletion of patches is not our principal concern with *Misumena*, it warrants careful attention, particularly in light of short-term changes in patch quality experienced by *Misumena*. Using Charnov's (1976) marginal-value theorem, the forager chooses how long it will remain in a patch by determining when it ceases to take up resources at a rate exceeding that available from the surrounding habitat (other patches in the vicinity). Although the marginal-value theorem contains flaws (discussed in Stephens and Krebs 1986), it retains interest and value as a result of its salutatory heuristic characteristics. In terms of our concerns about the tradeoffs between accuracy and generality, the inaccuracies projected from its use will usually be modest.

The intake rate in this model is an instantaneous function, so that when it drops below the average for the habitat, the forager should immediately seek another patch. Although the patch-choice model is deterministic, one should expect the intake rates to differ quite markedly over moderate periods of time (i.e., to exhibit a stochastic element [Kareiva, Morse, and Eccleston 1989]), but more so in some patches than others. How long should a forager remain in a nonproductive patch before moving; (that is, how coarsely does the forager divide up units of time?) Often, a forager might have several possible ways of calculating an appropriate giving-up time. The most direct measure is the time since the last capture. In a species such as *Misumena* that takes small numbers of very large prey, the differences in times between cap-

tures are sizable (Morse and Fritz 1982), and the resultant experimental error is large if it uses capture times as a criterion for giving-up
time. However, if the spider makes use of more frequently experienced
stimuli, such as numbers of visitors rather than numbers of captures, it
might make more accurate decisions, providing the spiders capture a
random proportion of these prey.

Some animals can use indirect cues in choosing patches for purposes
other than foraging sites (e.g., hibernation sites); therefore the use of
indirect cues in patch choice need not be out of the question. Problems
of this sort arise in claiming territories that simultaneously serve as
complete hunting sites for their holders, a type of territory common
among birds (Lack 1968). Since giving-up time requires longer-term
information for evaluating sites, it raises important questions about
how much information these foragers have at hand (the model assumes
complete information), how much of it is used, and the degree to
which progressively less recent information is discounted.

Since patch choice has an explicit spatial element, average patch
value takes on a distance component as well. Close-by patches with
lower availability of resources may therefore be as valuable to a forager
as more distant sites with richer resources (see Stephens and Krebs
1986). This concern increases the forager's difficulties in evaluating the
surrounding opportunities and identifying which ones are likely to decrease its success. The degree to which an individual retains information from when it was cruising for patches in the immediate and more
distant past will fine-tune the degree to which it can accommodate for
this factor. The more variable the distances among patches, the coarser
the accommodation to the predictions of patch choice will likely be, yet
the stronger the selective pressure to accommodate for it.

The simple models thus routinely require that a forager have ency-
clopedic knowledge of the environment about it and perfect recall of
events from the past. It has often been charged that these statements
are completely unrealistic attributes for any forager; however, these
complaints again miss the mark. If the models predict the observed
performance of a forager with reasonable accuracy, they raise the interesting questions of how that animal fulfills the stringent assumptions
or what regularities in the environment allow these assumptions to be
sustained. Answering such questions involves a robust research program that explores the extent to which the animal's own sensory system

and memory operate and the extent to which regularities in the environment account for success. Although major research programs of their own, they provide entrées into two questions about which we know far too little, and *Misumena* may provide the opportunity to address these questions. From what we know of experience and learning, we would expect their contributions to differ considerably with the animals in question. Our increasing insight into the learning abilities of some invertebrates leads us to believe that they incorporate considerably more such information into their decision making than was formerly believed (Papaj and Lewis 1993; Dukas 1998b), although they may use only a part of what is potentially available (Dukas 2002). The importance of environmental variation is likely to differ markedly with the habitat and will probably interact with the use of experience/ learning. Furthermore, the degree to which these factors interact may well differ with the group or species in question.

If patch-choice issues exceed those of prey choice, where an animal chooses to hunt becomes more important than its level of success in the areas that it does hunt. In many circumstances, this conflict will lead to tradeoffs between the two variables. Stephens and Krebs (1986) present models for situations in which both prey and patch factors are likely to be important. They make a variety of predictions, including one in which an individual that does not use information should treat a nondepleting patch in the same way as a prey item.

In our own work, we have made considerably greater use of patch choice as a variable than of prey choice, though I must reemphasize that these patches better fit the criteria of nondepleting than depleting patches. Early work (Morse 1979, 1981) demonstrated that adult female *Misumena* are opportunists in prey capture and that they take most potential prey they encounter, even though at times they pay a clear cost for doing so (Morse 1979). Rather, they appear to find patch-choice issues to be much more important. This is probably not surprising for a species that depends on finding often hard-to-locate hunting sites that attract many large, hard-to-capture insects.

By virtue of its tendency to eschew selectivity in diet choice and to conform relatively well to patch-choice variables, *Misumena* appears to compromise some opportunities favoring diet efficiency. If it discriminated in its prey choice, it should be able to confine its activities to the most profitable prey as well. Taking some of the smallest of the visitors to flowers, small hover flies, as well as the largest, bumble bees—even though they would lose mass on a strict hover fly diet and gain prodigiously on bumble bees—seems a poor compromise. However, the relative abundance of large and small prey may play a role in this decision making, such that if the small prey are relatively infrequently encountered, as they are on milkweed, the most profitable plant upon which to hunt (Morse 1981), such a generalist strategy will likely cost but little. And it should be noted that adult female *Misumena* do not often take truly tiny prey, including some that are taken by the early instars. There is a real scarcity of insects under 2 mm long in the captures of adult females, even where these tiny prey are abundant. Either the small items are perceived as unsatisfactory (though extremely valuable to early instars), or the spiders do not detect them. Whether the adult spiders can physically perceive such tiny prey (very small flies, thrips, etc.) or whether they simply ignore them is unclear. It seems questionable, however, whether the adults' forelimbs and chelicerae could efficiently handle extremely small prey, even if the adults did respond to them. Since the early instars take these prey readily, an answer to this question, and the basis for it, would be of considerable interest.

Sensory Capabilities

Our understanding of *Misumena*'s sensory capabilities and of the senses it uses in hunting and prey capture relies primarily on our behavioral observations. Where possible we have related these observations to studies of the sensory physiology of other species. Our observations suggest strongly that *Misumena* depend heavily on mechanoreceptors of one sort or another, which include tactile hairs, trichobothria, and slit sensilla. These organs are sensitive to touch, air currents, and mechanical stresses and vibrations, respectively (Foelix 1985). In terms of prey capture, it seems highly likely that *Misumena* make major use of both substrate vibrations and air currents. Changes in orientation make it clear that they respond to prey moving nearby, though out of sight; and that they also respond to prey flying nearby, but not in contact with the substrate.

The sense organs in question are extremely acute in species that have been tested. The sensilla permit the large ctenid spider *Cupiennius salei* to attack prey on the basis of the vibrations perceived from the plant on which it is situated. It can even use the difference in time of receiving stimuli on its different legs to obtain useful information (Barth 2002). *Cupiennius salei* has a body of 2 to 3 cm length and long legs, and it employs a sit-and-wait strategy. Whether a considerably smaller species like *Misumena* could use such cues as effectively is open to question. The trichobothria respond to the lightest air movements in *C. salei* (Barth 2002) and can even discriminate between predators and prey (Reissland and Görner 1985), simultaneously obtaining information about both distance and direction (Barth 2002). One would expect the tactile hairs to be equally responsive, though these crab spiders are notable for the modest numbers of sensory hairs present on their bodies (Gertsch 1979). However, these hairs should be more than adequate to provide the spiders with adequate sensory capabilities. Barth (2002) describes them as being extremely sensitive to the slightest of deflections.

In contrast to these senses, *Misumena* do not appear to depend heavily on simple visual cues to identify prey. In fact, if a prey item is still, the spider will usually walk right over it. However, although its visual capabilities are limited (Homann 1934), the probability of it attacking a moving target is much enhanced. These traits are very similar to those documented for *C. salei*, which will also typically walk over a still prey species, and for which moving visual stimuli are much more effective than still ones (Barth 2002). *Cupiennius salei* has visual adaptations that enhance its ability to detect movement (Barth 2002). Given *Misumena's* additional deficiency in detecting chemical stimuli (Anderson and Morse 2001; Leonard and Morse 2006), it is attractive to suggest that it detects motion similarly. Although *C. salei* responds to pheromones, which it perceives by both tarsal hairs and chemoreceptive hairs on the pedipalps (Barth 2002), pheromones apparently are not involved in the foraging behavior of *Misumena*. This is perhaps a reflection of its relative dearth of hairs.

Importance of Patch Choice and Prey Choice to *Misumena*

Depending on the circumstances, either the sites at which resources are obtained or procurement of the resources themselves may present difficulties to the forager. *Misumena* provides an excellent opportunity to compare these variables.

Patch Choice

Over most of their pre–egg-laying period, adult female *Misumena* are heavy and slow. They are well designed to capture large prey that come in close contact with them, but experience severe limitations in their ability to chase down these prey or to locate sites to wait for them. As gravid adults, females may even become so heavy that their ability to move about the vegetation on lines—which in other instances is a routine method of travel—is severely compromised. Shortly before egg-laying, their weight stretches these lines so severely that they sink to the substrate when attempting to cross the longer ones (Morse and Fritz 1982). They are thus classic sit-and-wait predators, and their patch-choice decisions are considerably more important to them than their choice of prey once they are on that site, or their level of success in prey capture. This priority results from the type of prey they typically capture, especially bumble bees, which are extremely large relative to the spiders, and most other insect visitors. Bumble bees are difficult to capture, even for formidable predators like large female *Misumena*, and the spiders are fortunate if they capture more than 3 percent of these large prey that they attack (Morse 1979; Morse and Fritz 1982). This low success rate necessitates finding exceptional sites that attract unusually high numbers of bumble bees and other large prey, thereby accounting for the primacy of patch choice in their lives. Their limited mobility makes finding these sites a challenge, however. Our conclusions on the importance of patch and prey choice for adult female *Misumena* are derived primarily from work with individuals hunting on common milkweed and pasture rose, two of the most important hunting sites of these spiders (Morse and Fritz 1982; Morse and Stephens 1996). Other types of hunting sites (e.g., those with much higher proportions of small prey, different overall abundances of prey) may skew the relative importance of patch and prey choice to these individuals and require study, but we have not as yet given them the attention they deserve.

The priority of patch choice over prey choice is best observed on common milkweed, the richest source of nectar in our study areas, and the most important hunting sites for bumble bees, shortly after most *Misumena* have molted into their adult stage. Milkweeds have large round inflorescences (umbels) of flowers that bloom sequentially from the bottom to the top of the stem. Hence, umbels on a stem differ in quality over much of the flowering season. The umbels are typically 3 to 5 cm apart and consequently function as discrete hunting sites or patches for the only modestly mobile adult female spiders, but patches that a spider can monitor simultaneously. Spiders leave little doubt about their ability to monitor all of these umbels simultaneously: if a prey item lands on an adjacent umbel, and the spider is in a hunting mode, it will rather quickly orient itself to the prey item, often even advancing slowly toward it. Umbels qualify as discrete patches because spiders cannot move between umbels rapidly enough to capture prey that they initially detect on an umbel adjacent to theirs before that potential prey item moves onto another umbel. However, information gained from observing the would-be prey item potentially provides the spiders with important information about the quality of the different patches (umbels) they are monitoring. As a result, most individuals shift between umbels of different quality relatively rapidly. They usually select the umbel most visited by prey within a few hours of arrival on the stem, which in turn is closely correlated with the number of nectar-producing flowers on an umbel (Figure 3.1: Morse and Fritz 1982; Morse 1988b). We have divided the umbels into three classes, based on the number of nectar-producing flowers on them: high-, middle-, and low-quality umbels, which have 25+, 5–10, and 0 nectar-producing flowers, respectively. Given the sequential flowering pattern of milkweeds (bottom to top), stems not infrequently have these numbers of active flowers in our study areas. Umbels with the most nectar-producing flowers attract the most insect prey.

Not all of the spiders, however, make the same decisions. Spiders that do not select the prime patch are of particular interest to us, for failing to exploit the umbels that attract the most insects appears to be suboptimal behavior. The response to prey on other umbels suggests that eschewing the prime umbel does not result from a perceptual inability to monitor the available sites effectively. It also does not appear to be a random event in the lives of certain spiders, for some individuals make this apparently suboptimal choice multiple times (Morse and Fritz 1982). We therefore obtain the impression of a population that is behaviorally dimorphic, with a distinct minority regularly choosing the suboptimal umbels. We have obtained the same distribution of for-

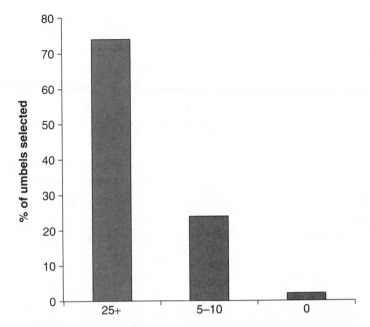

Figure 3.1. Patch use of milkweed umbels by adult female spiders. Umbels contained 25+, 5–10, or 0 nectar-producing flowers. Data from Morse and Fritz (1982); Morse and Stephens (1996).

agers whenever we have replicated this analysis in different studies (Morse and Stephens 1996).

Although milkweed is the most profitable hunting site available to adult female *Misumena* in our study region, it has a highly patchy distribution and does not grow in several of the sites occupied by the spiders. Other important flowers frequented by adult females include pasture rose and goldenrod *Solidago* spp. However, when given the opportunity to hunt on the latter two flowers, adult females do not gain biomass as rapidly as on milkweed (Morse 1981), though most individuals do capture enough prey to produce an egg mass. Accordingly, when milkweed and pasture rose were compared using sequential presentations, adult female spiders showed a stronger preference for milkweed than for pasture rose (Morse 2000c).

Choices among Umbels. The basis for umbel preference is of interest: it could be a response to the prey that are attracted to the flowers, or it might be a direct response to the flowers themselves. We tested this proposition using milkweed umbels of varying quality. Since we had earlier demonstrated that the spiders usually preferred umbels with many nectar-producing flowers to those with few nectar-producing flowers (Figure 3.1: Morse and Fritz 1982), this experiment provided the opportunity to establish the basis for their preference. Using walk-in cages ($1.8 \times 1.8 \times 1.8$ m) that covered several milkweed stems in the field, we placed adult female spiders randomly on umbels of different quality—those with 25 or more nectar-producing umbels, 5 to 10 such umbels, and 0 such umbels (high quality, middle quality, low quality). We then established how many of these individuals shifted from their assigned sites to higher-quality sites within two hours. Earlier it was shown that this period sufficed for most freely foraging individuals to make such decisions in the field. We first performed these tests with no bumble bees or other large insects in the cages. We then repeated the experiment with another set of spiders, but introduced enough bumble bees to equal the level of activity at similar sites outside the cage. Spiders in the presence of bumble bees were significantly more likely to improve their hunting sites than when bumble bees were absent. The test results with bumble bees did not differ significantly from those obtained outside of the cage in a "free-field control" (Figure 3.2). Whether individuals in the absence of bumble bees also obtained a slight advantage resulting from the direct assessment of flower quality remains an open question, for we found a nonsignificant trend for that shift, as we have in similar experiments. However, the latter results, taken one at a time, have never reached statistical significance. Satiated individuals (those that had captured prey within a day) responded similarly to those in the empty cage. Clearly, the direct presence of prey is the primary factor responsible for patch choice in *Misumena*, but the spiders did not respond to the prey if satiated.

Using the Australian crab spider *Thomisus spectabilis*, Heiling, Cheng, and Herberstein (2004) found that in the absence of prey, the spiders made choices between flowers that appeared to be based largely on olfaction. We have not obtained such a result from *Misumena*. Heiling, Cheng, and Herberstein's result followed the observations of Krell and Krämer (1998), who found that two species of African *Thomisus* were attracted to traps baited with eugenol, an essential oil that often is a component of flower fragrances. We have continually failed to get *Misumena* to recruit to traps baited with eugenol.

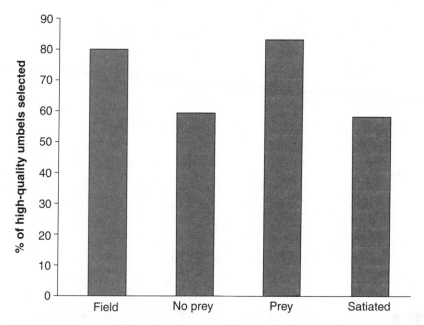

Figure 3.2. Choice of high-quality umbels with and without prey. Field = individuals foraging unrestrained in field, No prey = in cage with no prey, Prey = in cage with prey, Satiated = in cage but had fed during preceding day. Data from Morse (1988b).

Stimuli Used in Patch Choice. Although separation of umbels into three categories clearly sufficed for our analyses, the spiders' discrimination in selecting hunting sites is apparently even more acute than these results would suggest. As numbers of prey increased, we discovered a corresponding increase in the tendency of these spiders to leave high-quality umbels (Morse 1988b). We do not know the basis for this difference, but note that it matches a prediction from a risk-sensitivity model derived by Gillespie and Caraco (1987). In this instance, if average prey availability exceeds that required to lay a clutch of eggs, moving about might by chance position spiders at sites that will attract insects for a longer time into the future. This will enhance the probability of producing an outstandingly large clutch. Alternatively, this counterintuitive behavior might simply be a consequence of the large numbers of visiting prey creating so much activity on the hunting sites that satiated spiders sought other areas with less activity. Life on heavily visited umbels or flowers is definitely not a casual pursuit for a spider. She can count on being run over, run into, or knocked out of the flower by some of these bumble bees, which forage rapidly and often appear oblivious to the spiders (Morse 1986b; Dukas and Morse 2003)! A further possibility is that the physical contact with the bumble bees provides the cues that prompt the risk-spreading behavior proposed by Gillespie and Caraco (1987).

This assessment of patch quality works well for the spiders when the prey are common because it accommodates for the experience and decision making of the foraging bees themselves, rather than simply relying on the quality of the flowers. The bees themselves might not exploit the regional flower sources evenly for any one of a variety of reasons (newness of blooming, distance from nest, etc.). However, this foraging strategy is of little benefit to the spiders when environmental conditions, such as rainy periods, make it uneconomical for the prey to visit the flowers. During those times, potentially the optimal ones for changing sites, the spiders appear unable to make choices. This inability would thus seem somewhat of a constraint, even though the spiders clearly have better information for patch choice when prey are active. Then they can make decisions more quickly and from a greater distance. Prey may not visit these sites in the study area for considerable periods of time. Whenever it is rainy or wet, or overcast and cold, most insects do not visit the flowers. As a result, the sites do not provide cues for assessment. The costs of depending directly on the prey, rather than assessing the flowers, for choosing hunting sites would probably be higher for the spiders at poorer hunting sites than milkweed. Milkweed is likely to attract prey under less favorable environmental conditions than most other flowers. Milkweed's high resource levels allow the prey to make a profit on it but not on most other flower species at such times. Furthermore, milkweed routinely attracts large numbers of moths at night, which form an appreciable, though variable, part of *Misumena*'s resources on milkweed at times (25 percent or more, and 0 to 50 percent among individuals: Morse 1981).

The most common alternative hunting sites of adult female *Misumena* (pasture rose, goldenrod) do not attract nocturnal insect visitors (rose) or only occasional ones (goldenrod). In extensive nocturnal censusing we have found no insects on roses and a scant two moths on goldenrod. Pasture rose flowers typically close at night, preventing access to any remaining pollen, and the spiders often remain within them, while some spiders that hunted on goldenrod during the day moved down the stems into the litter at night (Morse 1981).

Levels of Patchiness. Umbel choice clearly represents an important level of patch choice, but it is not the only level of potential importance. In addition to choice among umbels, between-stem, and between-clone levels of patchiness, as well as between-flower species, differences are possible important variables. (Clones, as described here, are groups of genetically identical individuals. In common milkweed, individual stems are, or once were, connected by underground rhizomes. Usually, these milkweed clones form distinct clumps, though they sometimes expand until they interdigitate.)

Although adult females will remain on the umbels of a milkweed stem for considerable periods, at a point the last flowers on such a stem will senesce. In spite of considerable synchrony in flowering within a clone, some stems maintain flowers longer than others, often as a consequence of extra, late-flowering umbels. For those on senescing sites, the distances to the nearest still-flowering neighbors initially do not decrease significantly, though the number of such stems declines, thereby decreasing options. However, if their ability to respond to their prey is strong, their recruitment time to a favorable new site may not be greatly impaired. Eventually, however, usually within two to three days, the distance to the nearest flowering neighbor increases significantly (Figure 3.3: Morse and Fritz 1982). The spiders could select among stems at this time, though their perceptual abilities appeared inadequate to allow them to operate at the level of efficiency observed at the between-umbel level (Morse and Fritz 1982; Morse 1993c). Although they could monitor all of the active umbels of a stem at once, they could not accomplish this feat at the greater distances of the between-stem level. Consequently, they initially found favorable new stems at a random frequency and then selected or rejected these sites at a frequency comparable to that exhibited at the between-umbel level. They probably used the same sensory capabilities as in those choices (Morse and Fritz 1982). Here they were even faced with a simultaneous choice between two or more stems, but they lacked the sensory ability to make an a priori choice.

As a result of using different giving-up times for satisfactory and unsatisfactory sites, they were able to remain for nearly a week on stems with numbers of actively secreting flowers equal to those of the clone

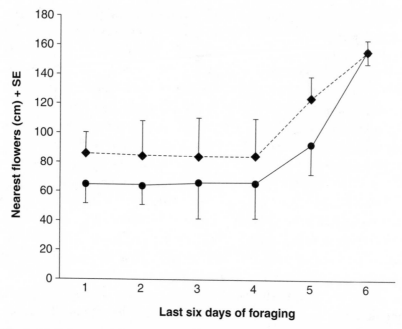

Figure 3.3. Relocation of spiders as milkweed flowers senesce. Distance from spider to nearest stem with nectar-producing flowers (solid line), nearest stem with more nectar-producing flowers than stem occupied by spider (dashed line) ± SE. Only plus SE portrayed on dashed line, minus SE portrayed on solid line, to minimize overlaps of SEs of the two lines. Modified from Morse and Fritz (1982). With permission from the Ecological Society of America.

as a whole, and with disproportionately larger numbers of prey, since the bees concentrated at the few sites that still provided foraging opportunities. The numbers of flowers on these sites significantly exceeded the number of active flowers on the stems that they vacated (Morse and Fritz 1982). However, as might be expected, given such a task, the variance in success at finding satisfactory sites was extremely high. Moreover, not all individuals on senesced umbels moved quickly. All of these factors probably combined to accelerate the rate of quitting the search and laying eggs (Morse 1984).

In contrast, movement over greater distances, such as to other clones of milkweed, is unlikely. We have not recorded any such instances during our work, which has involved well over 2000 individuals (Morse and Fritz 1982). The nearest clones were 20 to 30 m away, well in excess of movements at this stage. The longest single move noted by an adult female was 12 m, with moves averaging 2.2 m. Although two individuals moved over 25 m during their adult lifetimes (27, 26 m), this distance included periods before the critical stage at the end of the season. Thus, although the possibility of such shifts exists, especially where clones are closer to each other, they will not involve great distances. Furthermore, since the environmental conditions within such a circumscribed area are likely to be more uniform than they are over a wider area, synchronization of flowering in nearby clones will often be high enough that the window of opportunity generated from such a hypothetical move would seldom increase the foraging window significantly (Morse and Fritz 1982).

Similarly, adult females seldom recruited to other, later-blooming flower species at this time. We recorded no such movements between milkweed and other species while tracing marked individuals in this study. (Our sole instance involved an individual that moved from rose to goldenrod [Morse 1981].) All milkweeds in the study area were over 25 m from the nearest roses, and, although goldenrods were closer in many instances, milkweed flowers had senesced before the goldenrod began to flower. Adult females obviously do recruit to other species of flowers when their species senesces (we have observed this movement during other observations and at other sites), but the important point is that these flowering species are often significantly separated in both space and time.

Such physical separation between the flower species will seriously constrain the spiders' opportunities to squeeze in extra foraging days that increase their reproductive effort. No doubt this problem leads them toward a rather conservative tendency to lay their clutch if they do not succeed rather quickly in recruiting to a new site at this time (Morse 1981). This problem will be especially severe where individuals occupy especially patchy environments, such as the adult female crab spiders do. Highly volant animals, such as powerfully flying insects and birds, may experience less severe strictures of this sort than these spiders, all else being equal.

Temporal Issues. In addition to whether adult females will find milkweed, a problem arises of being in phase with the relatively brief flowering season of milkweed or other important species of flowers. If the spiders molt into the adult stage late in the season, they may have relatively little opportunity to exploit milkweed before it has completely senesced. The price for missing this opportunity may be great. Since spiders large enough to lay a clutch of eggs will gain considerable mass on each extra day they hunt on milkweed (Morse and Fritz 1982), the timing of their last molt and consequent number of days to forage on milkweed or a comparably rich site are critical. Although, as noted earlier, they will typically search for another site when their site senesces, they will not move far or for long. This is further suggested by their nests on average being only 3 to 4 m from their last hunting site (Morse 1989). Longer moves, punctuated by searches of several unfavorable stems, will probably provide the cues necessary to induce them to lay their single egg mass. Since over 20 percent of the individuals molting into the adult stage do not even accumulate enough mass to lay a clutch of eggs (Morse and Fritz 1982), time is a critical issue.

Patches and Fitness. All of these variables lead to the broader question of what constitutes a good patch in terms of the lifetime fitness of the spiders. This larger-scale projection involves further problems, such as what species of flowers are available at a site at different times of the year. With regard to the milkweed problem, this issue addresses whether other species of flowers blooming about the milkweed clone earlier in the season provide the spiders with resources adequate to molt into the adult stage early enough to exploit milkweed extensively. The matter is yet more complicated, for it requires that flowers be available at preceding stages as well. All of these assumptions require in turn that ample insects visit the flowers in question. Insect visitation rates are highly variable, but flower abundance and density affect these rates greatly (Vasquez and Simberloff 2003). So far, our impression is that the presence or absence of satisfactory flowers at each season is an even more critical, and more likely limiting, factor than the abundance of the flowers at a particular stage.

Since most or all of these flowers depend on pollinators, they will only survive where pollinator numbers are adequate. Certainly the long sequence of necessary conditions makes the issue of patch choice a longer-term proposition than one of response to the moment, though the spiders appear to respond to much shorter-term contingencies than those considered here. Anticipation of the future is a luxury that they probably cannot provision into their genome, and if dispersal is limited, only those populations that colonize sites with the appropriate community flowering phenology will harbor populations of this species. Most likely, regional *Misumena* populations share many traits of a metapopulation (Hanski and Simberloff 1997), differing from the standard in that sites for the local populations remain satisfactory for only limited periods.

Prey Choice

Although patch-choice variables play the dominant role in *Misumena's* resource exploitation, it is important to assess the effect of different prey in determining the value of these patches. Adult females appear to exercise few choices among prospective prey that visit the sites they occupy: their attacks largely or totally reflect the potential prey that visit the hunting sites. Consequently, they may routinely attack suboptimal prey, which in some situations they would probably profit from avoiding. On pasture rose, for instance, they attack most potential prey that visit the flower they occupy; prey that they do not attack for the most part visit the extreme opposite side of the flower occupied by the spider. As a result, they frequently capture the small hover fly *Toxomerus marginatus* and tiny halictid bees, upon which they lose mass, although bumble bees are regularly available and provide nearly their sole opportunity of registering a significant gain in mass (Morse 1979). Whereas the spiders register a 25-fold greater success rate in capturing small hover flies than bumble bees on rose (39 percent vs. 1.6 percent), one of these hover flies weighs only one-sixtieth as much as a bumble bee.

The exact basis for this failure to make choices is not clear, but is most likely a consequence of the rapidly changing patterns of prey availability over the course of a day. Correspondingly, these patterns cause the value of prey to change. As a result of the shifting patterns of prey visitation, *Misumena's* attack rate may rise to as high as 20/h in midmorning in extreme cases, only to fall rapidly back to nearly nil. With pollen normally dehiscing by midmorning and often exhausted of pollen by late morning, pasture rose fails to attract further visitors. The result is a corresponding lack of opportunities for the spiders occupying these flowers. Thus, although adult female *Misumena* would gain, on average, over 7 percent more biomass per day if they eschewed the hover flies, this cost shifts between 0 percent and 20 percent (midmorning) over short periods, thereby greatly complicating any solution that might be suggested by the mean value (Figure 3.4). This rapid fluctuation of resource levels, as well as the fact that pasture rose is only

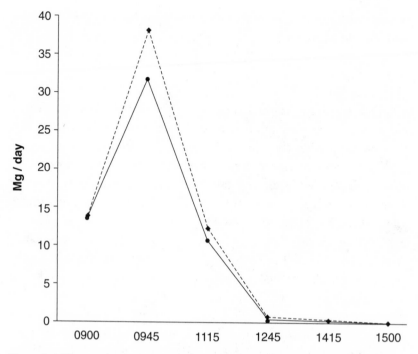

Figure 3.4. Biomass (mg) of prey obtained per day if capturing only bumble bees (dashed line), bumble bees and small hover flies (solid line). Generalist strategy yields less biomass than bumble bee specialist because some generalists capture hover flies and thus cannot capture profitable bumble bees when they are most abundant. Redrawn from Morse (1979). With kind permission of Springer Science and Business Media.

one of several flowers that adult female *Misumena* occupy, suggests why they do not specialize on a particular prey species. Many of these alternative flower sites differ markedly in both phenology and structure, and the visitors they attract differ accordingly.

Other possible explanations for this nonoptimal pattern of prey choice seem far less parsimonious. We have no evidence that the hover flies or other small prey provide special nutrients for the spiders because they struck at all other prey, including the small halictid bees and even male *Misumena*. Females known never to have fed on males experienced as high a success rate as populations from the field (Morse 1994, 2004). Few potential predators are present, and since bumble bee prey are often fed upon when the spiders are in less exposed positions than when feeding on small hover flies, predator avoidance appears unlikely to account for this behavior. The usual lack of other small insectivorous predators on the flowers obviates competition for bumble bees as a possible explanation. In general, the spiders show no sign of avoiding any of the visitors to the flowers, other than for a few huge queen bumble bees at the beginning of the season and some extremely large workers. Thus, avoidance of these formidable prey cannot account for the results, notwithstanding their low capture success (Morse 1979).

The best strategy for spiders hunting on pasture rose would be to catch bumble bees in the morning and feed on them in the afternoon when foraging opportunities are few. Processing small hover flies in the morning seriously cuts into possible time for capturing bees (Morse 1984). We saw no sign of the adults dropping already-captured hover flies at the approach of another prey item (especially a bumble bee). This strategy could possibly minimize loss from capturing hover flies. In contrast, earlier instars primarily capturing hover flies often dropped these flies to strike at another fly (Erickson and Morse 1997). At present, we cannot explain the basis for the difference in behavior between juveniles and adults.

This hover fly visits milkweed much less frequently than pasture rose, probably because in milkweed the pollen sought for their egg formation is locked up in the pollinia and thus is not available to them. With fewer unprofitable prey present on milkweed, bumble bees make up a higher proportion of the prey available to *Misumena*, and *Misumena* also capture bumble bees with higher frequency than on pasture rose. Furthermore, the availability of prey fluctuates much less on milkweed than on pasture rose. As a result, the spiders enjoy an enhanced level of gain in biomass on milkweed (Morse 1981). We do not have comparable data for earlier instars, but, with the exception of very large prey, such as the bumble bees, which they are most hesitant to attack and very seldom able to capture, they are likely to be similarly nondiscriminating.

Erickson and Morse (1997) illustrated how the value of the small hover flies changes for *Misumena* as they proceed through ontogeny. This hover fly is perhaps the only ubiquitous prey species in our study areas that is captured by all sizes of *Misumena*. Its value ranges from that of a rarely obtainable bonanza to second instars (Erickson and Morse 1997) to a low-value prey upon which adult females lose mass, at least on pasture rose on which they can usually capture these flies for only a limited period of the day (Morse 1979) (Figure 3.5). The flies

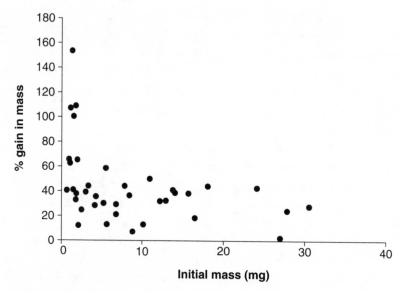

Figure 3.5. Relative gain in mass by *Misumena* of different size from feeding on small hover flies. An outlier 0.6-mg spiderling that gained nearly fourfold in mass is not plotted here in order to show the performance of other individuals in greater detail. Modified from Erickson and Morse (1997). With kind permission of Springer Science and Business Media.

are unlikely to be an important prey item for the youngest spiderlings, however, though second instars do occasionally capture them. We have quantified their gains in the laboratory, and we have also attempted to obtain a sense of their ability to capture these prey when maximally available. We have obtained a maximum of more than a threefold gain in mass of a 0.6 mg spiderling (!);the average gain of these spiderlings is considerably less, nevertheless averaging twofold (Erickson and Morse 1997). These gains relate to a single capture and do not clearly reflect the average success of a second instar, for the great majority of individuals did not capture a hover fly even when they were readily available.

In other experiments, spiderlings presented with hover flies throughout their second instar molted at a greater size than did those that had fed upon a single hover fly followed by *Drosophila* or those that were fed exclusively on *Drosophila* throughout the instar. However, it is greatly stretching the imagination to assume that spiderlings will feed exclusively on hover flies in the field, and the data thus far gathered do not suggest that a single hover fly will greatly enhance the spiderlings' success, anyway. However, it remains possible that if food were not provided virtually ad libitum, as it was in these experiments, a single hover fly would make a difference.

Larger *Misumena* experienced far more success in capturing hover flies than did the smallest spiderlings. Individuals weighing over 1.0 mg exhibited a significantly higher capture rate than the smaller spiderlings (Erickson and Morse 1997). All juveniles that captured hover flies showed a similar relative gain per unit time, a curve that remained remarkably flat over spiders that ranged from 0.6 to 40 mg. This similarity came about as a result of changes in foraging strategy that modified the frequency of capture, processing time, and biomass extracted. Larger spiders maintained this rate only by capturing multiple prey in numbers that increased with their mass. Unfortunately, we did not include adult females in our sample, but the results from the field study of adults (Morse 1979) make it seem questionable whether these large individuals would have been able to capture these prey frequently enough to maintain their body mass.

Thus, the number of prey experienced by these spiders affects the quality of their patches, but their value will differ greatly with the spider's size. Here we found that a small species of hover fly provided resources to a remarkably wide size range of individuals. However, there were marked limits to the advantages exhibited, both in the difficulties experienced by the smallest individuals in capturing the hover flies and in the multiple-prey strategy adopted by the later instars. Although the second instars gained mass at an average rate comparable to that of the later instars, they had the lowest probability of capturing hover flies as well as the greatest vulnerability to starvation. The second instars should be risk-averse at this stage, and hunting only hover flies in the field would be a risk-prone strategy. In contrast, adult females would profit from a risk-prone strategy where hover flies were the only common prey, supplemented by extremely difficult-to-capture, but much larger, prey. It would be useful to have information from other prey species to compare with the hover flies, although it is unlikely that other commonly encountered prey species would be available to a comparably wide range of spider sizes.

Yearly differences in prey available to early instars may curtail specialization in spite of the inordinate numbers of highly profitable dance flies (Empididae: Figure 3.6) available on their major hunting substrate, goldenrod, during some years. Young happening to coincide with such a pulse may enjoy enhanced survival and also contribute to

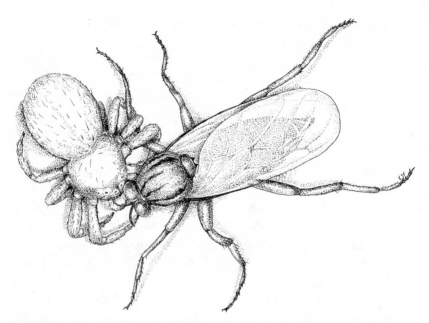

Figure 3.6. Misumena spiderling feeding on small dance fly (Empididae). Crab spiders usually feed first from the head of their prey. Illustration by Elizabeth Farnsworth.

the next generation far in excess of their actual numbers through superior weight gains, likely low predation, and enhanced success in overwintering. Some individuals increased in mass up to almost threefold during a dance fly gradation and proceeded through 1–2 molts over a 17-day period. Yet, large numbers of dance flies have only been present at the time of spiderling emergence in 4 years out of the last 17 (1988, 1998, 2001, 2004). This level of variation could hypothetically drive the behavioral patterns of an entire population.

Order of Presentation: Sequential and Simultaneous Choice

Whether sites are encountered sequentially or simultaneously may affect which one a spider will select. Studies by economists and cognitive psychologists have demonstrated the possible effect of order of presentation on choice—for instance, a shopper's selections (e.g., Gigerenzer et al. 1999; Read et al. 2001). Performance in simultaneous and sequential choices has been studied in detail among humans in as much as it assumes considerable importance in such practical affairs as predicting the purchasing strategies of shoppers (Simonson 1990; Ratner, Kahn, and Kahneman 1999). Order of presentation, for instance, plays an important role in influencing how and what they buy, particularly the impulse shopper. As a reflection of the perceived importance of choice order, Daniel Kahneman (see Kahneman and Tversky 1979; Kahneman 2003) received part of the 2002 Nobel Prize in Economics for his work on this subject, which integrated psychology and economics. The majority of current research on sequential and simultaneous choice appears in journals of marketing and organizational behavior, focusing on human-related issues. Animals exhibit many similarities, however. Tullock (1971), an economist, clearly grasped this point in his early paper, "The Coal Tit as a Careful Shopper."

Nevertheless, the recent behavioral literature usually is silent on this issue, implicitly treating decision making such as that of patch choice as a matter of sequential choice (Stephens and Krebs 1986), although many experiments assessing choices use simultaneous presentations (reviewed in Morse 2005). Only Stephens and Krebs (1986) and Stephens et al. (1986) explicitly refer to the issue as it relates to foraging repertoires. Sit-and-wait predators usually encounter potential new hunting sites in a sequential order (one at a time). A wandering spider will often be in a position to compare a new substrate with the one it occupies at that moment. Yet, often that site will not be a potential foraging substrate, and when it is, it may have been rejected earlier. The same problem should usually hold for opportunities at capturing prey. In relatively few instances does a predator confront two or more similar prospective prey in equal proximity, though certain variables could confound the issue (e.g., favored prey at greater distances than nonfavored prey). Fish that feed on planktonic organisms one at a time fit this category (O'Brien, Slade, and Vinyard 1976).

The sensory systems of predators may reflect the roles played by sequential and simultaneous choice in foraging. Different activities may require different abilities, and proficiency in one type of ability could affect one's potential in another. For instance, patterns of simultaneous and sequential choice may differ between hunting-patch choice and nest-site choice. The main difference between sequential and simultaneous choice is that in sequential choice they have no direct comparison. This lack of direct comparison should select for somewhat different powers of discrimination from those associated with simultaneous choice. Success should also vary with the configurations of the habitats exploited—distributions of satisfactory and unsatisfactory stimuli may vary widely. Simultaneous choice situations provide multiple opportunities for site selection or prey capture, even though success rate (per site or per attack) remains low. Selection for proficiency in sequential choice might be a consequence of the limiting conditions existing most often or at critical points.

We have identified preferences in hunting sites exhibited by young *Misumena* that provide an excellent opportunity to evaluate the roles of sequential and simultaneous choice in the foraging patterns of naive and conditioned individuals. In both simultaneous and sequential choice experiments, a majority of individuals selected goldenrod over aster, but the proportion choosing the favored resource, goldenrod, was higher in the sequential (93.3 percent) than in the simultaneous tests (77.0 percent) (Morse 2005). This pattern matches human tastes (Simonson 1990; Read et al. 2001).

Crab spiders, like humans, find the opportunity for choice a difficult one. A bee in the hand may be worth two in the bush! Adult female crab spiders hunting on pasture rose provide an excellent example. Since the rose flowers bloom for only a day, the spiders must choose a new flower each day. If they have to choose between two equidistant flowers, both attracting prey, they may expend considerable valuable time in making a choice. If only one such target is available, they make their selection more rapidly (Morse 1979). Since sequential and simultaneous choice produce somewhat different results, they may favor the selection of hunting sites in which sequential choice dominates. Failure to consider both possibilities may confound one's conclusions about behaviorally mediated constraints on foraging patterns. Yet these factors and their consequences are not routinely considered in relevant circumstances.

In our experiments, the efficiency of performance in sequential choice situations translated into greater periods of time spent foraging in favored hunting sites, which were associated with a sixfold difference in prey visitation rates. This advantage converts into greatly enhanced gains for the spiderlings at a particularly vulnerable stage, during which they are especially at risk of starvation and predation (Vogelei and Greissl 1989; Morse 1992a). Spiderlings in the favorable situations underwent extremely rapid growth, as much as nearly a threefold increase in 17 days, along with 1 to 2 molts (Morse 1992a), as opposed to little more than maintaining body mass over this period (Morse 2006).

Perceptual limitations curtail the conditions under which these advantages may be attained, however. The spiderlings cannot distinguish between goldenrod and aster inflorescences, more and less favored substrates, of similar size from a distance of 10 cm. Thus, movements by lines to potential hunting sites have a major random element to them. Since they depart from their natal nests on lines that average over 40 to 50 cm, their initial movements, at a highly critical period, have a large random element to them. Yet, the spiderlings will respond to gross differences in the size of targets at this distance, which must serve to focus their performance somewhat. However, we do not yet know the importance of the extent of this perceptual ability. Nor do we know whether moves shorter than 10 cm are more focused than those at 10 cm (Morse 2005). Yet, given the distance of most of the moves on lines (Morse 1993b), only a small minority of them fit into this short range.

Sites Providing Adequate Food

The previous sections suggest a number of limits to the sites that can support a successful *Misumena* population. What are these sites like, where do we find *Misumena*, and where do we not find it? Are food-related issues, such as those we have considered, central to *Misumena*'s distribution?

Types of Sites Occupied

Sites supporting *Misumena* populations are confined to areas that attract large numbers of prey. This constraint probably limits them to sites with flowers blooming through the season. Heavily forested areas do not provide that opportunity. Flowers occupying the forest floor usually precede the leafing-out of the trees in the spring—hence, their sobriquet of "spring ephemeral." With the closing of the canopy few opportunities exist for flowers, obviating the opportunity for these spiders, which depend on a season-long flush of flowers. Most of the sites favorable to *Misumena* today are varyingly artificial habitats—roadsides and old fields. In northeastern North America, these sites are patchy, though in heavily agricultural regions, patches occur at a larger scale than the ones at which we work. In the vicinity of our study area, satisfactory sites for *Misumena*—an old field, or a roadside open enough to support ruderal flowers—are widely scattered within a range dominated by forests and water bodies, which may provide a serious establishment problem in the first place.

Populations of *Misumena* are patchy even at sites that appear to provide all of the flower resources necessary to ensure success. The vacant sites deserve more attention, for we have not investigated them in sufficient detail to establish confidently whether they always provide the sequence of flowers necessary to sustain the members of a *Misumena* population through their entire life cycle. However, many of these sites appear to provide all necessary requirements. Vacant sites could also result from depredations of predators or parasitoids that make habitation impossible. In only a minority of instances does it seem possible that they cannot colonize one of these areas, small sites deep in the forest or islands that lie some distance from the mainland shore. Given the propensity of this species to balloon, however (Chapter 5), it seems likely that spiderlings have often reached these vacant sites, for some such areas do support populations. Since colonization of prospective sites is also required by various flowering species, their dispersal abilities and those of their pollinators may as likely prevent successful colonization as the ability of the spiders themselves. Species important to *Misumena* vary in their dispersal capabilities. Milkweed and goldenrod seeds are both widely wind-dispersed, but other species important to them (pasture rose, buttercup, oxeye daisy, cow vetch) have less impressive dispersal abilities. Their modest dispersal abilities may present a particularly important constraint preventing colonization of openings in the forest, which are generally of a successional nature and hence strongly time-limited.

In other parts of *Misumena*'s range, satisfactory environments may be more nearly continuous—where agriculture is more intensive. Even there, however, growth of flowers that will produce a continuously satisfactory environment may provide a landscape that is patchy at this larger scale. If these locales are less patchy for the spiders, opportunities for their predators and parasitoids, such as large mud-dauber wasps (Sphecidae), are probably less patchy as well, with a likely heavy impact that could account for the relatively low population densities of *Mis-*

umena across broad swaths of its range. However, predatory wasps are rare to nonexistent in our study areas, which appear to lie north of or more coastal than high concentrations of large mud-daubers (*Chaly-bion, Sceliphron, Trypoxylon* spp., etc.) that exact a heavy predatory pressure on crab spiders (Muma and Jeffers 1945; Dorris 1970; Rehnberg 1987). Our study areas are also largely deficient of spider wasps that hunt crab spiders. Parasitic wasps may, however, affect *Misumena*'s abundance and local distribution. They play an important role in some of our spider populations, with as many as 75 percent of the egg masses parasitized by the ichneumonid wasp, *Trychosis cyperia* (Morse 1988a, 1998c, unpublished data). It seems questionable whether these spiders could consistently withstand a 75 percent level of parasitism, and such sites may well be sinks.

Success at Different Sites

Population densities vary markedly at the sites that *Misumena* do occupy, and the basis for these differences could be accounted for in several ways. It is unclear just how great *Misumena*'s success must be for it to maintain a population over the long term. Some sites are clearly only available for short periods of time. Trees at several sites supporting high-density populations in my study areas in the early 1980s have subsequently shaded out important flower species, and the *Misumena* populations have disappeared with them.

Thus, local conditions unfavorable to *Misumena*'s foraging may play a more important role in limiting its numbers than overall climatic factors. At least, *Misumena*'s range extends far to the north and south of the sites at which I have worked (Gertsch 1939; Dondale and Redner 1978). In North America, Dondale and Redner (1978) recorded *Misumena* from Labrador, the Northwest Territories, and Alaska and south to the southern United States. To the best of my knowledge, no one has made relevant studies on *Misumena* in those marginal areas that would shed light on this issue.

Other workers, however, have reported briefly on various aspects of *Misumena*'s foraging activity. Lockley, Young, and Hayes (1989) found that *Misumena*'s nocturnal hunting activity on sunflowers in Mississippi considerably exceeded that in the daytime, a direct reflection of the differences in visits by insect prey. Nocturnal captures in Lockley,

Young, and Hayes's study also constituted a greater part of the spiders' overall foraging success than we have observed for more than occasional individuals in our study areas. The mean nocturnal contribution to the overall foraging success of our females on milkweed peaked at about 25 percent and usually fell well below that level (Morse 1981). The major nocturnal prey items in both areas were moths, especially owlet moths (Noctuidae). Differences between these two studies could result from high daytime temperatures depressing visits by the insect prey, though Lockley, Young, and Hayes did not make that suggestion. From Idaho, Robertson and Maguire (2005) reported the foraging behavior of *Misumena* on flowers that resembled that of our individuals.

The most probable source of potential climatic limitation in our study areas occurs in the winter, when these spiders apparently suffer substantial levels of mortality. Numbers of individuals in censuses immediately before and after winter suggest that in some years the mortality over that period reaches 50 percent or more (LeGrand and Morse 2000). These censuses also suggest that the youngest individuals and the males suffer the highest overwintering mortality. The age and size differential is in part a consequence of maternal success in obtaining prey early enough to produce clutches whose young have time to proceed through one or two molts before the end of autumn. By doing so, they may attain weight gains considerably in excess of later emerging young. This should provide them with vital resources for overwintering and enhance their probability of eventually laying their own clutches soon enough that their own offspring will get an early start. Thus, vital indirect factors play into these survivorship issues.

Capture and Handling of Prey

Misumena's distinctive morphology and physiology—squat, with large powerful forelimbs and fast-killing venom—play a major role in determining which prey it captures, how it captures them, and where it captures them. Most likely, the characteristic phenology and lifestyle of flower-hunting crab spiders are themselves a consequence of insects responding to the evolution of flowering plants and followed angiosperm radiation in the Cretaceous, though the fossil record provides no direct insight into this question.

Attack and Capture

Although I have somewhat minimized the importance of prey capture per se for adult female *Misumena* in this chapter, the spiders must be able to capture prey competently, even if they select the very best hunting sites. Adult females confine their activities almost exclusively to attacks from their perch site, though occasionally they will slowly move ("stalk") as much as a body length (Morse 1986b) in positioning themselves to attack a prey item. Their tendency to stalk is in part a consequence of how rapidly the prospective prey item moves. They move more often in the process of attacking slow-moving honey bees than the faster-moving bumble bees, and their far higher rate of success in capturing honey bees (Morse 1986b) probably results in part from their ability to adjust their orientation to that prey more effectively in the extra time available. Thus, they launch fewer attacks from an oblique angle, a position that minimizes the extent to which forelimbs can function to pin the prey long enough to strike it effectively. Probably a major factor accounting for the adult females' far lower success rates in attacking large prey on pasture rose than milkweed and goldenrod (only 25 percent as accurate as on milkweed and goldenrod: Morse 1981) (Figure 3.7) results from inappropriate orientation to their prey when they strike. This is partly a consequence of the shape and size of the flowers. On the other hand, the spiders have more opportunities to attack prey on rose than on the other two species, which, however, does not make up for their poor success rate in attacks there.

The spiders attacked the different prey species on milkweed at similar distances, although they oriented themselves toward bumble bees at greater distances than toward honey bees. Attack distances were similar for high-quality and middle-quality umbels (Morse 1986b). These results do not follow predictions from diet theory (Stephens and Krebs 1986) for attack patterns. Orientation distances are probably a consequence of the conspicuousness of different prey species. Theory predicts longer attack distances, and presumably orientation distances, for profitable prey. It also predicts that middle-quality umbels, because they attract fewer insects, will result in longer attack distances than high-quality umbels. Thus, the results fit the patch use model rather than the diet model, consistent with the predominating role of patch choice in these circumstances (Morse and Stephens 1996).

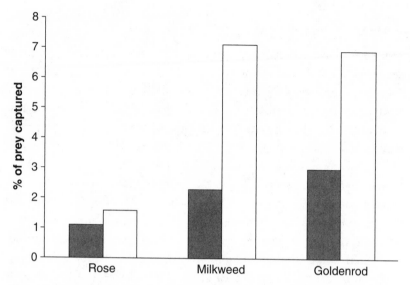

Figure 3.7. Percentage of prey captured by adult female *Misumena* on various flowers. Dark bars = % of prey within one body length captured, light bars = % of attacked prey captured. Data from Morse (1981, 1986b).

Minimum distance to prey during a visit (2–4 cm), rather than mean distance (4–7 cm) or time of prey on an umbel, was the best predictor of an attack (Morse 1986b). The longer the visit of the prey item, the lower its minimum distance to the predator became, presenting the spiders with enhanced capture opportunities.

In addition to exhibiting partial preferences, the spiders did not attack a higher proportion of insects visiting middle-quality umbels than high-quality umbels. This indicates a level of inflexibility not anticipated by theory. This result held, in spite of a visitation rate five times as great on high-quality umbels as that on middle-quality umbels (Morse 1986b). Capture rates closely followed the attack rates, with five times as many prey caught on the high-quality umbels as the middle-quality ones. The results thus portray adult female crab spiders as responding in a rather insensitive all-or-nothing way to the cues experienced on these substrates.

Our studies of *Misumena*'s hunting behavior in Maine differ strikingly from those of Chittka (2001) on the hunting behavior of adult female *Misumena* he studied in Germany. Chittka's adult females were considerably more mobile than those in any of our studies. As classical sit-and-wait predators, our spiders seek out quality sites and move slowly on them, if at all, when hunting, virtually never attacking prey farther than two body lengths away (Morse 1986b). Although our adult females will occasionally stalk prey for these short distances (1–2 cm), these efforts seldom if ever yield prey. It is thus crucial for them to occupy the appropriate site in the first place, and no more than minor adjustments in position yield prey. Chittka (2001) described his spiders as sometimes moving rapidly within inflorescences to capture prey, though he does not quantify these statements. This behavior does not characterize adult female *Misumena* in our experience, although fifth- and sixth-instar females (Morse 1986b) sometimes do move in a way more similar to Chittka's animals than do our adult females. (Similarly to our *Misumena* populations, both Beck and Connor [1992] and Schmalhofer and Casey [1999] comment that the very similar *Misumenoides formosipes* must be in close proximity to their prey if they are to catch it.) Further comparisons between the *Misumena* populations would be profitable.

Surprisingly little information details strikes at prey made by spiders, especially the parts of the prey's body struck (Foelix 1996a, 1996b), notwithstanding the oft-cited "neck-bites" referred to in general sources (e.g., Bristowe 1958; Main 1976). We have performed several experiments on this question, focusing on the performance of second and middle instars (Morse 1999a). Just emerged, naive second instars struck nonselectively at different parts of *Drosophila melanogaster* prey, which are slightly larger than the spiderlings. Thus, more successful strikes were made to the abdomen than the head or thorax, consistent with the different surface areas of these body parts (Figure 3.8a). These individuals were run in the same way several successive times, but their patterns of attack did not change with experience. Strike locations did not result from different individuals specializing in attacks to different parts of the body: of those run two or more times ($N = 29$), not one confined its attacks to a single body area. In contrast, similar presentations of small hover flies to middle-instar *Misumena* of similar size to the flies resulted in attacks that focused on the heads of these flies (Figure 3.8b). Unfortunately, we do not have similar data for middle instars attacking *D. melanogaster* or second instars attacking the hover flies.

Neither do we have equally detailed data for the adults. This is a crucial lack, since their success in capturing very large prey, their primary resource, is so low. However, observations suggest that they do depend

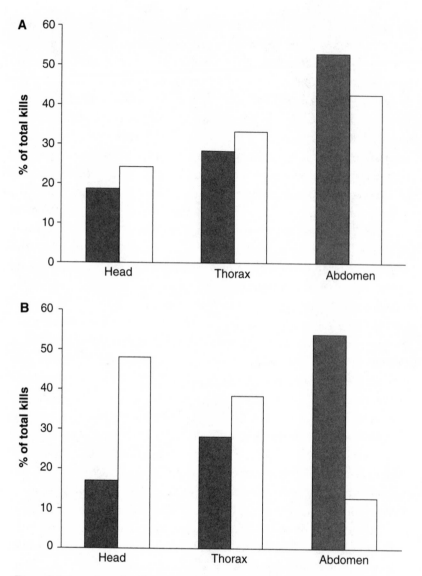

Figure 3.8. Location of strikes on prey: (a) second instars on *Drosophila melanogaster*; (b) middle instars on small hover flies. Dark bars = % of strikes expected on basis of total area, light bars = % of strikes observed. Data from Morse (1999a).

on strikes comparable to the widely reported "neck-bites" for their largest prey; although they do not accurately direct their strikes due to the rapid motion of the bumble bees, almost all successful strikes do land between the head and thorax (probably the so-called neck-bites) or between the thorax and abdomen. Strikes in other areas appear to be unsuccessful, probably because the spiders do not obtain effective purchase on these large prey unless they strike in these two areas. They appear to use their raptorial forelimbs to hold these prey down temporarily while they strike. In contrast, they often capture small prey such as the small hover flies entirely with their chelicerae.

Both the characteristics of the hunting substrate and the nature of the prey affect the spiders' attack patterns. They experience markedly different opportunities on milkweed and pasture rose, two substrates on which adult females capture a large proportion of the bumble bees taken. The normal positioning of adult females within the cup-shaped rose flowers should enhance the number of opportunities, although, as noted, the shape of the rose flowers often results in fast-moving prey approaching from a relatively unsatisfactory orientation. However, on milkweed, where the spiders often are partly concealed among the flowers and thus below the level at which the bees forage, movements of the bees exhibit less directionality relative to the spiders than on rose. A spider's hunting may therefore attain a more effective "stealth" component than on rose. Analogous differences probably occur on all of the substrates the spiders occupy. One could anticipate comparable differences accruing to second instars hunting on two of their most frequent substrates, goldenrod and aster.

Many observers have remarked about how quickly crab spiders dispatch their prey. Crab spider venom is thus generally assumed to be particularly potent, although currently no actual evidence specifically supports this supposition (Nentwig 1986; Pollard 1990). Pollard determined that his *Diaea* completed transfer of venom to their fly prey within 15 seconds of striking them, and that they adjusted the amount of venom to the size of the prey attacked. The effects of this venom may be even stronger in some situations. Adult female *Misumena* frequently rendered bumble bees matching their own mass virtually motionless within 5 to 10 seconds, consistent with a similarly rapid transference of venom. This rapid knockdown, as well as the tendency to drop off on a line at contact, minimized the prey's opportunity to obtain purchase on the substrate, which might allow it to escape.

These opportunities and difficulties will affect the quality of
different hunting sites, though the number of prey visiting the sites, a
consequence of the rewards available to the prey, probably constitutes
the largest single factor determining a site's attractiveness. However,
the flowers dispense their nectar and/or pollen rewards in different
ways, as well as the size of rewards. Furthermore, the spiders experience limits in their ability to adapt from one substrate to another
(Morse 1999b, 2000c). Further work is needed to establish how these
factors interact.

Handling

Although second instars struck at flies nonselectively, in every instance
they commenced feeding on the head, shifting the fly around within a
few minutes after killing it. Middle instars also commenced feeding on
the heads of hover flies. The preference for feeding first on the head
appears to be widespread. Pollard (1990) also reported that *Diaea* sp.,
crab spiders from New Zealand, processed their prey initially from the
head.

The basis for starting to feed from the head region is not clear,
though Pollard (1990) noted that his crab spiders *Diaea* sp. removed
food from the head region more efficiently than the abdomen. That
difference, however, may only be a consequence of which area is processed first. The spiders frequently fed initially from the eyes, although
since the eyes make up much of the head, this apparent preference may
represent no more than a consequence of the area occupied by the
eyes. *Diaea* sp. with excess available prey confine their feeding to the
head region, discarding their prey after feeding from the head and then
capturing another prey, rather than feeding from the abdomen (Pollard
1989). The contents of the visual and/or nervous systems might also
provide especially rich sources of critical elements, though we did not
explore this possibility.

Soon after killing its prey, the spider begins to feed, making contact
with its mouthparts to one of the wounds made by the chelicerae. Initially it withdraws fluid from the prey, using suction from its muscular
pharynx and stomach. It then refluxes this material, mixed with its own
stomach fluid, back into the prey and continues this alternation for
some period. By this time relatively large amounts of this liquid are

passing back and forth between the spider and the prey (Pollard 1989, 1990). The stomach fluid has a powerful effect, reducing soft tissues of the prey to a liquid consistency—essential, since the spider cannot ingest particles larger than about 1 micron (Bartels 1930). No evidence exists to suggest that the venom itself, though enzymatic, plays a role in the breakdown of tissues (Nentwig 1986).

Draining versus Dismembering Prey

Although the crab spider strategy of draining the exoskeletons of their prey (Foelix 1996b), rather than dismembering them into an often unrecognizable mass, differs from that of most spiders, it is a trait shared with cobweb weavers (Theridiidae) and some sheet-web weavers (Linyphiidae). A number of true bugs (Hemiptera) also use this strategy (Cohen 1995). Treatment of prey is associated with the presence or absence of tooth-like structures on the chelicerae and gnathocoxae (basal segments of pedipalps); those of crab spiders and cobweb weavers are largely or totally untoothed, as opposed to groups that break up their prey (Collatz 1987). Breaking apart prey permits close access to the food, though these spiders must still cover the broken-up prey with enzyme-containing liquids. Arguably, crab spiders, though denied this close access, do not suffer the extent of desiccation losses from their prey that the dismemberers do, although the desiccation loss of intact prey may nevertheless be significant (Pollard 1990). One would expect the absence of teeth to be a derived trait that has developed independently within both crab spiders and cobweb weavers. Prey characteristics, moisture conditions, and ability to produce enzymes of both adequate strength and volume all potentially play a role in determining the conditions under which the crab spider strategy might develop (see Chapter 5).

Physical Factors

Several physical factors may modify the foraging demands and patterns of the spiders. Although they act primarily as constraints, their effect on their prey may differ markedly from their effects on the spiders.

Water Balance

Misumena normally obtain most of their required moisture from their prey, though their needs become apparent in the laboratory if precautions are not taken to provide a source of moisture. Provisioned adult and late-instar females can routinely be kept under most temperature and humidity regimes likely to occur in an unregulated laboratory. Early instars and males, however, are vulnerable to low-humidity conditions. This vulnerability can be seen from the increase in their mortality rates during periods of drought (characterized by sustained low humidity and, frequently, elevated temperatures), unless we take careful precautions. Thus, we routinely provide the approximately 0.6-mg second instars with moistened strips of paper toweling. Failure to do so will result in rather quick deaths of the young under dry conditions, often in less than five days (Morse 1992a).

As expected from their differences in mass, the ca. 4-mg males are not as vulnerable to desiccation as the smaller second instars. However, during one particularly dry and hot summer, we experienced unprecedented mortality among the males, which we immediately alleviated by providing them with the aforementioned moistened toweling. During usual summer conditions we do not need to take this precaution with males, as long as they are regularly fed.

However, body water loss can be readily observed in the field among females guarding their egg sacs (Morse 1987). These individuals usually do not take prey and thus do not regularly gain body mass from this source. As a result, their mass typically declines slowly, up to nearly 20 percent over a 30-day period (Figure 3.9). This decrease in mass seemed unlikely to cause mortality, however, for members of a comparable group maintained in the laboratory lost an average of 39.8 percent of their mass before succumbing (Morse 1987). During periods of rain, the field-based individuals may increase in mass as much as 10 percent. This increase is the result of apparent drinking in some instances. However, the prevailing atmosphere at such times, during which the spiders may become wet, permits them to take in moisture through the integument as well. At such times they may emerge from their normal guarding position on the underside of their nests and expose themselves directly to the rain, becoming thoroughly wet in the process. We do not believe that this behavior is confined to the postpartum females. Yet the apparent mass gains resulting from it can be most easily mea-

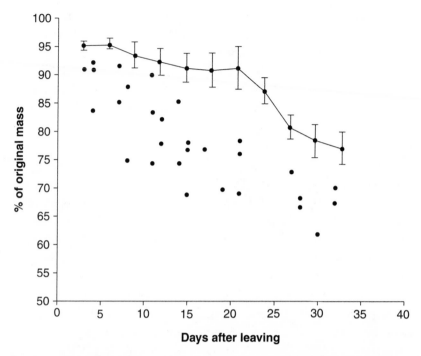

Figure 3.9. Loss of mass by post-laying adults in the field and laboratory. Points with error estimates (SE) = % loss in mass of spiders in field. Points without error estimates are percentage losses of individual spiders without food or water in the laboratory. Modified from Morse (1987).

sured there, owing to their usual failure to feed during this period. Gains of individuals enclosed in nylon tricot bags cannot be attributed to feeding. Under conditions experienced regularly by *Misumena*, dew droplets will routinely be available in the early morning, and a majority of their drinking may occur then, since we have made few such observations.

In addition, the ability of postpartum females to absorb moisture can be readily demonstrated by placing severely desiccated individuals into containers with moistened paper towels that they cannot contact. In that saturated humidity, these individuals will rapidly take up most or all of the weight deficit that they have encountered since laying their clutch of eggs, sometimes up to 30 percent of their former body mass (Morse 1987).

Pollard, Beck, and Dodson (1995) have reported that nectar feeding will provide the moisture male *Misumenoides formosipes* need to maintain their water balance. Under most circumstances, *Misumena* will have access to the dew that forms on plants in the early morning, and nectar would, for those purposes, not offer a superior source of moisture to these droplets. In fact, under drought conditions, nectar will often become extremely concentrated. Although this change might increase its potential food value per unit volume, it would decrease its value as a source of moisture.

Temperature

Although adult female *Misumena* are sensitive to extreme temperatures, they remain active over a markedly wider range of temperatures than their most important prey—bees, flies, and moths. These activity patterns are quantified in the closely related *Misumenoides formosipes* and *Misumenops asperatus* by Schmalhofer (1999) and Schmalhofer and Casey (1999). Their hunting success may thus vary widely as a consequence of the prey's resultant activity. Because of the heterothermal existence of their major prey (Heinrich 1979), bumble bees, temperature assumes great importance for adult female *Misumena*. To provide an idea of the consequences of bumble bees' thermal life for *Misumena*, I first present some aspects of their thermal sensitivity.

Low Temperature and Its Effects on Prey. The large prey upon which adult female *Misumena* primarily depend typically fly only under favorable weather conditions: those in which the temperature is high, the sun is shining, and/or unusually rich nectar sources are available. Bumble bees, for example, must maintain a thoracic temperature of 30°C or more in order to fly (Heinrich 1979). Although flight muscles generate much of this heat, thermal demands dictate the conditions under which the bees find it energetically economical to forage. Thus, on overcast or rainy days, bumble bee visits to flowers greatly decline or even cease. Heat loss will be acute under rainy conditions, since the insulatory quality of the bees' pile is often lost at this time. These effects leave the spiders without prey, and at such times their growth rate naturally declines.

When foraging on relatively modest resources, such as goldenrod in the late summer, these bees are also extremely sensitive to dips in temperature and radiant energy. When bumble bees forage on goldenrod under a partly cloudy sky with an air temperature of roughly 22°C, if a large cloud temporarily blocks the sun, resulting in a drop in ambient temperature of no more than 2°C, the bees become ectothermic and unable to fly. Although they continue foraging, they do so at a much reduced rate (see Heinrich 1979) and would thus seem vulnerable to the spiders.

These factors are potentially important to adult female *Misumena*. Under these circumstances, one might predict that the spiders' ability to function effectively at lower temperatures than their prey (Schmalhofer 1999; Schmalhofer and Casey 1999) would provide extra rewards for them. Because these bees are difficult for the spiders to capture, and they procure only a very low percentage of those that they attack, slowly moving, semitorpid individuals would appear to be especially vulnerable targets for the spiders. Yet, we have not observed any such bonanza. Since the spiders respond most strongly to the movement of prey, the slow movements of the semitorpid bumble bees apparently do not attract as much attention as when they are more active. Active bumble bees move over the flowers at a much greater rate, vibrating the flowers considerably more than when they are semitorpid. Moving more slowly, semitorpid bees obviously do not traverse as much space as the highly active ones; thus, they do not come into contact with the sit-and-wait spiders as often as do highly active ones.

A similar situation likely exists at night where *Misumena* have access to nocturnally active prey on milkweed. (Most other flowers used commonly by *Misumena* attract few if any moths at night.) Night hunting provides a sometimes important supplemental food supply for the spiders (Fritz and Morse 1985), since milkweed flowers attract myriads of moths. Noctuids and geometrids are the most important prey to the spiders as a result of these moths' abundance and size. Moths forage much more slowly over the flowers than do bumble bees and spend much longer periods at individual flowers than do bumble bees. At the higher nighttime temperatures experienced in our study areas (ca. 15°C and above), most noctuids and geometrids can fly instantly, yet their relatively slow foraging motions are more closely akin to semitorpid than highly active bumble bees. However, when temperatures fall below approximately 15°C, the common noctuids (and other moths as well) become semitorpid, and their rate of movement progressively declines with the temperature. Clearly, they should become especially easy targets at low temperatures, though their escape response is simply to drop like a stone into the litter if accosted, so a spider must strike accurately at this time. We do not know under which condition the moths are more vulnerable to the spiders, but the results from the bumble bees suggest that the spiders gain little advantage from the moths' behavioral shift.

Maximum Temperatures. At the opposite end of the spectrum, *Misumena* likely experience upper boundaries to their heat balance on hot days, especially when they are in cup-shaped flowers that focus radiation on the area they usually occupy (Morse 1979), and act as miniature solar reflectors (Hocking 1968; Kevan 1972). Schmalhofer (1999) points out that the temperature of a flower surface may exceed that of the air by as much as 15°C, such that these surfaces may at times reach 45°C or more. This temperature approaches the critical thermal maximum (highest body temperature at which a subject can escape a hotter site) of *M. formosipes* (and probably *Misumena* as well), a remarkably high temperature of 48.2°C. At such times they alternate their activity between hunting sites and nearby shaded areas. Often this shift simply requires them to move to the opposite side of a flower petal, as can be seen most clearly when they hunt on pasture rose flowers. However, this behavior places them in an inconvenient position to capture prey during part of the time and must decrease their rate of capture somewhat. If a large prey item such as a bumble bee visits the flower, however, the spider may creep over the edge of the petal to place itself in a hunting position. Such behavior is unlikely to be predator avoidance, since the spiders do not exhibit this behavior when hunting on pasture rose unless the temperature and exposure are excessive. The spiders may also position themselves similarly on ox-eye daisies, although most daisies in our study area flower before maximum summertime temperatures materialize. Furthermore, daisy flowers probably do not function like solar ovens as pasture roses likely do, because they are flat rather than cup-shaped and because their color (white vs. pink but with heavy UV-range reflection in our analyses) seems less likely to produce elevated temperatures. Schmalhofer (1999) suggests that the exceedingly high critical thermal maximum of *M. formosipes*, relative to most spiders, is an adaptation to hunting on flower surfaces, presumably shared with *Misumena*.

Bumble bees generally continue to fly at a high rate through the heat of an average clear day. Their activity may decline somewhat during the hottest part, which could, however, also be a consequence of declining nectar or pollen volumes available for them. On exceptionally hot days, their activity will cease almost entirely during the middle of the day. At this point the temperature has presumably risen to a point at which it is unsafe to fly, owing to the heat generated in the process, which will raise their thoracic temperatures to a lethal level (Heinrich 1979). Because the spiders are likely to be in partial hiding at such times as well, this decline in prey activity may not greatly lower their potential rate of prey capture. That is, when the spiders would be least efficient at capturing prey, prey availability has declined most significantly. Such extremes in temperature occur infrequently in our northerly coastal study area, however. We have observed only one day over 30 years in which high temperatures apparently totally curtailed bumble bee activity, at that time exceeding the normal summer maximum by 8°C. Thus, it is questionable whether constraints on resource exploitation at high temperatures affect adult female *Misumena* as severely as those under low temperatures. Young *Misumena*, which should be more vulnerable to temperature change than adults, usually hunt in goldenrod and can burrow deeply among the flower heads, which lowers their exposure greatly.

Consequences of Changing Sites. The low availability of prey during unfavorable temperature conditions would appear to provide a favorable and relatively cost-free time for the spiders to change sites (see above). Movement could thus occur when cost was minimal, and other dangers should not exceed those at other times and might even decline. In particular, if in danger of predatory wasps, the spiders are probably less vulnerable than when those insects are in flight. The theory of risk-sensitive foraging further predicts that with lowered prey availability, it should prove advantageous to shift to another site (Gillespie and Caraco 1987). Yet, patch choice of adult females remains primarily a function of prey visitation rates (Morse and Fritz 1982), so that spiders lack the critical information necessary to make accurate shifts at such times. Given the absence of prey, any site should appear as favorable as any other. Under those circumstances, the best strategy might be simply to sit tight on a site that has already yielded prey, and therefore may yield prey in the future. Furthermore, flowers will not senesce as rapidly under cold and rainy conditions as at other times. (However, if the spider has already exploited a foraging site for an extended period, the flowers at that site will probably soon senesce.)

Choices made in the absence of prey also run the risk of yielding low prey visitation rates when conditions improve. Although flower number and prey visitation are highly significantly correlated, considerable variation occurs about the mean, even within a clone of milkweed (Morse 1986c), because plants are on the edge of a clone or otherwise obscured. The possibility of wandering totally out of the clone could further lower the putative gain from shifting sites in the absence of prey activity.

Nectar and Pollen Feeding

Potential alternative sources of sustenance for flower-frequenting spiders include nectar and possibly pollen. We have relatively little information on how much they are used, though we have observed apparent nectar feeding by both *Misumena* spiderlings and adult females, during which they crowded head first into goldenrod flower heads (Morse 1987). However, they do not regularly feed on nectar, even while it is being secreted heavily by the flowers. We have made few records of the behavior, although as a result of regular inquiries from several colleagues we have continually been on the lookout for it. This impression differs considerably from that of Pollard, Beck, and Dodson (1995), who reported nectar feeding to be an important means of sustenance, as well as moisture, for adult male *Misumenoides formosipes* on flowers such as wild carrot *Daucus carota*. They also noted that female-guarding male spiders on wild carrot preferred 30 percent sugar solutions to pure water.

We have not observed nectar feeding in male *Misumena*, which routinely feed on small hover flies and other insects that roughly equal the males in size. Wild carrot is not yet in bloom when most female *Misumena* are hunting, and male *Misumena* in our populations perform only modest amounts of female guarding. It thus remains questionable whether the moderate periods of guarding recorded (1 to 2 days: Holdsworth and Morse 2000) severely constrain their nutrient or water balance. Furthermore, the food value of nectar to these carnivores is limited. Males with access to nectar in Pollard, Beck, and Dodson's (1995) study exhibited a trend toward higher survival rates than those without nectar. Vogelei and Greissl (1989) found that juvenile *Thomisus onustus* (Thomisidae) did not molt when provided only with nectar.

Pollard, Beck, and Dodson (1995) also suggested that the spiders obtain some sustenance from pollen. If so, the grains would have to be soaked in nectar or water because of the apparent inability of spiders to imbibe solid objects larger than 1 micron in diameter (Bartels 1930). Vogelei and Greissl (1989) have reported pollen feeding by *Thomisus onustus* in the laboratory, suggesting that the spiders ingested the pollen as proposed by Pollard, Beck, and Dodson (1995). Vogelei and Greissl found that spiderlings survived better with sucrose solutions than pollen. They suggest that nectar would provide better results than a pollen solution because nectar may contain sugars, amino acids, proteins, lipids, vitamins, minerals, and pigments, while the pollen amino acids are deficient in critical nutrients, especially tyrosine. Others have obtained mixed results in tests on the use of pollen with species from several other spider families (Smith and Mommsen 1984; Carrel, Burgess, and Shoemaker 2000).

Some Potential Bonanzas

Under certain circumstances, predator and prey may meet in conditions that present particular advantages to the predator in relation to the prey. We would not expect them to be frequent or stable situations, for long-term consequences would otherwise dictate adaptation or extinction of the prey. Next I present two potential instances of such relationships between *Misumena* and its prospective prey.

Insects Burdened by Milkweed Pollinia

Whenever unusually rich resources become available, they should provide extra opportunities for the spiders, though the spiders' ability to respond efficiently to them may be related to the predictability of such bonanzas. Spiders might respond innately to predictable resources, but innate responses may not suffice to permit efficient exploitation of infrequently or unpredictably available resources. Under these circumstances, behavioral plasticity would serve them well. We have not systematically investigated this issue, but the spiders' weak response to a seemingly excellent source of food—insects trapped in the stigmatic slits of milkweed flowers—raises questions about the extent of their ability to respond to opportunities outside of the ordinary.

Milkweeds, in common with other members of their group (formerly the family Asclepiadaceae), present their pollen in packets called pollinia, which are dispersed by visiting insects whose feet or legs become entangled in the stigmatic slits of the flowers. As insects move over the flowers, one of their legs may occasionally drop into a stigmatic slit. The leg may slide along the stigmatic slit until it contacts the pollinarium, made up of a central "grasping" corpusculum or clip and two pollinia. This spring-like mechanism of the corpusculum attaches to the insect's leg, foot, or perhaps a hair that has become similarly entrained (illustrated in Morse 1985a). If that insect is large enough (bumble bees, etc.), it pulls the pollinarium out of the flower with little impedance, although honey bees are not infrequently temporarily entangled (4.7 percent, but < 1 percent for bumble bees observed visiting flowers: Morse and Fritz 1989). However, smaller visitors may have to struggle to extricate themselves, and even small bumble bees and honey bees may be temporarily trapped (several seconds) before breaking free. Yet smaller visitors, especially those in the 15 to 40 mg range, including flies and geometrid moths, may become permanently lodged in these flowers, often struggling for hours before expiring from heat or water loss. These actions provide the spiders with potentially important extra opportunities, since they experience low success rates in capturing large prey, especially the largest, and most rewarding, ones.

Yet, the spiders did not capture these individuals at a significantly higher rate than nonentangled prey. Entangled prey made up only 3 percent of *Misumena*'s captures on milkweed, and entanglement did not increase predation rate. The spiders did not even capture permanently entangled flies and geometrids significantly more often than freely moving individuals of these groups (Morse and Fritz 1989). It perhaps is surprising that the spiders did not respond more strongly to these disadvantaged potential prey, but the failure to exploit more permanently trapped insects may be a consequence of *Misumena*'s classic sit-and-wait hunting strategy. Thus, unless an insect became entangled very close to a spider's hunting site, the spider did not attack it. Adult female *Misumena* readily captured similarly struggling bumble bees presented to them by a gloved hand. However, the spiders may not have perceived the vibrations produced by the periodically struggling entangled insects to result from prey.

The spiders have further opportunities to capture prey species that have escaped the stigmatic slits with pollinia and corpuscula on their appendages. Some of these insects, especially honey bees, accumulate large numbers of pollinia and corpuscula, often in strings, on their legs and even their mouthparts. Those that carry pollinia attached to their feet both forage more slowly (Morse 1982a) and slip from the flowers more frequently while foraging than do those without these impediments. In a word, they are clumsy! Honey bees captured by the spiders carried more pollinia and corpuscula than did a random sample, but no such differences emerged for bumble bees (Morse and Fritz 1989). The failure to improve their capture rate on these impaired bumble bees implies that even at their slowed rate these bees still moved too rapidly to present easy targets for the spiders.

Insects Tripped by Draglines

In common with most other spiders, *Misumena* lay down draglines almost everywhere they go. As a result, sites occupied for several days often contain considerable numbers of these lines festooned about their hunting sites. Such constructions bear a superficial resemblance to some of the simplest webs produced by spiders, which alert the spider and slow the prey down enough that the spider has time to attack it, thereby enhancing its effective range of capture. These accumulations of lines may form an analogue to a simple web, although the crab spiders themselves almost certainly arose from web-building forms. The general absence of webs is thus a derived condition (see Shear 1986). As I noted earlier, the sole webs thus far known from crab spiders (*Diaea* sp.—Jackson et al. 1995) may have been derived from this type of construction. We have observed that bees occasionally are tripped by these lines, which are not sticky. By chance alone, one would predict that these lines would increase the success rate of prey capture by the spiders. To date, however, we have not found any difference in capture rates of tripped and untripped bees (Morse and Fritz 1989). Indeed, we made only four observations of entanglement by lines that exceeded 5 seconds in 285 hours of observations, three of honey bees, and one of a bumble bee. In none of these instances was a spider close enough to the bee that it would have had an opportunity to capture it. These lines do not present a measurable danger to the bees, and our data provide no support for such an origin of webs.

These constantly laid lines provide the spiders with an anchor, such that if they fall from a site, they will not become measurably displaced from a favorable area. They serve an especially valuable role when a spider contacts a prey item large enough to bowl the spider out of its hunting site into the litter below. Such an encounter would greatly decrease the probability of the spider retaining contact with the prey, cause it to lose its favorable hunting site, and expose it to the danger of predators in the litter layer. Not only does the dragline prevent these unfavorable outcomes, but the flexibility of the lines is so great that a large insect such as a bumble bee will merely displace the spider and its bee prey a few cm below the hunting site suspended in the air. Under these circumstances, the prey item cannot gain any purchase on the substrate that would help it to bolt free of the spider, and the spider can quickly dispatch it with its extremely potent toxin.

Synthesis

Studies of foraging have usually concentrated on animals that were difficult to follow over significant periods of a life cycle. For that reason few relatively direct estimates of lifetime fitness consequences of foraging behavior exist. *Misumena* provides one of the best opportunities yet available for fitness-related studies because of its relatively simple, sit-and-wait hunting regime focused on flowers and large insects that can both be readily measured and manipulated over large parts of a life cycle. Patch-choice decisions assume far greater importance than diet decisions for *Misumena*, a consequence of the importance of selecting sites that attract many relatively large prey (e.g., bumble bees for adult females).

Advantages of the patch-choice patterns of some adult female *Misumena* are starkly superior to others. This raises the question of how this suboptimal behavior is retained within the population, given the intensity of selection apparently acting against such individuals. This apparent paradox focuses attention on other parts of the life cycle, including the early instars. Success in prey capture also differs there, with the most successful individuals increasing in size some 2.5 orders of magnitude over periods of less than three weeks. Whether experiences in the early instars strongly affect adult behavior is a key question explored in detail in Chapter 6. The value of prey species shifts strikingly over ontogeny, as best seen by changes in profitability of small hover flies, one of the few species taken by all stages of *Misumena*.

During ontogeny the basis for patch-choice changes radically from one in which innate traits prevail, resulting in a preference of the young for goldenrod flowers, to one in which choices of hunting sites are dictated primarily by prey visitation. In contrast, no striking differences in prey selectivity take place: diet choice is a much less important foraging variable for *Misumena* than patch choice. Throughout their life cycle these spiders do not exhibit strong prey choices, although adults fail to respond to some of the very smallest prey used by early instars (thrips, etc.), whether by choice or inability to detect them. In contrast, early instars cannot cope with the large prey of the adults, such as bumble bees, but will readily attack prey at the outside limits of their abilities.

The picture for males does not differ measurably from that of females during the first half of ontogeny as far as we can yet tell. In fact, its similarity is revealing. However, as adults, the males exhibit a far more active time-use profile than either adult females or females of their same instar. The patch choices of the males resemble those of the females, with the result that they often occupy the same sites as the females.

ℐ 4

Fitness Payoffs

$\mathcal{I}$F ONE THINKS OF LIFE history tactics as the ways by which organisms allocate the available resource pool to different parts of the life cycle, and foraging tactics as the ways by which they set the size of the resource pool via acquisition, the relationship between the two areas becomes transparent (Morse and Stephens 1996). Then it becomes logical to ask how foraging decisions and resulting success influence basic life history variables of fecundity and survival. Although the link between foraging behavior and fitness is the central assumption of optimal foraging theory, as I have noted it seldom is adequately tested in the field. A comprehensive analysis will contain both life history variables measured in the field and experimental manipulation of the variables necessary to establish the causative basis for selection (Grafen 1988; Wade and Kalisz 1990). Suitable species are required for investigating and integrating foraging behavior and its life history consequences. *Misumena* is a remarkably good species for this type of analysis because of the relative ease of both gathering information on its basic life history parameters and its tractability for performing critical experiments in the field.

Lifetime Fitness Payoffs versus Partial Fitness Payoffs

In this book I am particularly interested in the consequences of lifetime fitness as opposed to partial fitness payoffs. Partial fitness might, for instance, refer to factors facilitating success during a certain part of the life cycle, such as the reproductive season and the overwintering period. Investigators working on partial fitness components sometimes tout their results as providing the information key to fitness considerations of a particular trait. Some workers may argue that such partial fitness measures will provide the information necessary to evaluate a particular component of fitness. One cannot be confident, however, of its broader significance without extending the analysis to incorporate the rest of the life cycle; otherwise, such analyses are vulnerable to counteracting tradeoffs. More often than not, partial fitness components involve species from which it would be extremely difficult or impossible to obtain more comprehensive estimates of lifetime fitness.

Rich in information and detail as iteroparous species (those that breed more than once) may be, this very wealth of variation stands ready to thwart most attempts to measure lifetime fitness, in the present instance as it relates to foraging variables. Lifetime fitness studies have been conducted on laboratory populations (e.g., Lemon 1991), thereby skirting the difficulties of field studies, but necessarily conducting them in an artificial environment. I believe that our field-focused study of *Misumena* (Morse and Stephens 1996) presents the most comprehensive effort to estimate the lifetime fitness consequences of differential success in foraging behavior that incorporates most or all of the major life history factors.

Since optimality thinking entered behavioral ecology through studies of foraging, not surprisingly this work dominated early studies and thinking related to lifetime fitness. Important concerns thus included the fitness consequences of a particular foraging strategy, which were typically evaluated in terms of energy gain/time. Though a reasonable estimate (see Chapter 3), that measure only provides an extremely indirect estimate of lifetime fitness, and the opportunities for countering tradeoffs seem legion. Of greater concern, these measures of energy gain/time came to be viewed as appropriate for the purpose (see Houston and McNamara 1999).

Other Studies of Foraging and Fitness

Laboratory experiments can often provide the way through difficult problems in studies of behavioral and evolutionary ecology. However, since the basis for such studies is to understand the animal in its natural world, they must be carefully conceived and interpreted, and ideally be rooted with observations from the field. By virtue of the unnatural conditions they necessarily introduce, laboratory experiments may otherwise produce results irrelevant to the setting in which the subject evolved.

Laboratory studies attempt to control variables to a workable degree. Since birds and mammals, favorite subjects of life history studies, typically present unworkable problems in the field, by virtue of their potentially long lifespan, multiple breeding episodes, and resource exploitation patterns that require harvesting widely dispersed resources, laboratory studies offer only possible solutions. Such controlled studies bring these and other variables within the range that they can be monitored and measured. Obviously, however, they are highly artificial because some of the most critical variables (e.g., dispersal) have been grossly restrained. Although these studies are useful in providing explicit numbers for otherwise impossible-to-obtain variables, it remains unclear exactly how to interpret them. Such studies would be highly informative if performed to compare with the natural situation, but the fact remains that they have probably been performed because a field approach is not feasible. Comparison with partial fitness measures from the wild is another possibility, as long as these partial measurements are not taken as definitive answers to the basic fitness problem at hand. Thus, such studies are best considered in the context of the insights they may bring to bear upon lifetime fitness issues, rather than any estimate of patterns exhibited in the field.

Lemon (1991, 1993) conducted an aviary study of lifetime fitness on captive zebra finches *Taeniopygia guttata*. He controlled the energy intake of the birds and concluded that his results supported the foraging theory assumption that a direct relationship exists between population growth and net rate of energy gain. Although this relationship, in which net rate of energy gain is maximized, would hold in an ideal situation, efficiency maximization in the wild does not guarantee success. Fitness relates to the end result, rather than the way it is achieved. An animal could maximize its fitness either by maximizing its input of resources or by directing aggression at its competitors. Furthermore, tight regimes such as those run on these birds present a far more predictable world than the one in which zebra finches evolved. In the field zebra finches are among the most highly nomadic of bird species and are explosive breeders that concentrate on flushes of resources brought by rain in the Australian Outback. These outbursts are almost singularly unpredictable in both space and time and generate what are perhaps the most opportunistic lifestyles known among higher vertebrates (Serventy 1971).

Lemon (1993) did not indicate the origin of his birds, but he noted that they had been bred in captivity for several generations. Thus, the degree to which artificial selection for aviary conditions modified critical aspects of their behavior can only be surmised. The birds exhibited a significant heritability for foraging patch choice (Lemon 1993), consistent with fitness advantages for individuals that make patch-choice decisions similar to those of their mother. However, it is critical to know that these differences do not come about primarily as a consequence of maternal effects, which Lemon could not eliminate. Also, the consequences of selection on animals that have evolved under some of the likely most plastic selective regimes possible may obscure the laboratory results. Thus, Lemon's study provides valuable insights and supports certain assumptions of foraging theory, but does not permit a critical evaluation of the lifetime fitness consequences of this behavior.

A field study of lifetime fitness on Columbian ground squirrels *Spermophilus columbianus* by Ritchie (1988, 1990, 1991) evaluated the ability of these squirrels to achieve an optimal diet, a measure of the proportion of monocots and dicots taken. Squirrels that came closest to the predicted diet gained more mass than others, had higher survivorship, and produced larger litters. By the third year they had achieved six times the reproductive output of the deviators, strongly supporting a relationship between predicted optimal diet and lifetime fitness, although Ritchie (1990) could not test experimentally most of the key variables or estimate recruitment into the next generation.

To measure the inheritance of this behavior, Ritchie (1991) then compared the foraging performances of mothers and offspring—both those he separated at weaning and those allowed to remain with their mother. Results for the two groups were relatively similar, except that the separated squirrels foraged more variably than did those not separated. He could not eliminate possible maternal effects, so it remains unclear whether the behavior has a genetic basis. Neither could he determine how the deviators are maintained in the face of strong selection (Ritchie 1990).

Both studies are thus notable in realistically addressing the central issue of lifetime fitness as it relates to key foraging variables. At the same time, however, they illustrate the difficulties involved in carrying out this type of study.

Misumena—Lifetime Fitness

One of the most attractive aspects of working with *Misumena* is that it permits relatively direct and accurate estimates of lifetime fitness in relation to the most important potentially limiting variables, and under realistic conditions. Obtaining these measures requires far more work than making simple calculations of energy gain/time, instead incorporating a wide range of evaluations made at key points throughout an animal's life cycle. As I have already noted, it is unrealistic—indeed, practically impossible—to monitor individuals of long-lived, multiparous generalist species in the detail necessary to obtain all of the information critical to lifetime fitness. *Misumena* has a far simpler life cycle than those species, even allowing us to treat several of these critical variables as constants.

I now outline aspects of *Misumena*'s life cycle critical to its lifetime fitness. Since I am particularly interested in the relationship between prey capture of adult females and recruitment of resultant young into the next reproductive cohort, I begin with the foraging activities of the adult females. I consider the bases for success in capture and how they translate into gain in mass and allocation to eggs. Placement of eggs may strongly affect success, both of the eggs and the emerging young. These factors include danger of predation and opportunities for dispersal. Condition of the female may play an important role here, maternal effects continuing well into the next generation. I then proceed to a brief discussion of the costs incurred by juveniles before evaluating the relative importance of the various factors covered in this section.

Prey Caught

Highly effective prey capture is a basic aspect of fitness for female *Misumena*, though by no means is it the only issue of importance. However, as noted in Chapter 3, although prey capture dictates the size of the energy pool available to adult females, in *Misumena* it depends on choosing the sites that attract the most large insects. These sites provide the spiders with the greatest number of opportunities to attack their extremely difficult-to-capture principal prey. Prey capture of the adult females focuses strongly on large prey, for without them, the spiders probably cannot accumulate sufficient biomass to lay a clutch of eggs. It is also questionable whether penultimate females can accumulate enough biomass to molt into the adult stage without taking relatively large prey. However, penultimates, as far as we can ascertain, cannot subdue a bumble bee—at least, we have only once seen them do so in 30 years of intensive field work. At the very least, these events are rare enough to exert little selective pressure on the system. The matter of particular concern here is whether the penultimates can make kills of maximum possible size often enough to reach adulthood.

Biomass Accumulated

Rates of biomass accumulation vary markedly among adult females within any given year (much more than between years) in accordance with the sites they occupy (Morse and Fritz 1982). Increases in mass thus differ greatly among adult females depending on whether they hunt on prime-condition milkweed, pasture rose, or goldenrod. Suboptimal sites (those that attract few prey because of the minimal rewards obtained from hunting on them) of all of these species yield poor results (Morse 1981). Although we do not know whether the spiders would choose poor- or middle-quality milkweed sites over high-quality pasture rose sites, we predict that they would select roses in preference to the poor milkweed sites, at least while pollen remains, solely as a consequence of the number of prey that the different flowers attract.

In the best of situations adult females hunting on high-quality milkweed umbels rapidly increase in mass when they capture large bumble bees, with the greatest gains reaching 50 mg/d for three days, though most gain at a much slower rate (Figure 4.1). The largest bees are al-

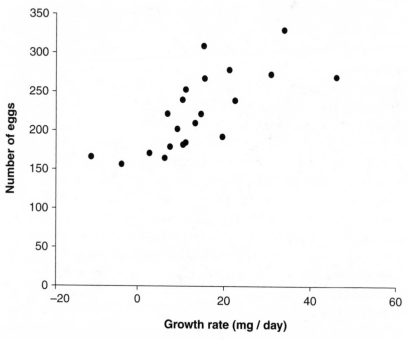

Figure 4.1. Gains in mass of adult females followed for several days in the field and weighed periodically, and their relationship to eventual egg production. Redrawn from Fritz and Morse (1985). With kind permission from Springer Science and Business Media.

most invariably *B. terricola,* and the spiders gain significantly more mass when feeding on this species than on any of their other prey, including *B. vagans* (Fritz and Morse 1985). Bumble bees provide such large resources that they temporarily satiate the spiders. The spiders usually will not capture more than one bumble bee every other day, though some may take an additional bee on the following day. An exceptional individual processed three large bumble bees in two days, increasing in mass by over 150 mg in just over 24 hours, but it died shortly after! Though a single observation, when combined with the large number of observations of adult females taking prey at lower rates, it suggests a clear limit to the rate at which the spiders can process prey, beyond which they cannot proceed without catastrophic consequences. This hypothetical limit brings to attention the report of Higgins and Rankin (2001), who found that unusually high intake rates by earlier instars of the orb-weaving spider, *Nephila clavipes,* resulted in disproportionately high mortality at their next molt.

At the other end of the spectrum, individuals that make poor for-aging choices may even lose mass while hunting on milkweed. In our study 17 percent of these adult females lost weight over the several days they were monitored. Over 20 percent of the individuals closely monitored failed to lay a clutch, and none of them reached 114 mg. This is the lowest immediately pre-laying mass we have yet encoun-tered in the field (Fritz and Morse 1985).

Even at the rate of one bumble bee every other day, the spiders put on surprisingly large, rapid weight gains. The earlier maturing indi-viduals may reach a pre-laying mass of 400 mg under ideal circum-stances, though the standard mass at laying is much lower, in the vicinity of 220 mg. They add as much as 80 to 85 percent of their total mass as adults, the vast majority of it invested in egg production (Fritz and Morse 1985). More often than not, the relatively low average mass at laying is a consequence of flower senescence. Even on the richest hunting site, milkweed, senescence of the flowers often plays a central role in limiting the spiders' mass and eventual clutch size. Milkweed blooms for several days before bumble bees recruit to its flowers, so the bees miss a substantial proportion of the total nectar production. On average they did not recruit to milkweed until 7 to 10 days after the first milkweeds came into bloom, though their numbers then built up quickly (Morse 1982a). Responding directly to the bees as they do, rather than to the flowers, the spiders thus do not regularly position themselves on site when the bees first arrive on the milkweed. It takes the spiders yet more time to recruit to the milkweed in re-sponse to the bees. At the same time, the spiders do not tarry on flowers that attract no prey, as they might if they recruited directly to flower quality.

Egg Mass

Female *Misumena* in the study area lay a single egg mass in the field, whose size and timing depend primarily on their success in capturing prey. The egg mass itself typically makes up a remarkably consistent proportion of the female's total body mass, exhibiting but slight ten-dencies to veer away from linearity at the very lower end of the size range (Figure 4.2). Egg masses are extremely large, running at nearly two-thirds (mean = 65.5 percent) of the total pre-laying biomass. With a coefficient of variation of only 10.7 percent, reproductive effort is

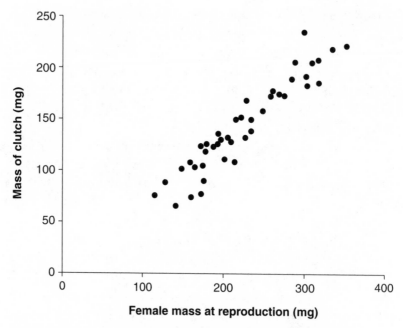

Figure 4.2. Mass of female spiders at reproduction versus mass of clutch. Redrawn from Fritz and Morse (1985). With kind permission of Springer Science and Business Media.

thus by far the least variable reproductive trait (Fritz and Morse 1985). The spiders lay down virtually all of the resources for this egg mass during the adult stage, reemphasizing the importance to the adult females of finding good hunting sites. Although the mass of individual eggs differs greatly within the population (0.40–0.81 mg, mean = 0.63 mg), eggs within a brood differ very little in mass (Fritz and Morse 1985). Egg number is not correlated with the mass of individual eggs ($r^2 = 0.015$); hence, this common tradeoff does not occur in *Misumena* (Figure 4.3). Almost all of the eggs not preyed upon hatch (94.5 percent), and level of success does not differ with the size of the clutch.

Although females produce but a single brood under normal circumstances, several factors suggest that they are capable of laying a second brood. Some individuals that have produced and guarded a brood will subsequently abandon their nest and begin to feed again, increasing measurably in size during the process, though not proceeding past this point. A small proportion of these individuals (0.8 percent: Morse 1994) actually make a second nest, which they then guard, though

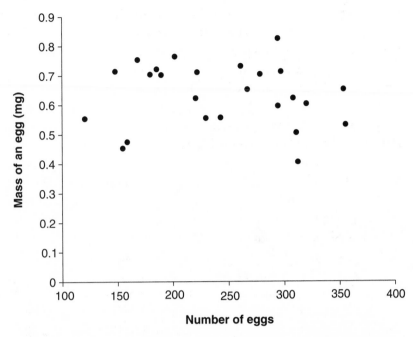

Figure 4.3. Mass of individual eggs versus number of eggs. Redrawn from Fritz and Morse (1985). With kind permission of Springer Science and Business Media.

these nests ("pseudonests") contain no eggs. During one particularly early, warm summer, two individuals we had previously provisioned apparently laid a second brood, though predators attacked these nests, preventing us from evaluating them further. Thus, we wished to determine whether *Misumena* were routinely physiologically capable of producing more than one brood, since many species of spiders lay multiple broods (Foelix 1996a).

We therefore captured the largest penultimate females at the very beginning of the season, fed them as much food as they would take, and mated them as soon as they molted into the adult stage. In this way we could get the spiders to lay egg masses significantly earlier than they usually produced them in the field. Ten days after they laid, we removed the mothers from their broods, remated half of them, and again fed large prey to all of them. Under this regime a majority of these individuals laid second broods. All of the broods from remated individuals were completely fertile. In contrast, half of the group that we did not remate produced completely fertile broods, but the other half produced only partially fertile broods. Thus, the spiders would not always depend on the presence of males at this time to produce a viable second brood, though some would profit from a second mating. Relatively few males remain in the field at the time that females might remate, however (LeGrand and Morse 2000).

Thus, the spiders are physiologically capable of producing more than one brood, but the short season at our northern study area ordinarily prevents them from producing a second one. Under exceptional conditions, combined with an early spring, an individual might produce two successful broods in a summer. However, such second broods would necessarily hatch extremely late in the season. Late broods suffer disproportionately high mortality in overwintering, not to mention the possibility that they, too, will often mature at the late end of the yearly cycle, if they survive. Such offspring will probably lay (single) late clutches themselves, which will experience high mortality. All in all, though of considerable evolutionary interest, second broods, ecologically speaking, make an insignificant contribution to the population. Thus, it serves these spiders best to guard their nests assiduously. In this way they minimize the mortality of their clutch, which in fact they normally do (Morse 1987, 1988a).

The ability to produce a second clutch is nevertheless of considerable interest in light of the high reproductive effort of the first brood, which considerably exceeds that of similarly nesting species in the study area (Morse 1994) and seems appropriate for a semelparous (one reproductive output) species. However, the low metabolic rate of spiders may facilitate survival of the females well past their first brood. The characteristics of the second brood are particularly interesting, in that the mass of the females was similar for both of their broods (marginally and nonsignificantly lighter for the second brood: Morse 1994). The spiders did not compromise on Brood 1 for Brood 2, since over 95 percent of the resources for the second brood were gathered between broods. Furthermore, the reproductive effort in the first brood played no role in determining whether a spider laid a second brood. Thus, spiders made no accommodations for Brood 2, although the ideal conditions presented them could have facilitated their success.

These results suggest that female *Misumena* should regularly produce two clutches south of the study areas, where the season is longer. The much smaller *Misumenops tricuspidatus* regularly produces two or more clutches in Japan (Hukusima and Miyafuji 1970) and China (Zhao, Liu, and Chen 1980; Li and Zhao 1991). Fink (1986) found that the green lynx spider *Peucetia viridans* (Oxyopidae), also a sit-and-wait predator on flowers, typically produced two broods south of a line running across central Florida. The lynx spiders maximized their chances of a second brood by laying their clutch in an area attracting substantial numbers of prey. Thus the female could guard and hunt at the same time, thereby minimizing the hiatus between broods. This strategy would not work for the nest sites chosen by our crab spiders, since they select sites that are often maximally advantageous for the foraging of their young (near goldenrod), but are less favorable sites for any foraging they might undertake. Alternatively, the earliest individuals might be able to lay on flowering milkweed, though they seldom adopt that strategy. Instead they usually move to the periphery (Morse 1993a).

Densities of *Misumena* are low in many areas to the south of our study area. Consequently, we have not investigated the possibility of double-broodedness there. We would expect such individuals to reduce their care (guarding) of the first brood in order to telescope the season as much as possible. We would predict little change from the typical behavior we have outlined in any populations to the north of our study areas, though it is of interest to ask how long they can guard at the northern edge of their range—Labrador, the Yukon, and Alaska (Dondale and Redner 1978).

Eggs are laid at night, which probably minimizes both water balance problems and dangers of predation or parasitism by day-flying insects. We observed two broods being laid at 23:30 and 01:30 (Morse 1985b). The eggs, which adhere in a cluster, are milky white when laid, but within a day they turn leaden gray (= plumbeous—Smithe 1975). They are placed in a nest made by folding the end of a living leaf under the rest of the leaf (Chapter 2). After laying the eggs, the mothers secure the sides of the nest with sheetlike silk. The nests themselves probably prevent excessive desiccation. The sides of the nest leaf are usually gathered rather tightly by early morning, though a substantial proportion has not yet completed the task, which they have usually completed by the end of the morning. Within a few days the spiders often bind their nests to surrounding leaves and stems with strands of silk, which provides the female with cover and often additionally provides cover for the nest. Other lines may extend to nearby leaves, which is the likely consequence of the limited movements made by the mothers, though they spend most of their time on the underside of the nest leaf in a guarding position.

Nest Sites

Not only do the nests face danger from predation or parasitism, but poor nest location could easily discount or negate earlier success by providing little opportunity for the young to recruit to nearby hunting sites. However, sites favorable for offspring over a month in the future will probably not have reached favorable condition when females select their site, forcing them to rely on indirect information (or to make random decisions). Furthermore, since gravid females are slow and heavy, they are less mobile than if they produced smaller clutches, thereby shortening the likely cruising range for selecting a site. Although moving on lines even at this time, the lines often slump to the understory from their weight. They moved only 4 to 10 m between their last foraging site and the site at which they placed their egg sac (Morse 1985b), potentially limiting the options open to them.

Nest location is further tied to the mothers' foraging strategy. If females do not find another hunting site rather quickly after quitting a senescing one, they "give up" and produce an egg sac with the resources they have on hand. Presumably this strategy trumps a risky game of betting they will eventually find another site that allows them to enhance their foraging success. Progressively later clutches likely experience lowered survivorship. As time passes, nest-searching spiders lose mass, not to mention their ongoing vulnerability to predation. During the peak year of a meadow vole, *Microtus pennsylvanicus*, outbreak, twice as many spiders disappeared at the time of nest-site search as in any other year during which we measured this variable. This finding tempted the speculation that vole predation accounted for that difference in loss (Morse 1985b). These rodents are well known to feed extensively on a wide range of invertebrates, as well as on vegetation (Johnson and Johnson 1982).

The spiders nested more often on nonflowering than on flowering milkweed stems. This was probably a consequence of their movements away from their last foraging site, which often took them to the periphery of the clone, where most of the nonflowering stems grew. Those laying on nonflowering stems moved more than twice as far from their last foraging site as those laying on flowering stems (mean = 9.6 vs. 4.2 m), which may be a consequence of those moving shorter distances not traveling far enough to leave the area of flowering stems (Morse 1985b). Movements from the last hunting site to the eventual nest site may represent an effort to reach sites less vulnerable to parasitism, or may simply be the consequence of an unsuccessful effort to find another hunting site.

Contrary to prediction, we found no relationship between type of location used for a nest site and size of the individual involved (Morse 1993a). In general, the spiders selected sites with a higher density of vegetation than that of the immediate vicinity. Goldenrod is the densest, bushiest substrate in the vicinity of optimal *Misumena* foraging sites. When combined with the high probability of using a milkweed leaf away from the flowering milkweed, gravitating to maximum-density substrate in these habitats resulted in the spiders nesting near goldenrod (within 1 m) four times as frequently as predicted by chance (Morse 1993a). This choice produced the added benefit that goldenrod, the preferred foraging substrate of just-emerged young (Morse 2005), will have started to bloom by the time the young emerge. Nevertheless, the short dispersal distance interacts with substrate choice, since only about half of the spiders placed their nests within 1 m of goldenrod. This is a likely consequence of goldenrod's patchy distribution, which, though the dominant flowering plant when the spiderlings emerge, does not totally blanket the habitat. Success of young at nests near and away from goldenrod did not differ up to emergence (Morse 1993a), though dispersal patterns at these two locations differed markedly (see below).

Misumena also located their nests nonrandomly within the plants they frequented. Those placing their nests on the favored substrate, milkweed, typically used leaves smaller than most of the leaves on the stem, frequently the third set from the top (Morse 1985b). These elliptical leaves (mean = 8 cm long, 3.5 cm wide) were large enough that a simple fold provided adequate space to surround the egg mass with flocculent silk and to secure the edges of the leaf tightly with silk, minimizing the space covered only by silk. The spiders cannot manipulate larger leaves in such a precise way, and nests placed on them usually contained considerably larger expanses protected only by silk (Morse 1985b). They also used similar-sized small leaves on other parts of the plant, such as the axillary leaves sometimes produced when an original leaf is lost. Thus, they appeared to respond primarily to the characteristics of the leaf rather than to its position on the stem.

Survivorship of Egg Masses

Female *Misumena* prefer milkweed leaves to any other substrates we have tested (Morse 1985b). It is by far the most frequently chosen nest site in the field, relative to the availability of substrates. Milkweed leaves are pliable, and the spiders can fashion them into tight nests that leave relatively little space between the parts of the leaves that they have folded together. Ichneumonid parasitoids *Trychosis cyperia* attempt to place their eggs in *Misumena*'s nests by attacking the silken areas that bind together the different parts of the leaf, rather than by inserting their ovipositor through the leaf itself (Morse 1988c). Although the spiders do use leaves of other species for nests, they choose alternatives infrequently when milkweed leaves are available (Morse 1990), and those nests suffer considerably higher levels of parasitism than those on milkweed. For example, chokecherry leaves, a substrate seldom used, though shaped very much like milkweed leaves, are stiff and difficult to fashion, with the result that the spiders cannot bend them with the precision seen for nests on milkweeds, leaving substantial parts of the nest covered only by their silk. Not surprisingly, the wasps parasitized experimental chokecherry nests at a considerably higher rate than milkweed nests (Morse 1990).

The timing of parasitism was not random, for the success of nests left by the parents did not differ from those in which they remained to the end (Morse 1988a). Neither did these mothers differ in size from those that remained (Morse 1987). However, the mothers leaving the nest remained, on average, well over one-half of the time to emergence of the spiderlings. Since nests from which we immediately removed the mothers after laying suffered high rates of loss, the majority of nest losses occur over the first half of the nest period (Morse 1988a). Consistent with this point, observations of *Trychosis* at *Misumena* nests (Figure 4.4) suggest a considerably greater sustained interest in nests with recently laid eggs than those in an advanced stage (Morse 1988a).

Figure 4.4. Ichneumonid parasitoid *Trychosis cyperia* confronted by female *Misumena* on the top of her nest. Illustration by Elizabeth Farnsworth.

The wasps may need to inspect the nest in order to determine whether it is suitable for ovipositing. Presumably the parasitoids cannot successfully parasitize the older egg masses, though we do not know whether or not they attempt to do so. However, mothers should not increase their vulnerability to parasitoids by leaving their nests well before the young emerge. Leaving would hypothetically enhance their opportunity to produce a second brood, though, as noted earlier, they appear unable to produce a second brood naturally in our populations owing to serious time constraints. The continued presence of females may also enhance survival against spiderling predators at emergence from the nest (see below). Many females leave their nests when their young do, or immediately prior to them (Morse 1987).

Large egg masses should represent a major boon for maximizing fitness; however, they could also be especially vulnerable to tradeoffs. Large egg masses should provide maximally rewarding targets for would-be predators and parasitoids. In fact, *Trychosis*, the most common nest parasitoid of *Misumena*, and the major source of its egg mortality, lays a single egg in a *Misumena* egg mass, and the resulting offspring consumes the entire egg mass of all but the very largest clutches. (Since a single parasitoid larva attacks the entire brood, it often is referred to as an egg predator.) Thus, only those spiders laying the largest broods have the possibility of attaining any reproductive success if parasitized. Since parasitoids well fed as larvae have the highest productivity themselves (Godfray 1994), if egg-limited, they should concentrate their search on the largest *Misumena* clutches. However, if they spend most of their time searching, they may exploit any nest they find.

Contrary to prediction, large egg masses of *Misumena* suffer lower levels of parasitism from *Trychosis* than do small clutches and from other less common sources of predation as well (Morse 1988a). This difference probably occurs because the large egg masses are guarded by large, active mothers. Recall the nearly linear relationship between size of the egg mass and size of the mother. Thus, rather than a tradeoff penalizing large clutches, large clutches significantly enhance fitness at this stage of the life cycle.

Nevertheless, even large egg masses suffer some mortality from the wasps, as well as from other sources. It is therefore important to ask why *Misumena* literally "puts all of its eggs in one basket." Laying two or more small clutches rather than one large one might appear to be a preferable strategy if the spiders face a substantial chance of losing a clutch. This strategy would be especially favorable if multiple brooding resulted in some of the eggs being laid earlier in the season. This strategy is followed by garden spiders *Argiope aurantia* in the study area rather than after the mother waits until she has amassed enough resources to lay the largest clutch possible. Certain bases for *Misumena*'s strategy of single-brooding might be imagined, however. Variation in nest production could make an important difference. Although *Misumena* place their nest in a leaf, garden spiders hang their egg sacs on the edge of their web or adjacent to it. In this way garden spiders may provide some protection even while they continue to feed at their web. In contrast, *Misumena* produces a nest separated from its foraging area and thus could not provide effective physical protection to a nest and hunt at the same time. As noted, their guarding does improve success.

Misumena and its parasitoids sustain a more complex relationship than suggested here, for only in three years out of six in the principal study area did large size result in a significant advantage, though the overall six-year result was highly significant (Morse 1988a). Over three of those years we monitored two additional sites, permitting us to test this factor further. We obtained the same results each year for all three sites—large females at all three sites were significantly more successful in one of the three years. These year-to-year differences probably resulted from differences in the relative numbers of spiders and parasitoids, which could have a cyclical basis, although our data set is not long enough to verify that point.

As with *Misumena*, the presence of *Trychosis* should depend on the size of the habitat, to the degree that it reflects the size of the host population. According to this argument, the larger the site, the larger the population of parasitoids and the higher the probability of their continued survival. But our data do not suffice to determine whether population size directs the level of parasitism. If *Misumena* nests are usually at a premium, one would expect the parasitoids to expend large amounts of energy and time hunting for these nests. Under these circumstances, the searching ranges of parasitoids should increase and lead to dispersal, consistent with the similarity in parasitism levels at nearby sites from year to year. The next-most-important parasitoid, the scuttle fly *Megaselia* sp. (Diptera: Phoridae), appears to be present in only part of the sites tested, as opposed to the ubiquitous *Trychosis*. The impact of any other parasitoids was clearly negligible.

The interactions between female *Misumena* and *Trychosis* are impressive affairs in which the spiders exhibit a heightened pattern of activity in the presence of the wasps, vigorously attacking when one lands on the nest, often knocking it off in the process (Morse 1988c). Oddly, however, the spiders do not normally kill the wasps—in this entire study we have recorded only one kill of a *Trychosis* by a spider. We do not know what constrains the spiders, though the wasps might produce pheromones, or simply be unpalatable. *Misumena* do not hunt actively while guarding their nests, but they will attack, kill, and consume insects that inadvertently visit the nest (Morse 1987). Although with rare exceptions these prey do not suffice to maintain their body mass (Morse 1987), they do demonstrate the spiders' ability to attack and kill other insects at this time. (The overall frequency of prey capture at nests is 1 per 51 days [Morse 1987], twice the tenure of eggs and young in a nest.) However, if a wasp persists in attacking the nest, the spider's aggressive behavior may wane. Although persistent wasps may thus succeed in parasitizing a clutch, the larger and more vigorous spider mothers experience a lower level of parasitism from the wasps. Thus, even if the defense is only partial, it exerts a measurable and significant effect (Morse 1988a). When engaging the wasps physically, the guarding spiders may knock the wasps off the nests such that they do not always continue their attack (Morse 1988c). The wasps do not appear adept at finding the spiders' sites once they are displaced from them, such as after a spider engages them and knocks them off its nest. Such a displacement may usually suffice to protect the nest from the wasp, something that the large spiders accomplish more effectively than do the smaller ones. Large spiders may, however, simply be in better physical condition, and thereby more active, as a result of their superior size (Morse 1988a).

Little is known about the biology of *Trychosis*, so we do not know what other factors dictate its success. It is currently unknown whether *Trychosis* has alternative hosts (Townes and Townes 1962; H. Townes, personal communication). So far, we have reared *Trychosis* only from *Misumena*. Other members of the genus *Trychosis* are specialist parasitoids (egg predators) of spiders (Townes and Townes 1962; H. Townes, personal communication), and *T. cyperia* probably is so as well. *Trychosis* seems unlikely to attack the other common crab spiders in the area, members of the genus *Xysticus*. In our sampling to date, *Xysticus* spp. appear to support a totally different group of parasitoids/nest predators from *Misumena*, perhaps to be anticipated for the species that place their egg sacs on or near the surface of the ground or litter. Other spiders in the study area with comparably positioned and constructed nests (*Pelegrina insignis, Enoplognatha ovata*) have also exhibited different sets of parasitoids in the limited sampling conducted to date. This finding is not surprising in light of the usual modest host range of spider-attacking parasitoid wasps (Gauld and Bolton 1988). It seems most likely that *Trychosis* would also attack nests of closely related crab spiders, such as members of the genera *Misumenoides and Misumenops*, but they are too rare in our study area to make these comparisons. Given their rarity, we would not expect them to contribute significantly to *Trychosis*'s welfare in our study areas.

The scuttle fly *Megaselia* sp. typically lays several eggs in a *Misumena* egg sac, each appearing to attack a single spider egg (per *Megaselia* egg). The attacked spider eggs "leak" their contents, however, with the result that entire clutches are usually lost to ubiquitous molds. The flies account for only 10 to 15 percent as many attacks on *Misumena* egg masses as does *Trychosis*, and we have not found them in all of the study areas in which we have worked.

Ants (especially *Lasius neoniger*) are the most important conventional nest predators. However, they typically are so aggressive that female *Misumena* do not place egg masses at locations where they will encounter these ants. Where that occurs, either under experimental conditions or when the ants occupy a site after the spider has laid its egg mass, the ants may so persistently attack the guarding female that she abandons the nest. If she has not woven her nest tightly, the ants will find the openings in the nest after she abandons it, enter, and consume the nest contents. They usually do not secure entry to tightly sealed nests and thereby will probably provide some protection from other would-be predators under these circumstances. However, when the young are about to emerge, they make small holes through the silk, and at this time the ants invade the nest and consume the young. We only learned about the potential impact of ants as a result of experimental manipulations of nest sites we ran to learn about other effects. In the process, we inadvertently set up some of the experimental nest sites in areas frequented by ants. The ants probably do not exert a major effect on naturally placed nests, but their overall ubiquity may severely constrain the use of otherwise satisfactory nest sites. Their extensive nectar robbing may further decrease the number of potential prey for the adult female spiders earlier in the season.

Mortality of Emerging Young

Young emerge from their nests approximately 26 days after egg-laying, though the time will differ substantially with the ambient temperature (and with relationships among the young—cannibals: see below). Somewhat over half (ca. 55 percent) of the females guard their nests until the young begin to emerge; many of these females remain until all the young have left the nests. Of the mothers that do not remain until this point, about half have died and the other half have abandoned, the latter remaining for an average of 16 days (Figure 4.5: Morse 1987).

Young whose mothers remain until they emerge enjoy a further advantage of protection at this time as well (Morse 1992a). The major source of danger at emergence comes from jumping spiders, primarily

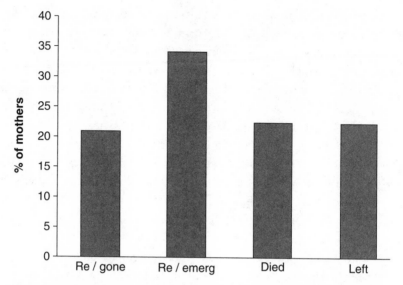

Figure 4.5. Time that females guard their nests. Re/gone—remained until all young left nest, Re/emerg = remained until young emerged, Died = died before young emerged, Left = left before young emerged. Data from Morse (1987).

Pelegrina insignis. Most of the jumping spiders have reached their middle instars when young *Misumena* emerge and will even attack the spiderlings as they emerge from their nests, if their mother is not present. If the spiderlings' mother is present, the jumping spiders capture few if any of the spiderlings—even in her emaciated state she can easily attack and ward off the jumping spiders. In experimental introductions of the jumping spiders to nests guarded by *Misumena* mothers, the mothers not only effectively defended their sites, but occasionally even captured, killed, and consumed the jumping spiders, all of which were considerably smaller than the mothers. Although we did not observe any such attacks under natural conditions, and only once found a mother feeding on a jumping spider, the frequency with which it occurred under experimental conditions made it clear that the mothers offer significant protection at this time.

Loss at Dispersal

Offspring initially disperse from the nest on foot and then either on lines between strands of vegetation or by ballooning (Figure 4.6) (Morse 1993b). Although it is difficult to quantify mortality associated

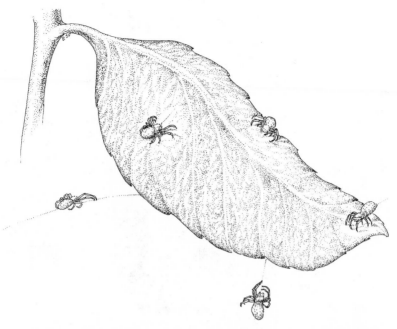

Figure 4.6. Misumena spiderlings dispersing from vicinity of nest. Spiderling at right letting out line in preparation to balloon, one at lower left on end of line let out prior to ballooning, one at lower right dropped on line prior to ballooning. Illustration by Elizabeth Farnsworth.

with dispersal, most observations suggest that it is extensive and largely random in our *Misumena* populations. Differential losses during this period, resulting from variables ranging from offspring size to brood number and placement of the nest, could dwarf the effects occurring at other stages of the life cycle. However, we have no suggestion that these losses deviate from random in relation to foraging patch choice, the behavioral trait in question. Probably nest placement plays the greatest role in dispersal, and nest location is independent of earlier success (Morse 1993a). However, since recruitment to adjacent flowers plays the greatest role in determining whether the young remain at their natal site or balloon away, location of their nest is crucial.

The spiderlings exhibit a rather stereotyped behavioral pattern upon leaving their nests. They typically move upward on their nest plant, usually to the very top, and from there they play out lines to nearby vegetation. If they are on an isolated site, from which their line does not make contact, they often balloon at that point. However, if the parent places the nest near surrounding vegetation, the offsprings'

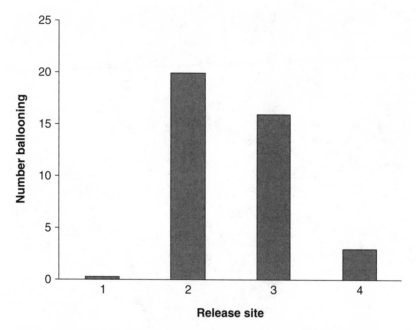

Figure 4.7. Number of spiderlings dispersing by ballooning from various substrates (Morse 1993a). 1 = goldenrod flower, 2 = milkweed leaf, 3 = aster flower, 4 = goldenrod buds. With permission from the Ecological Society of America.

lines will typically contact it. If there is a satisfactory hunting site, often goldenrod in bloom, the spiderlings usually remain to hunt (Figure 4.7). If successful, as when large numbers of dance flies (Empididae) occupy the flowers, the spiderlings often remain for long periods. Typically they stay until the flowers senesce—in one instance for two and one-half weeks. Under these circumstances, the probability of ballooning away from the vicinity of the birth site declines progressively, and spiderlings gaining large amounts of mass will probably remain in the immediate vicinity throughout their lives.

Not infrequently, the spiderlings will repeatedly put out lines without contacting a flowering stem; under these circumstances the spiderlings subsequently balloon. Typically, they will move unsuccessfully on two to three lines before ballooning. Probably the ballooning act ideally serves merely to waft the spiderling a few meters away to a new site, where its chances of finding a profitable site will be enhanced. After all, the general vicinity of where the spiderling finds itself must be suitable, or its parents could not have reproduced there in the first place. The inner fringe of the ballooning frequency distribution ranged from 2–3 to 10 m (Morse 1993b), and some of these spiderlings landed in a profitable patch of goldenrod. However, we have only been able to measure the inner 10 percent of this frequency distribution, with one exception in which we succeeded in measuring a ballooning distance of slightly over 50 m. Most individuals moving beyond 10 m had already ascended so high that they would not land in other parts of the study area, but would have been wafted over the forest canopy or water. Whether or not these spiderlings control their ascent or descent once aloft is unclear (see Suter 1999), though theoretically they should be able to do so (R. B. Suter, pers. comm.). Those that land in the forest canopy presumably could balloon again, potentially into a favorable area.

Tolbert (1977) reported that garden spiderlings *Argiope aurantia* sometimes ballooned as many as six different times. However, once spiderlings have left the vicinity of their takeoff point in the region of our study areas, the probability of landing on a satisfactory hunting site, even after six ballooning events, is very low. More extensive open areas should result in higher "reballooning" opportunities. The frequency of spiderlings in the aerial "plankton" is routinely high, suggesting that ballooning is a strategy fraught with danger. Furthermore, these samples are usually well stocked with crab spiders (Yeargan 1975; Salmon and Horner 1977; Greenstone 1990). Thus, although these aeronauts constitute the propagules that will produce new populations, their likely contribution to the dynamics of an established population is small (Weyman, Sunderland, and Jepson 2002). Bonte et al. (2003) have found that habitat specialist species, which would experience a far patchier landscape (for their purposes) than generalists, show a lower propensity to balloon than do the generalists. This relationship bears some resemblance to the forces that would act on *Misumena* in open habitat and in small patches like those of our study areas.

Costs for Juveniles

Juveniles encounter a wide variety of costs, which we are only now beginning to uncover and quantify. One factor that exacts particularly high costs, but that we have not considered in analyses of the adults, is molting, a dangerous act that frequently ends in mortality. Dangers range from an inability to exit from the previous exoskeleton, often the

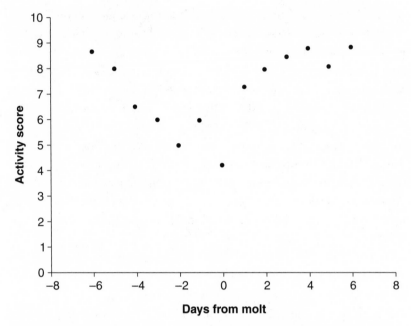

Figure 4.8. Activity levels of juvenile female spiders at different stages of molt cycle. Activity measured in terms of whether an individual moved over a five-minute period. Tests ran for one hour, so the maximum possible score was 12. Modified from Sullivan and Morse (2004).

consequence of an excessively dry environment, to the post-molt period in which the new exoskeleton is still soft, leaving the spider virtually helpless and vulnerable to a wide variety of potential predators, including those that it would normally prey upon itself! In addition, changes associated with molting require large amounts of time during which individuals remain relatively inactive, from roughly five days before the molt to one day after it in *Misumena* (Figure 4.8: Sullivan and Morse 2004), a period comparable to that of many other spiders (Foelix 1996a). Middle-instar female *Misumena* thus spent roughly one-third of their time in the molt phase, during which they took few if any prey, and during a small part of which they were totally helpless and vulnerable (Sullivan and Morse 2004). Because spiders pass through considerably more instars than do many holometabolous insects, it is of interest to ask why they do not also decrease their molts, which would greatly decrease their inactive and vulnerable time. However, failure to do so may well be a fundamental constraint of spiders, in that the eventual size attained appears to bear a close relationship to the number of instars experienced. Thus, although male *Misumena* may have decreased their number of instars, it has probably come about in the process of diminishing their size (Higgins and Rankin 2001).

Which Are the Most Important Variables?

The combined work required to measure the likely variables involved in the selective process—12 summers of work—far exceeded what we could conduct simultaneously. Arnold (1983) pointed out the virtues of path analysis for combining studies that would be infeasible to perform as single efforts in their own right, such as this one. Path analysis (Wright 1968; Li 1975) provides an excellent way of integrating the experimental and observational studies outlined in the preceding sections. It permits us to establish the magnitude, correlations, and causal relationships of key foraging and life history variables and their roles in lifetime fitness. In the process, it allows us to establish the strengths of different causal pathways between the dependent and independent variables.

A chain of causal events (path) proceeding from maternal hunting patch choice through maternal mass, clutch mass, and number of dispersal-age young was the dominant element for *Misumena* and explained a large part of the variation (Figure 4.9: Morse and Stephens 1996). Other paths that incorporated parasitism of the egg mass and predation of young leaving the nests made only small impacts on the results, though these results should not be taken to suggest that parasitism and predation play no role in *Misumena*'s life. To the contrary, they are potentially potent selective factors, rather than factors that normally dominate the dynamics of these populations. No significant tradeoffs were found, primarily because a single factor, maternal mass (a maternal effect) resulting from foraging success, provided major benefits for successive life history stages. Since differences in the numbers of eggs, egg loss, and mortality at dispersal resulted almost entirely from differences in maternal mass, they are controlled by the maternal generation and thus are appropriately attributed to the lifetime fitnesses of the mothers rather than those of the offspring. The results

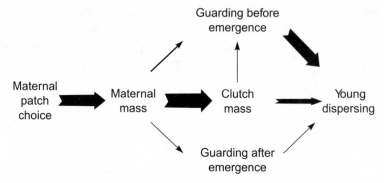

Figure 4.9. Path analysis of *Misumena's* key life history events (Morse and Stephens 1996). Path coefficients convey the direct effect of one variable on the variation in another, holding other variables constant. Here I present them as correlations whose strength is denoted by the width of the arrows. The correlations themselves are standardized partial regression coefficients obtained from multiple regression on standardized variables of each of the dependent variables. Kingsolver and Schemske (1991) present a helpful summary of the uses of path analyses, from which I have drawn here.

demonstrate how a single important factor—in this instance foraging patch choice—can dictate subsequent success through many subsequent stages of a life cycle. Grafen (1988) appropriately termed this ripple effect the "silver spoon effect."

Effects at Subsequent Stages

The overall impression that emerges from the preceding sections is that selection for quantity dominates selection for quality, at least in terms of traits manifesting themselves in the young. Characteristics of the mother appear to dictate the differential success of the young. Whether this advantage affects traits that manifest themselves later in ontogeny, either in the subsequent instars or adult stage, needs to be ascertained, however. The female's ability to enhance the number of her surviving offspring is a consequence of her quality, in turn derived largely from her foraging success as an adult, and, to a lesser degree, earlier in life. Thus, maternal effects, generated largely through the mothers' differential ability to protect their young (Morse and Stephens 1996), play a very important role in the success of the spiderlings.

The provisional conclusion that offspring quantity rather than quality constitutes the primary benefit resulting from foraging success does not constitute a complete answer in itself; otherwise it would be difficult to imagine selection acting on these centrally important traits. However, since the young remain under the control of their mother to various degrees during these earliest stages, this force would minimize selection for traits that distinguish among them. Thus, traits differentiating individuals would more likely occur subsequent to dispersal. However, between-brood differences do occur in both size and activity that could affect success of the second instars.

We are currently exploring in detail whether factors affecting offspring "quality" arise largely after the spiderlings disperse. Young of different broods varied systematically in substrate selection and in their tendency to shift these traits in response to environmental variables (Morse 2000a). Yet, they exhibited low variation in their response to small insect prey at this time, though they improved their responses to a variety of stimuli, given the opportunity (Morse 2000b). Presented with a standard diet *(Drosophila melanogaster)*, they retained differences in mass exhibited at birth (Morse 2000b) through the second and into the third instar, at which point we terminated the study. These differences could have important consequences for them in the future, for both survival (through the winter) and fecundity (for the size of the brood eventually produced).

A largely unexplored aspect involves predation on the newly independent young. In the absence of continual recruitment, their numbers at a site decline fairly rapidly. These sites abound with potential predators of the spiderlings, including their own sibs, or other conspecifics, especially somewhat older ones. Since these young will even take prey considerably larger than themselves (Erickson and Morse 1997; Morse 1998), predation events are likely to be a consequence of who grabs whom first! However, among the most likely predators at this time are jumping spiders, especially *Pelegrina insignis*, the same species that captured spiderlings as they left their nests. We attempted to quantify the effect of jumping spiders by releasing dyed spiderling *Misumena* and subsequently capturing jumping spiders at these sites to survey which ones' mouthparts were covered with dye. (If spiderlings are dusted with powdered micronite dye before release, particles of this dye will remain on them until they molt. In the meantime their predators will accumulate this dye on their bodies as well, especially their mouthparts.) Preliminary results suggest that the jumping spiders capture these spiderlings at a rate that may account for much of the decline in their numbers observed at a site. Other possible predators of spiderlings on these flowers include damsel bugs (Nabidae), assassin bugs (Reduviidae), various other hunting spiders, and yellowjackets (*Vespula* spp.).

Thus, selection appears to act regularly at some stages of the life cycle but not at others, or if so at a level too weak to measure accurately. Furthermore, the strongest secondary effect we obtained, differential success of female size in enhancing nesting success, held in some years but not in others, and was stronger at some sites than at others (Morse 1988a; Morse and Stephens 1996). These results are perhaps not startling, but explicit information is seldom available for multiple traits in fitness studies, so that they cannot be readily evaluated. Some of the traits for which we found no differences could yield significant differences in another year, or under different circumstances. It would be surprising if these traits did not yield significant differences somewhere in space or time, for we assume that they have at one time or another been under either direct or indirect selection. These results further demonstrate the danger of drawing conclusions about lifetime fitness based on partial fitness estimates, especially if not adequately replicated.

Cannibalistic Broods

A small proportion of *Misumena* broods in our populations exhibit a strong tendency toward cannibalism. Because of their relative rarity, we have largely ignored them thus far, incorporating them in their appropriate mortality considerations. However, they provide important opportunities to investigate relationships between number and size of eggs and offspring.

In a few broods, relatively small numbers of spiderlings emerge from the nest well after the normal time of departure. These individuals are noticeably larger than those from other broods, and in some instances the graded weights of the young even make it possible to infer how many sibs that individual has consumed! Since early success is problematic among the emerging young, and initial size advantages were likely retained under many circumstances (Morse 2000b), cannibalism could sometimes prove adaptive. We do not know whether the eggs (and consequent emerging young) from these broods differ in size from the mean of noncannibalistic broods. However, since the eggs in *Misumena* broods are remarkably similar in size (Morse 1993b), we suspect that the cannibalistic individuals come from ordinary-sized individuals as well. (If the cannibalistic individuals came from broods with especially small eggs, this habit could serve as a rather inefficient means of size compensation, though information to date does not suggest that the cannibals have this sort of origin.) These consumed young serve their sibs in a manner equivalent to trophic eggs.

Some circumstances might favor large emerging young, especially if chances are high that a first meal will not be forthcoming. When newly emerged young do not quickly find satisfactory foraging substrates, they commonly balloon, but the larger these individuals become, the lower the probability of ballooning (Morse 1993b). Cannibalism could thus serve as an alternative sedentary strategy, which enhances the chance of finding a satisfactory hunting site near the emergence site. If the site permitted their parents to reach maturity, and if their mother found enough food to produce the clutch from which they emerged, a sedentary strategy should prevail. Egg size is totally independent of clutch mass in *Misumena* (Fritz and Morse 1985; Morse and Stephens 1996) and presumably an attribute of the individual female. However, since these cannibalistic broods are relatively uncommon, they may not have been represented in the samples we have analyzed for this variable.

Since young from large eggs maintain their initial weight advantage over those that emerge from smaller eggs when offered similar diets (Morse 2000b), cannibalistic young would presumably acquire some of these same advantages. They would also emerge from their nests with some feeding experience, although we do not know whether such experience would advance their success in capturing other prey. They might even reach a large enough size as cannibals to be able to molt into the third instar without further sustenance. In this case they might be able to handle larger prey than before by virtue of their enlarged mouthparts and raptorial forelimbs. Given the usual abundance of small hover flies on goldenrod, a prey item that second instars experience great difficulty in capturing (Chapter 3), these third instars might profit initially on prey not easily captured if they emerged as small second instars.

If these cannibalistic broods have a genetic basis (which we have not demonstrated), cannibalism may act as a selective factor in the population, favoring large young under some circumstances. Then, one might expect to see it eventually suppressed and larger eggs favored. However, eggs of most clutches do not currently approach the maximum egg size found in our populations, suggesting that selection toward larger eggs is not strong. Furthermore, several of these cannibalistic young reach body masses at least twice those of the largest young to emerge directly from eggs (> 1.5 mg vs. 0.8 mg, respectively). That result suggests that cannibalism allows a quantum leap in size over that possible through modification of egg size, and that any gradual selection toward larger egg size would fail due to other pressures within the population.

We have not been able to attribute cannibalism in these broods to any environmental factors, in spite of its likely utility in some environmental conditions. Hence, our inclination is to suggest a direct genetic basis. Mayntz and Toft (2006) suggested that wolf spiders *Pardosa prativaga* were dimorphic for a cannibalistic propensity, and some individuals given a cannibalistic diet fared better than those on a *Drosophila* diet. The rarity of cannibalism in *Misumena* populations necessitates a strong observational effort in order to answer these questions. With rare exception we have only noted this behavior during years in which we monitored large numbers of broods, a full-time endeavor in its own right. Nevertheless, such an investigation might richly reward the effort, since it would likely provide broader insight into the issue of cannibalism itself, the conditions under which the predisposition evolves, and its relationship to trophic eggs.

Synthesis

Although by definition lifetime fitness plays the central role in determining the success of individuals, comprehensive measures of this proposition that incorporate both survival and fecundity components are rare in the literature, especially as they relate to key foraging variables. The standard procedure has been to assume that the critical conditions are met if results are consistent with such an interpretation, rather than to measure and manipulate multiple variables.

Misumena provides an excellent model for studies of lifetime fitness in a natural population, largely because of life cycle traits that permit a considerable number of simplifying conditions. For instance, semelparity and the habit of taking small numbers of large prey make it possible to control two of the biggest problems facing investigators studying this phenomenon. The most important life history event of a female *Misumena*, from the viewpoint of fitness consequences, is its success in selecting sites at which to hunt as an adult. Making good decisions at this point will maximize its chances of obtaining abundant food that ultimately results in a large clutch of eggs. Tradeoffs, in the form of disproportionate rates of mortality on large clutches, do not occur at later stages of the life cycle and sometimes even modestly enhance the advantages gained from hunting patch choice (Morse and Stephens 1996). These subsequent stages are enhanced as byproducts of the original gains; the enhanced size of the adult female secures gains for the subsequent stages, which in this instance make up an aspect of maternal effects.

Tradeoffs are not inevitable components of all life history events (Bell 1984; Scheiner, Caplan, and Lyman 1989; Arnold 1992), and the silver spoon effect facilitates the retention of multiple advantages passed on along the life cycle. Noordwijk and de Jong (1986) and Houle (1991) have pointed out that if resource acquisition exceeds variation in allocation of resources to a trait associated with a potential tradeoff, such tradeoffs may be erased or even be replaced by positive correlations. *Misumena*'s high variation in foraging success, combined with its high and relatively invariable reproductive effort, could account for the absence of tradeoffs in successive stages of the life cycle (Morse and Stephens 1996). One point deserving more credit is that some individuals are more successful than others, at least under the conditions experienced, and that a good boost in success at one stage or another may provide all that is needed for the enhanced success of subsequent stages. At present, it is not clear how frequently such tradeoffs occur. However, the apparent absence of negative correlations between life history traits makes *Misumena* an ideal species for such an investigation.

Misumena's life history traits and the accompanying environmental variables suggest important limitations that compromise its response, but simplify its study. These traits and environmental factors include a high, constant reproductive effort; absence of a correlation between egg size and clutch size; an inability to overwinter as adults; and lack of time to produce more than one brood. In particular, the low variance in reproductive effort does not suggest a high genetic propensity to shift from its facultatively semelparous condition. At the same time, since *Misumena* can produce a second brood under ideal experimental conditions—conditions unlikely to occur in our study area—the currently observed set of traits likely serves it better than any alternative. The causal relationships among seasonality, semelparity, and reproductive effort certainly demand further attention.

More investigations of this sort are needed in order to obtain a comprehensive understanding of how the interacting variables we have discussed work under natural circumstances (Kamil, Krebs, and Pulliam 1987). Not only will enhanced information on other systems provide this understanding, but it will offer important insights into the role of genetically based tradeoffs. This issue has proved a contentious one in discussions of the evolution of life histories. Particularly needed is information about the respective contributions of genetic and environmental information in tempering decision making of the sort required here.

✄ 5

Constraints on Success

𝒯HE BASIC FORAGING MODELS MAKE predictions that
sometimes are impossible to achieve. Nevertheless, these predictions
set goals that foragers should emulate as closely as possible. The difference between the predicted and the observed intake provides a convenient scale for comparing the success of individuals. Their performances often are much lower than the predictions, as a consequence of
the ubiquity and importance of constraints in the natural world. Several factors may account for why individuals do not closely match the
predictions; I will focus on four of them (competitors, predators, physical limitations, and stochastic variables), which interact in most circumstances. The relative importance of these factors may differ with
the instar, sex, and situation in which an individual finds itself. They
will all be subject to environmental and genetic variables, which jointly
act to dictate the responses exhibited.

The Degree to Which Predicted Success Is Achieved

Foraging may thus impose many limitations on individuals, which are
clearly illustrated by adult *Misumena* females. Since foraging dominates
their activities, in accordance with the importance of acquiring resources for reproduction, these constraints may assume great importance for their success and eventual fitness. Foraging success also as-

sumes a high premium in other parts of *Misumena*'s life cycle. Thus, I will also evaluate the role of constraints in limiting the success of these stages and their implications for the adults.

The degree to which an individual conforms to simple predictions is in part a consequence of the principle of allocation (see Stephens and Krebs 1986), an evaluation of the effort put into the several factors noted earlier. If the focus is on foraging, it is easy to forget that other demands simultaneously compete for an animal's "attention." However, these demands may precipitate behavior inappropriate for maximizing foraging efficiency—if a small foraging bird must scan the sky for predators, its time and attention for foraging will decline. Traditionally, it has been assumed that a complex of genes controls the predisposition for a suite of such interdependent traits, with the animal exhibiting varying amounts of phenotypic plasticity, itself likely to be under genetic control. However, examples now exist of single genes controlling at least significant aspects of foraging behavior. Notable among them is the foraging gene *(for)* that controls the "rover" and "sitter" genotypes in *Drosophila melanogaster* (Sokolowski 1980, 1998) and likely in other species as well (see Chapter 6).

Here it is most appropriate to focus on the bases for nonconformity to simple predictions. Identifying and quantifying these nonconformities provides useful information for evaluating the nature and magnitude of constraints that control efficiency at any task. Gould and Lewontin (1979) made the important point that imperfect fits to predictions, often treated as tradeoffs, may alternatively be a consequence of organisms being integrated wholes that cannot be cleanly broken down into independent and separately optimized parts (i.e., adaptation has limits that cannot extend beyond the genetic information available to it). This is, in effect, a plea for thoughtful alternative hypothesis testing. Under these circumstances, a response could be an expression of a genotype incapable of reacting directly to a contingency, rather than a compromise between two flexible variables. In this situation phenotypic plasticity is unable to ease a constraint. Genetic drift could even have eliminated genetic variance that otherwise would have permitted an adaptive response. Thus, although I consider the concept of tradeoffs to be useful, it is inappropriate simply to assume their presence. They may well provide a plausible explanation of an observed relationship, but independent confirmation is required to verify a relationship. It is important to keep this point in mind during the following discussion, which assumes adequate genetic variation.

The question of whether a phenotypic approach produces valid measurements has generated discussion and controversy for some time, as is illustrated in papers written two decades ago by Reznick (1985) and Partridge and Harvey (1988). At issue are the answers to questions about the frequency and importance of genetic and environmental factors in natural systems and how they differ with spatial and temporal variables, how constraints resulting from their effects operate, and how they contribute to the "tempo and mode" (Simpson 1944) of evolutionary change.

Constraints may result in tradeoffs where maximizing efficiency of productivity comes at maximum risk, with predation usually being the risk under consideration. Much of the attention paid to predator avoidance has focused on birds and mammals (see Lima and Dill 1990). The sparrow-brushpile paradox (Pulliam and Mills 1977; Grubb and Greenwald 1982) provides a stark example of the conditions likely to generate such a tradeoff: as it ventures outward from the brushpile, the sparrow may encounter increasingly rich concentrations of seeds, but in exploiting the seeds, this sparrow exposes itself to progressively greater danger from bird-hunting hawks the farther it ventures from the brushpile. (Interestingly, this condition may have been progressively exacerbated by the sparrow's past foraging efforts, which selectively depleted the seeds closest to the brushpile.) Such birds typically exhibit high levels of vigilance under this condition, frequently scanning for predators and often deigning to leave their safe haven. Thus, pressure from predators may constrain foraging opportunities as a result of facilitating avoidance behavior, with consequent problems for survival of the prey species.

The costs of these avoidance responses, which are likely to depress food intake well below an optimal rate, obviously have a strong selective nature, imposed on them by the predator. Consistent with this argument, the predator-avoidance behavior of some flock-forming birds appeared to decline in the 1960s (Morse 1970), when numbers of hawks declined catastrophically as a result of high DDT concentrations. These birds frequently ignored stimuli that inevitably produced

alarms, such as flyovers at low elevation from forms superficially resembling *Accipiter* hawks (e.g., doves, cuckoos) in areas with normal-
sized hawk populations (Morse 1973).

These ideas have recently enjoyed much greater attention from ecologists, with the recognition that often they may exact demographic effects on populations that even exceed those of direct predation (Lima
1986; Lima and Dill 1990). Such factors, now often masquerading
under the euphemistic term *trait-mediated indirect interactions* (TMII)
are contrasted with *density-mediated indirect interactions* (DMII), classical predation, whose impact is taken to be density-dependent. Trait-
mediated interactions, strategic changes in prey phenotype, behavior,
and the like, in response to the presence of a predator (Preisser, Bolnick, and Benard 2005), may have a developmental, morphological,
physiological, or behavioral basis (Werner and Peacor 2003). The behavioral aspect primarily concerns us here, which may involve foraging
effort, resulting resource income, conversion of resources to progeny,
and consequent vulnerability to other predators (reviewed by Preisser,
Bolnick, and Benard 2005).

It is now becoming clear that predator avoidance behavior is ubiquitous. Ydenberg (1998) has stated that many animals are always at some
predation risk, even if never attacked. Such behaviors as group foraging that decreases uptake efficiency presumably exist because of their
role in decreasing vulnerability to predators, thereby generating a
tradeoff. Since many of these studies have focused on small birds and
mammals, do other groups follow this pattern as well? Do animals with
major foraging demands and minimal contact with prey exhibit
predation-related constraints, and if so, are they subject to change, as
with the bird flocks noted earlier?

Major Types of Constraints

In the context laid out above, I will consider the role of several likely
constraints on foraging, especially as they relate to *Misumena*. They
should not be taken as a complete list of constraints likely to befall foragers, but they provide a sense of the problems an average foraging
animal likely faces. In that sense, *Misumena* is thought to differ from

other species only in the ease with which I can access several of these constraints.

Competitors

Competition may be either intraspecific or interspecific. In defining competition as vying for resources in limiting supply, intraspecific competition should exert a stronger effect on foraging capabilities than interspecific competition because members of the same species will normally overlap in their requirements most heavily. Thus, unless they are heavily outnumbered by heterospecifics, their cumulative effect on each other will be strongest.

Competition, however, occurs when resources are in limited supply; if the resources (prey items, hunting sites, etc.) are not in limited supply, competition cannot be assumed. Ecological theory (Miller 1967) predicts that intraspecific competition will result in exploitation of a broader resource base that includes suboptimal opportunities for some members of the population. That partitioning could come about through physical combat, formation of a dominance hierarchy, or differentially efficient exploitation. Under some circumstances, social hierarchies might exist in the temporary absence of resource limitation. Interspecific competition will cause a niche shift, with the less effective competitor squeezed into suboptimal space, which if inadequate, results in the disappearance of that species.

Intraspecific Competition. Intraspecific competition is unlikely to be a frequent problem for adult female *Misumena*, owing to their generally low densities. Satisfactory foraging sites, defined by the visitation rates of insect prey, greatly exceed numbers of adult *Misumena*, even though the spiders often experience difficulty in accessing these sites. Although relatively sedentary, adult females do move about, especially as the flowers in their hunting sites senesce. Thus, even if they are moving randomly, they should occasionally come in contact with each other. We seldom observe these events in the field, and the reason for that difficulty can be easily demonstrated experimentally. When they come in contact, adult females respond aggressively to each other, with the result that one, usually the smaller, quickly withdraws. No more than a touch of the forelimbs usually suffices to produce this separation. We have not observed active fights or cannibalism in these situations, though it would be hard to imagine that it never occurs. Over the years of this study, we have only twice found one adult feeding on another. However, if one inadvertently places a second individual in a small collection vial (as inevitably occurs in the field, in spite of best intentions!), one always kills the other.

We have no evidence that adult females respond specifically to conspecific draglines under these circumstances, although these lines could provide the cues allowing individuals to remain separated. Work to date (Chapter 9) suggests that either these females do not produce pheromones or the males, at least, do not respond to them (Anderson and Morse 2001; Leonard and Morse 2006). However, we do not know whether this restriction holds for female-female interactions. Females could potentially obtain information from structural cues on the lines, but again we currently have no direct evidence to support this possibility. Perhaps the myriad of lines inevitably present in the environment, usually made by a wide variety of spiders, makes sorting out such potential stimuli a difficult proposition.

The issue of competition among earlier instars is not as clear as that for the adult females. Sometimes early instars are extremely crowded, especially at the foraging sites closest to their nests, usually goldenrod, to which a sizable proportion of a brood may have recruited. Although the first individuals to depart the nest often face difficult decisions in finding hunting sites, those that lay out lines leading to rich foraging sources may provide ready pathways for their sibs to join them. Thus, concentrations of over 100 individuals sometimes accumulate at such sites. However, numbers of spiderlings at these sites usually decrease rather rapidly over time (Morse 1993b). This decrease may take place by random movement, dispersal in response to sites perceived deficient of prey or to contain too many competitors, or it could result from predation on the spiderlings themselves from a number of species, including conspecifics—even sibs. Even at these sites, however, high-quality hunting locations probably are seldom at a premium, though finding high-quality sites in the first place may place a formidable burden on the spiderlings. These sites can absorb large numbers of individuals, and it is questionable whether the spiderlings seriously depress the number of flower visitors to their most common substrate, goldenrod inflorescences. Flowering goldenrod attracts large numbers of small insects, sometimes in swarms that provide clearly superabundant foraging opportunities for the spiderlings. In years of dance fly abundance, several dozen may occupy a goldenrod inflorescence at once.

Subsequent instars are considerably less abundant than the newly emerged second instars; this is a likely consequence of both dispersal and mortality. Overwintering mortality is also high, judging from differences in numbers in autumn and the following spring, which will further lower the numbers of older instars, and the resulting potential for competition.

Interspecific Competition. Misumena's most likely interspecific competitors are other sit-and-wait predators on flowers. They use similar hunting sites, but none are abundant. Given the large number of such superior sites, their impact thus seems likely to be inconsequential. The most frequent include other crab spiders, primarily *Xysticus emertoni*; jumping spiders, primarily *Pelegrina insignis* of a wide variety of instars; ambush bugs (Phymatidae); assassin bugs (Reduviidae); damsel bugs (Nabidae); and preying mantids (Mantidae). They all take prey similar to those of *Misumena*, but combined they seldom achieve a density comparable to *Misumena*, at least of the juveniles. Thus, interactions are likely to be incidental and confined to the occasional mutual claiming of hunting spots. These hunting spots can usually be easily substituted if two individuals reach a common site in a typical area containing many satisfactory sites. In the absence of limiting resources, it is inappropriate to invoke competition. The more realistic possibility is that some of these predators also prey upon *Misumena* itself. This danger probably occurs considerably more frequently among early instars than adult females.

In sum, it appears highly unlikely that competition plays an important role in preventing *Misumena* from achieving strong fits to foraging theory predictions. Although we can cite instances in which other species have usurped *Misumena* from its foraging site, these events take place infrequently. Moreover, their importance is likely to be swamped by other factors that might depress a good fit to the prediction. This conclusion concurs with that of Wise (1984, 1993) for some other spider groups. Thus, competitors are unlikely to modify the general conclusions relating to patch choice and prey choice drawn from models discussed in Chapter 3.

Predators

Predators could prompt the spiders to adopt a defensive strategy that decreases their hunting efficiency. Here I outline some of the predatory pressures experienced by different members of a population and the likelihood that they will constrain foraging behavior. The spiders clearly do exhibit predator-avoidance behavior. For instance, they will quickly dart to cover in response to the rapid movement of a large object, such as a human hand, and they will often partly or completely "bury" themselves within the flowers of an inflorescence. However, some of these behaviors may warrant an alternative interpretation, such as ambushing prey or escaping unfavorable microenvironmental conditions, or they might function in more than one context. As voracious predators themselves, it is also of interest to observe whether *Misumena* in turn engender predator-avoidance behavior from their prey.

Intraspecific Predation (cannibalism). Cannibalism is often treated as an entity separate from predation, but in the present context it seems appropriate to include it under the rubric of predation (the capture and consumption of another animal) by considering it as intraspecific predation (Nishimura and Isoda 2004). Inadvertent cannibalism presents a potential source of mortality among the early instars, though its frequency in the field requires closer attention. Laboratory experiments that we have run with closely confined second instars demonstrate that most broods exhibit very little, if any, cannibalism among either sibs or similar-sized groups of mixed parentage, even under starving conditions. These results resemble those for wolf spiders *Hogna helluo* (Roberts, Taylor, and Uetz 2003) and may be a consequence of nonsibs often being closely related, hence discounting the advantages of nonsib cannibalism (Dong and Polis 1992; Pfennig 1997). Thus, in common with a number of arthropod species (e.g., Dickinson 1992; Schausberger and Croft 2001), *Misumena* appear to exhibit early-instar kin recognition. Yet, these newly emerged individuals will readily take prey at this time, both in laboratory situations and in the wild. In fact, cannibalism may play a central role in density regulation of wolf spiders (Wagner and Wise 1996).

Dispersal is usually thought to mitigate cannibalism, and in its absence cannibalism is considered a normal outcome (in the context of intraspecific predation). Given the ability of these spiderlings to capture prey as large or even larger than they are (Erickson and Morse 1997), the capture of sibs or other conspecifics presents no physical problem. How often this event occurs and under what circumstances are the matters in question. Why spiderlings might largely eschew sibs or other similar-sized conspecifics under confined laboratory conditions, yet take them in the wild, is an unanswered question.

This discrimination could simply be a consequence of constantly renewed recognition. That is, if the spiderlings frequently come in contact with each other, perhaps they will not cannibalize each other, but if they do not remain in close continual contact, they will engage in cannibalism. Many instances of invertebrate learning exhibit rapid decay if they are not frequently reinforced (e.g., Turlings et al. 1993; Abramson 1994).

Such a behavioral pattern would serve the spiderlings well as second instars still in their natal nests. Barth (2002) noted that *Cupiennius salei* spiderlings did not cannibalize sibs until they reached nine days of age, at which point they had exhausted their yolk sac. *Hogna helluo* became intensely cannibalistic after molting into their third instar, a stage at which they have usually dispersed (Roberts, Taylor, and Uetz 2003). However, a few *Misumena* broods are cannibalistic within the nests (Chapter 4), such that undersized broods of varying-sized spiderlings eventually emerge, at times even differing so strikingly in mass that one can confidently estimate the number of sibs they have eaten!

Interspecific Predation. Adult females and late instars. *Misumena* are likely targets of other predators in turn, including the potential competitors of *Misumena* noted earlier. Many of these species take large prey, and hence most *Misumena* are at risk, although large females are considerably less vulnerable than the earlier instars.

Most evidence for predation on adult females is anecdotal, consisting of either single or occasional observations of predation or possible predation. The literature strongly suggests that mud-dauber and spider wasps (Sphecidae, Pompilidae) play an important role in the dynamics of many spider groups, with crab spiders among the most affected. In contrast to our northerly study sites, predatory wasps likely play a major role in the welfare of *Misumena* in many areas. To the south, Muma and Jeffers (1945) and Dorris (1970) found that mud-daubers, especially *Sceliphron caementarium*, took large numbers of late-instar *Misumena*, *Misumenoides*, and *Misumenops*. *Misumena* are much less common in seemingly satisfactory sites south of our study areas (Rhode Island, Maryland), probably because of the regular presence of mud-daubers and spider wasps that focus on crab spiders, though we have not been able to test this supposition. However, mud-daubers are uncommon or absent from our study areas, and we have been unable to document any impact of their presence on *Misumena*. The coastal location of our study sites in Maine probably contributes to their northerly appearing aspect, as numbers of mud-daubers increase immediately inland from the relatively cool coastal aspect. In fact, we have never observed mud-daubers in a study area with an easterly exposure that frequently receives fogs directly from the cold ocean, while we see them infrequently at a westerly exposed and more sheltered site, 10 km away. Although spider wasps often occur in our study areas, almost without exception the species observed concentrate on different families of spiders. Over the 30 years of this study, we have made only two observations of a spider wasp, *Dipogon sayi*, capturing small adult female *Misumena*.

Yellowjackets and bald-faced hornets (*Vespula* spp., *Dolichovespula* spp.: Vespidae) frequently visit the flowers occupied by *Misumena*, usually confining their activities to searching for nectar, though they capture insects there upon occasion. Thus, they might prey upon *Misumena* as well, especially smaller individuals, although we have made no observations of any such captures. Adult *Misumena* actually prey on the wasps under these circumstances, sometimes taking yellowjackets in numbers and occasionally even capturing bald-faced hornets. Certainly these hornets are one of the most formidable potential prey in the study area!

The popular literature reports that vertebrate predators, especially birds, represent a key source of predation on *Misumena* and similar species. Yet we have not documented a single instance of successful or attempted bird predation on these spiders during the entire period of this study, involving thousands of hours of careful observations in the field. If bird predation were an important consideration, we should have observed it over this period. That the colors of *Misumena* are consistent with being cryptic to arthropod, rather than vertebrate, predation at some flowers (see Chapter 2) further suggests the relative unimportance of vertebrates as predators of *Misumena*. Observations in the field make it quite clear why birds are unlikely to play a significant role in predation on *Misumena*. For the most part *Misumena* occupy herbaceous plants that offer little in the way of perch sites for birds, making it difficult for them to obtain easy access to the spiders. Furthermore, most of the sites occupied by these spiders are in open areas, eliminating the presence of some of the most common insectivores, including those that take large numbers of spiders from the vegetation. These insectivores include the most likely bird predators in these circumstances, species that forage while hovering at a source (flycatchers, etc.). In the infrequent instances that such species land on the vegetation frequented by *Misumena*, they do so with difficulty, bending the vegetation over in the process. The spiders respond very quickly to any vibration of their substrate, almost instantly darting out of sight behind the closest available vegetation.

The most promising opportunity for such an attack occurred in one of our study areas in which a power line stretched directly over the middle of a large milkweed clone. A small flycatcher (eastern phoebe: *Sayornis phoebe*) that nested nearby often perched on this line, and when milkweeds were in bloom, they occasionally ventured down to perch on the umbels at the top of the largest milkweed stems (one year only of 18 years we worked at this site). However, the birds experienced considerable difficulty in maintaining their balance on these umbels. Since we were conducting intensive work with all of the adult females on this site and did not lose any of them over this period, we are confident that the phoebes took no *Misumena*. Neither did the birds' behavior suggest that they hunted for spiders there.

Other spider populations may be far more affected by bird predation than our *Misumena* populations, however. Birds regularly attacked members of some populations of the funnel-web spider *Agelenopsis aperta* and appeared to affect their prey capture strategy (Riechert and Hedrick 1990; Riechert 1991). Birds also regularly attack orb-weavers in their webs.

During the meadow vole *Microtus pennsylvanicus* outbreak in the study area, resighting rates of marked adult females dropped to one-half that of other years (Morse 1985b). Since gravid female spiders, though regularly moving on lines, often are so heavy that these lines sag to the ground when the spiders attempt to cross them, the gravid spiders temporarily descend to the substrate upon occasion. This renders them especially vulnerable to terrestrial predators, such as the voles. Although generally thought of as herbivores, these rodents regularly feed on small invertebrates, such as *Misumena* (Johnson and Johnson 1982). I must emphasize, however, that we had no direct basis for inferring predation by the meadow voles, though the correlation was striking. We have not obtained this result again, so even if it could be attributed to the voles, it is likely an infrequent phenomenon.

These observations and extrapolations strongly suggest that adult female *Misumena* are occasionally preyed upon, but that these acts are unlikely to result in significant mortality. Nevertheless, the question remains of whether these would-be predators alter *Misumena*'s behavior in a way that decreases their foraging efficiency.

The low level of mortality suffered by the adult females, despite their apparent vulnerability, suggests that they may not exhibit strong antipredatory strategies. As Dukas (2001) has argued, however, success may well be based on vigilance that is not apparent under natural conditions. For instance, their inevitable tendency to move upward, both when on the ground and in vegetation, may distance them from depredations of meadow voles and small snakes, among their most likely terrestrial predators (Morse 1985b). At the same time, this behavior conveniently positions them in the most favorable hunting sites of their plant.

Although I argue that predation on adult female *Misumena* is not common, this condition does not carry over to other life history stages.

Early Instars. Early-instar *Misumena* are probably most vulnerable to small jumping spiders, which occur frequently in the goldenrod inflorescences. They include the species that are most likely to attack the young as they emerge from their nests (*Pelegrina insignis*, etc.: Chapter 4) (Morse 1992a). These jumping spiders, intermediate instars themselves, weighing 1.5 to 10 g, appear to capture significant numbers of second-instar young foraging on goldenrod, as we have demonstrated by releasing *Misumena* spiderlings marked with powdered micronite dye and then measuring the proportion of jumping spiders containing this dye on their mouthparts. Although a crude technique, it has established beyond doubt that *P. insignis* is a major predator of the spiderlings. We have also begun to manipulate numbers of marked jumping spiders on the goldenrod inflorescences to obtain a clearer picture of their behavior and dynamics at this time. Preliminary results indicate that these spiders are highly mobile and that they quickly colonize empty sites that attract insects. The moderate numbers of jumping spiders present on goldenrod clones, plus the high frequency of clones occupied, suggests that their numbers on a clone are controlled by a density-dependent mechanism, probably through aggressive behavior. We have not tested that hypothesis, however.

In addition, a wide variety of other predators on the flowers, including several other spiders (middle-instar sac spiders [Clubionidae], early instars of the nursery-web spider *Pisaurina mira* [Pisauridae], and occasional brown crab spiders *Xysticus* spp. [several species, especially *X. emertoni*]) doubtless take these young as well. However, the largest sit-and-wait predators on these flowers are unlikely to capture significant numbers of newly emerged young. An adult female *Misumena* will rarely take a prey item of less than 2 mm, and spiders in general experience difficulty in capturing prey that are unprofitably small (Nentwig 1986; Bartos 2004).

These data suggest that predation on young *Misumena* considerably exceeds that on adult females. If so, the young should exhibit antipredatory strategies. Our best evidence for antipredatory behavior comes from the differences in performance of these young on goldenrod inflorescences with different densities of flower heads (Morse 2006). The density of these flower heads varies markedly, both within and between species, even at a local site. We discovered that naive, just-emerged spiderlings on some of these clones quickly captured abundant dance fly prey, but that on other inflorescences attracting similar numbers of flies they initially captured almost none. A subsequent analysis of the flower-head densities of goldenrod inflorescences revealed that initially the spiderlings captured prey on clones with relatively widely spaced flower heads, but not on those with tightly spaced flower heads. Instead, within several minutes the spiderlings on the tightly spaced heads hid under the dense thatch of the flower heads. The sparser inflorescences nevertheless provided cover, as well as easy access to the surface of the inflorescence, and the spiderlings on these sites quickly adopted an aggressive prey-capture routine.

Given the ubiquity of small jumping spiders on goldenrod inflorescences, it is tempting to suggest that the "hiding" *Misumena* spiderlings were responding (instinctively) to predation danger, while their sibs, placed on more open but protected sites, were adopting a high-gain strategy at locations that provided both hiding and hunting opportunities. Although the spiderlings hid rather than hunted on the dense inflorescences, those on sparse inflorescences took advantage of the opportunity to hunt.

We followed up these results with two experimental studies comparing spiderling responses to thinned and "normal" inflorescences. One of these studies ran longer than the two-hour period of the initial tests. Initially, we obtained results similar to those from the original observations (Morse 2006). However, the difference in behavior at dense and sparse sites largely disappeared some time after two hours. Performances on the two types of sites remained similar over the 24 hours of the experiment, suggesting that the initial difference in response to site characteristics had only a short-term effect. However, in the process the group on sparse sites gained an early advantage that might have long-term significance, since it involved early foraging success. Furthermore, spiderlings foraging on the dense sites probably did so at considerable risk because of their exposure on top of the flower heads. One of these experiments was conducted in a year when dance flies were not recruiting to the flower heads; the other experiment took place during a period of major recruitment. We obtained similar results from these two experiments over the period that the times overlapped. This suggests that the spiderlings responded primarily to their substrate rather than the prey.

Another potentially important predator of spiderlings is the harvestman *Phalangium opilio* (Phalangiidae), a common inhabitant in the study area and one that often ventures into flowers (Morse 2001), including the goldenrod inflorescences frequented by *Misumena* spiderlings. In common with other harvestmen, *P. opilio* have serious water-balance limitations (Cloudsley-Thompson 1968) and seldom occupy the floral canopy during most of the day, including the period during which the spiderlings are most active. However, the harvestmen commonly occupy these sites at night and, as generalized predators or omnivores, would likely consume any spiderlings encountered. This proposition could be readily tested with dyed young *Misumena*, since other studies (Morse 2001) have demonstrated that, similar to jumping spiders, *P. opilio* feeding on dyed food retain this dye about their mouthparts for at least a day following feeding.

Adult Males. The literature abounds with statements that adult males of many different groups experience extremely high mortality rates associated with their search for females (e.g., Andersson 1994). This impression seems almost ubiquitous with regard to mortality in male spiders (see Vollrath and Parker 1992). Most references also suggest that male spiders eat little or nothing during this period, although these sources often lack information supporting this assertion. Since male web-building spiders fail to produce webs as adults, they lack their previous source of food (from their webs). However, if they occupy females' webs, possibilities of obtaining food exist. A lifestyle of totally eschewing available food should decrease their bodily condition and thereby increase their vulnerability to predation, disease, and parasitism.

In contrast to this commonly perceived pattern, male *Misumena* suffer low mortality over the majority of their adult lives, capture prey regularly, and retain their original mass (see Chapter 8). Vigorous censusing of males over their adult stage (June–August) revealed insignificant changes in numbers through most of this period, despite the lack of recruits to the populations. (All antepenultimate and penultimate males have molted into the adult stage by then [LeGrand and Morse 2000].) Only near the end of their lifetimes, as they senesced, did their numbers decline. These males remained highly active, extremely fast, and difficult to capture during all periods but their senescence (Sullivan and Morse 2004). This difficulty of capture holds for both human experimenters and, presumably, their would-be predators, probably mostly other small predaceous invertebrates that are considerably less cursorial than the male *Misumena* themselves. Though vulnerable to aggressive females with which they attempt to mate (often already-mated individuals), only the oldest individuals late in the summer seem to experience cannibalism of this sort regularly (Morse and Hu 2004).

Late in the summer, however, we recorded a steep rate of decline in the number of males of these populations, which roughly coincided with the clearly increasing senescence of wild-captured laboratory individuals at this time. Numbers in the field, though remaining nearly constant from the time that all penultimates had recruited into the adult cohort (LeGrand and Morse 2000), decreased rapidly at this time, one to two weeks before heavy mortality began in the wild-caught laboratory individuals. Other work at this time (Morse and Hu 2004) demonstrated that the activity of males did decline as they reached old age. Thus, the disappearance of individuals from the field censuses could come about as a result of either a decrease in activity (in which case they might become less visible in even a carefully conducted census) or increasing vulnerability to predators. At any rate, the longer-term significance of this decline in numbers of males in the field is clear by extrapolation with the laboratory individuals—death. By the time males reach this stage, only occasional virgin females remain in the field; the apparent programming of senescence would therefore only affect their fitness under unusual circumstances. Although it is relatively easy to provide virgin females in the laboratory for mating at these times, they are by then virtually nonexistent in the field. However, since these females might not manage to lay a clutch if they came to maturity this late in the season, any matings with them probably would not yield any fitness benefits to the male in question.

General Comments. Thus, we have obtained relatively little information on predator-avoidance behavior in *Misumena*. As adults, the spiders exhibit few traits that can be unequivocally identified as predator-avoidance behavior. They will periodically move between the front and back sides of rose petals when hunting there, but the heightened frequency of this behavior on particularly hot days suggests that this activity has a primarily thermoregulatory basis. We have made no observations of bird predation over the long tenure of our field program. Yet as Ydenberg (1998) notes, this negative information cannot be taken as definitive evidence for the absence of predation. And elsewhere, wasp predation on spiders may far exceed that seen in our study areas. Many spiders have an instinctive fear of these wasps (Foelix 1996a), and it would seem unlikely that they would lack a method to minimize their contact with these deadly predators. Unfortunately, owing to the scarcity of these wasps in our study areas, we can add little to what must be an important factor for many *Misumena* populations. However, as discussed in Chapter 4, parasitoids may have a major effect on *M. vatia*.

Misumena as the Predator. When observing *Misumena*'s predatory acts, it is hard to avoid inquiring about its own effect in generating antipredatory behavior on the part of its prey. Predator-avoidance of *Misumena*'s bumble bee prey bears considerable similarity to the spiders' own predator-avoidance problems; it also warrants attention because of its possible consequences for *Misumena*'s success.

Earlier (Morse 1986a) we found that bumble bee workers encountered *Misumena* only infrequently when foraging on milkweed, their favored foraging site as well as that of the spiders. Given the numbers of spiders and the length of the worker bees' foraging lifetime, any individual had only a modest probability of ever encountering a spider. Attacks by the spiders, even if traumatic, resulted in only a 3 to 5 percent probability of capture. Given that the bees were members of social groups, and probably resource limited, adopting an avoidance strategy that decreased foraging efficiency might appear unlikely. Each social insect worker contributes to the overall resource base of the nest and the fitness of the queen. If by ignoring the spiders they brought more resources back to the nest than by engaging in avoidance behavior while foraging, an occasional capture would be far more than balanced by their higher foraging accomplishments, which could be channeled into a larger number of reproductives than at a nest whose members practiced avoidance. And we observed little sign of behavior suggesting that the bees avoided the spiders. On pasture rose, where the spiders hunted in the middle of the cuplike flowers, the bees, here foraging for pollen, frequently literally ran over the spiders, totally ignoring them as far as we could tell. The spiders often avoided those bees and at other times were bowled out of the flowers by them, only maintaining contact with the flower by their dragline. These interactions seem to bear little relationship to Ydenberg's statement. Yet these bees were the spiders' most important prey.

Given that the data set published in 1986 (Morse 1986a) far exceeded any data sets available to hypotheses predicting predator-avoidance behavior (Clark and Dukas 1994), Reuven Dukas and I revisited this question, using the bumble bee-*Misumena* system (Dukas and Morse 2003; Dukas, Morse, and Myles 2005). In experiments using relatively high densities of *Misumena* (upper densities recorded in censuses of natural populations—one *Misumena* on every three to four flowering milkweed stems) and control patches with no spiders, we found that the smallest of the bumble bees at our sites, *Bombus ternarius*, visited sites with *Misumena* less frequently than predicted by chance, as did the similar-sized honey bees *Apis mellifera*. These results are consistent with an indirect effect impacting the bees; the mere presence of predators generated avoidance behavior by the prey, negatively affecting their foraging proficiency (reviewed by Schmitz, Krivan, and Ovadia 2004). Previously, we (Morse 1986c) had demonstrated that the spiders, not surprisingly, captured small bumble bees more successfully than larger ones. However, we did not find a significant avoidance response from the two larger bumble bees (*B. vagans* and *B. terricola*), the species making up our earlier data set (Morse 1986c). Thus, these experiments demonstrated predator avoidance in the most vulnerable species but not in the others.

If the results came about from the bees learning to avoid the spiders, the high proportion of unsuccessful attacks could provide excellent opportunities to learn avoidance behavior. In one of our experiments, we monitored the behavior of individually marked, attacked bumble bees, allowing us to determine whether they would return to that patch less often than predicted by chance. We did not find such an avoidance effect, even among the most experienced workers. However, of greatest interest was the large number of unmarked bees that continually entered the area, which presumably were naive to the earlier interactions of conspecifics at these sites. Looking at the bee-spider relationships from the viewpoint of the spiders, we find that any possible avoidance behavior on the part of the bees will have vanishingly little significance for the spiders' capture rates. Even if experienced bees did avoid spider sites, they were but a small fraction of the large throng of bees visiting the flowers.

Attacks thus did not appear to modify the behavior of the bees, at least at the level of the experimental patches. So if learning is taking place, it does so at the level of the plant or inflorescence occupied by the spider, which we did not investigate. The bees may not respond to attacks because they are so accustomed to being struck by the vegetation during windy days that they do not distinguish unsuccessful attacks by the spiders from the surrounding wind-blown vegetation—or, perhaps the not infrequent jostling on the inflorescences that occurs among the bees when at high density at rich sites. Lastly, it simply may not pay to distinguish among these contacts. Even in this relatively high-density experimental spider population, we estimated that an attack would only occur once every second day (Dukas, Morse, and Myles 2005). However, our speculations may be inaccurate, since Dukas (2005) subsequently found that beewolves (*Philanthus bicinctus*)—large sphecid wasps that specialize on bumble bees—may decimate local bumble bee populations, but that the bumble bees fail to avoid these far more formidable predators.

Interestingly, Clark and Dukas (1994) predicted that members of social insect colonies should exhibit avoidance behavior; Dukas (2001) found such behavior in captive honey bee colonies. However, we (Dukas, Morse, and Myles 2005) did not find such behavior among members of wild colonies of bumble bees, even when subject to heavy predation. This relatively small sample suggests that predator-avoidance behavior does not have a heavy effect on natural field populations, but the question requires much more empirical work. Even if the situation from the field turns out to be commonplace, the presence of such behavior in the controlled honey bee populations invites explanation.

Finding a Hunting Site

Predator avoidance is only one of several constraints that might modify foraging success. A second potentially important constraint involves finding food in patchy environments in which the sparrow in the brushpile alluded to earlier might opt to concentrate on finding large, rich, rare patches of food, or to go for the potentially less profitable but low-risk option of sparser, less patchy food supplies that will assure it a maximum probability of survival. (I present the problem as if it were independent of the predator-avoidance issue discussed earlier. A sparrow's frequent dilemma under these circumstances may well be whether to court starvation with a conservative foraging regime, to court predation with foraging that concentrates on the richest food sources available, or to disperse to another area, which likely will entail high starvation and predation risks as well, not to mention risks associated with dispersal itself.)

Finding a hunting site appears to be a hit-or-miss proposition for *Misumena* of all ages. Following the occupation of a substrate, they then assess it and remain or move on. Their constraints may be either short term (e.g., starvation) or long term (meeting deadlines imposed by the season, whether by completing reproduction before the end of the season or reaching a favorable size for overwintering).

Spiderlings. As young leave their natal nests, they typically disperse on lines to other vegetation (Chapter 3). At such times they possess no more than a limited ability to select sites from a distance, at least to se-lect between "targets" of similar size, as illustrated by their failure to choose between hunting sites only 10 cm away, far less than the lines by which they usually leave their nests (Chapter 3). Thus, the initial part of recruitment appears to be random. However, they may make subtle decisions based on differences in the size of the targets they select at that distance (Morse 2005). Most likely, however, they follow the standard predictions of giving-up time (Chapter 3)—assessing the site they have reached, and, then depending on their assessment, remaining or moving on.

We have not yet explicitly determined what cues (Chapter 3) the spiderlings use to select a hunting site, but following experiments on their foraging behavior, we believe that substrate plays a major role. At this stage, previous experience does not appear to make more than a limited difference in their decisions.

The contingencies acting on the largest of the newly emerged spiderlings should not be as severe as those on smaller ones, since large young survive a starvation regime longer than do small ones (Morse 1993b). Thus, large individuals should be the most conservative in decision making at this stage. However, we found no size-related differences in the dispersal behavior of newly emerged spiderlings, which may simply reflect the importance for all individuals, large or small, to begin to feed as rapidly as possible. Even though the small may not survive as long as the large in the worst of situations, they all probably experience extreme pressure to feed, which would gain them a size advantage that lowered predation rates, lowered overwintering mortality, and gave them a head start toward a large, early reproductive effort.

Adult Females. Most females must shift their foraging sites during their adult stage. This requirement may demand considerable movement relative to their limited mobility. At times they may travel on lines or walk through the understory. From this position they must make decisions about where to find a hunting site, which almost certainly will be located in the canopy, usually well above them. Individuals facing this situation climb stems randomly (Morse 1993c), acting as if they have a sense of the insect activity above them, but are unable to localize it. They also exhibit a strong tendency to move upward at nearly all times. The sites they choose are no more likely to be satisfactory than if picked at random. Once they have expended the time and energy on the climb to the canopy, if the site turns out to be unsatisfactory, they must either climb back down in order to move on to the next site or put out a line to adjacent vegetation. Closer up, they appear to orient primarily or entirely to the presence of large prey (Chapter 7).

The characteristics of the substrate itself appear to assume little importance in their own right.

Physical Constraints

In addition to sensory limitations and constraints dictated by organisms of the same and other species, the spiders experience a variety of
physical constraints that involve conditions beyond which they cannot
function. I briefly consider a few of the most important here.

Desiccation of the Spiders. *Misumena* of all ages and both sexes are
probably subject to desiccation in the field, although they are normally
quite resistant to excessive losses. The smaller individuals (early instars,
males) should be the most vulnerable, judging from laboratory studies
(Chapter 3). How often these conditions are lethal is unclear. Probably
more important are the constraints that these conditions impose on the
areas the spiders can occupy and exploit.

Desiccation of Prey. Use of external digestion may limit intake to some
degree, depending on the size of the prey and the rate at which the
spider can consume it. Prey lose moisture quickly under the conditions
that *Misumena* usually capture them, and the resulting desiccation of
the prey will cause the body contents, now broken down by the spiders'
enzymes, to increase in viscosity, perhaps to levels at which the spiders
cannot extract them. Pollard (1989) noted that this evaporative process
could significantly decrease the amount of resources that the small
New Zealand crab spider *Diaea* sp., a species similar in size to middle-
instar *Misumena*, could extract. Tests run on the hover fly *Toxomerus
marginatus*, a major prey of middle-instar *Misumena*, revealed that
these flies lost about half of their mass as a result of desiccation when in
the full sun for three hours, roughly the period that *Misumena* of this
size require to process them (Figure 5.1: Morse 1998). Presumably the
water loss resulted in the spiders' inability to remove as much nutrition
from the flies as if the flies had not desiccated (see Pollard 1989). If
placed totally in the shade, these flies lost only 12.5 percent of their
mass over three hours. Though a much more modest decrease, it could
still cause problems for the spiders. In the field these spiders alternated
their time in and out of the sun, though often consuming their prey in
the shade under a flower. This behavior probably lowers the viscosity

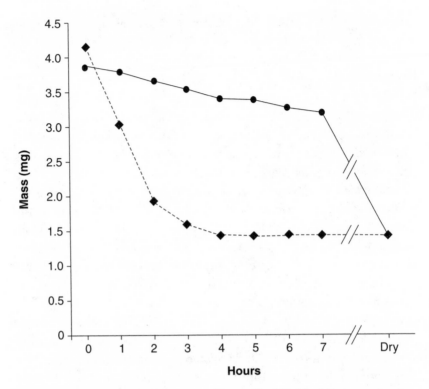

Figure 5.1. Loss in mass by desiccation of hover flies killed by *Misumena* when in the shade (solid line) or sun (dashed line) (Morse 1998). Variance largely confined to differences in size of individuals and thus not shown here. Prey in the shade required well over seven hours to reach constant mass. Prey taken from spiders immediately after kills. With kind permission of Springer Science and Business Media.

problem, though their behavior at this time could also be mediated by other factors such as their efficiency of digestion at different temperatures, problems of their own possible thermal overload, and vulnerability to their own predators.

The spiders exhibit a characteristic pattern of feeding on their prey. Typically they commence to feed from the head, even though they may not have initially struck the prey in that region. Thus, the prey may lose additional moisture as a result of these extra wounds. However, in experiments testing this additional possible difficulty, the presence of these extra wounds did not increase the rate of moisture loss beyond that occurring in the unwounded individuals (Morse 1998). Indeed, hemolymph at the wound sites appeared to coagulate rapidly (Morse

1998, 1999a), which may be of considerable importance, since S. D. Pollard (pers. comm.) notes that such extra holes might limit the spider's ability to generate the vacuum necessary to reflux the digestive fluids and imbibe the contents of the prey. Foelix (1996b) shows a scanning electron micrograph of a puncture wound made by a *Xysticus cristatus* in a blowfly that has coagulated in this way. These holes measure only approximately 30 μm in diameter and thus should coagulate rapidly.

Loss from evaporation will differ in response to several other variables. Even within a habitat, the importance of this factor should differ with the size of the prey and the nature of its exoskeleton. Adult female *Misumena* depend heavily on bumble bees for the majority of their prey intake under most natural circumstances, and honey bees are the most important prey nearer agricultural sites (Morse and Fritz 1982). Merely as a consequence of their size, rates of evaporation should be much slower in these large insects than in small syrphid flies. However, since adult female *Misumena* on average feed on medium-sized bumble bee workers for five to six hours (Morse and Fritz 1982), the actual length of this period should have an important effect on the amount of resources that could eventually be extracted. These differences might also affect the probability that a spider will take its prey out of direct exposure to the sun.

Since adult female *Misumena* capture considerable numbers of medium-sized moths (primarily Noctuidae and Geometridae) during the night when hunting on milkweed, differences in desiccation rates could affect the profitability of these prey as opposed to diurnal prey subject to greater rates of desiccation. However, the spiders' own metabolic rates will be lower at night than during the daytime, so their gain from nocturnal prey could even exceed that of diurnal prey. The diurnal-nocturnal issue is not an important option for most flowers occupied by *Misumena* because, other than for milkweed, they do not attract substantial numbers of prey at night. The other most important hunting sites for adult female *Misumena* in my study areas, pasture rose and goldenrod, attract no more than an occasional moth at night (Morse 1981).

By tearing apart their prey before abandoning them, most spiders somewhat ameliorate these problems of viscosity. However, these prey remains will desiccate rapidly on exposure. Since spiders feeding in this way still have to feed through external digestion, spitting enzymes onto their food, they, too, must compromise the profitability of certain prey.

Seasonal Constraints

These spiders face not only a variety of physical and biological constraints, but also temporal ones. I have already emphasized the costs from the heightened overwintering mortality of late broods. It is also important to maximize overlap with flowering at important hunting sites. In Chapter 9, I will also discuss the consequences of being out of phase with optimal opportunities for mating.

Timing of the final molt may constitute a significant temporal constraint, for our *Misumena* do not overwinter in the adult stage (Morse 1995). Up to now we have failed to find a single marked female adult that has overwintered out of a sample of over 2000 marked individuals (adult males have died well before this time). Nor have we observed a successful overwintering among individuals experimentally confined to wintering sites favorable for penultimates.

The inability to overwinter as an adult seriously constrains the life cycle of some members of a population. The closely related *Misumenops tricuspidatus* can overwinter in both juvenile and adult stages (Hukusima and Miyafuji 1970; Li and Zhao 1991), which should save these individuals from undertaking such activity patterns as those described for *Misumena*. It would be of considerable interest to know if this species undergoes any seasonal modifications of hunting behavior similar to that of *Misumena*.

Although slow growth normally is counterproductive, late in the season penultimate females may decrease their rate of prey capture, thereby preventing them from molting into the adult stage before winter. At this time, penultimates gain mass at only one-eighth the rate of adults at the same sites, though patch-use strategies of the two groups do not differ.

We take this striking difference between penultimates and adults to be a consequence of the penultimates slowing their intake rate at the same time that the adults are increasing it as much as possible. The adults are faced with the problem of gathering enough resources to lay a clutch as soon as possible. Despite comments to the contrary for the closely related *M. formosipes* (Gertsch 1939), the clutch must hatch before the first frost, and even then will probably experience high overwintering mortality as a result of their small size. If the penultimates gained mass at a higher rate, they might reach a size at which they would molt into the adult stage, with virtually no chance of producing a brood.

It is also unlikely that any adult males will have survived until this time (LeGrand and Morse 2000), so mating opportunities will be nil. Their only viable strategy therefore is to wait until the next season to produce an early brood, whose young should have a maximally high success rate. Size, nutrition and growth rate all play a role in regulating insect molt (Blakely 1981; Forrest 1987; Nijhout 1994). In spiders, a low availability of food delays maturity in species that do not overwinter as adults (Miyashita 1969; Wise 1975). (The low rate of intake by late-season penultimates may simulate the effects of a food shortage.) Thus, a suboptimal feeding strategy may enhance fitness, even though such penultimates will suffer the danger of mortality over an extra winter.

The hunting strategy of antepenultimate females (Morse 1995) is also consistent with this interpretation of the penultimates' behavior. Antepenultimate females also profit from gaining mass at a maximum rate, for this increase will provide them with extra resources for overwintering and will speed them along toward egg-laying sooner in the following summer than if they adopted a less active foraging strategy at this time. They already lag behind the current penultimates and are at risk of maturing so late in the following year that their young will not emerge in time to maximize their feeding opportunities. The antepenultimates are under no danger of molting into the adult stage in the autumn, and they may enhance their overwintering success if they manage to increase their mass before winter ensues.

Leg Loss and Other Infirmities

Along with many other arthropods, *Misumena* frequently lose one or more legs in the course of a lifetime. Leg loss may occur in a variety of contexts, including molting, interactions with conspecifics and predators, and wedging a leg in the vegetation or substrate. Shedding a leg may provide an escape from predators that grasp their prey, though not those that directly attack the body (Formanowicz 1990; Punzo 1997). Although its overall significance differs with the species, it clearly carries potential costs. It is generally conceded that multiple leg loss will impede locomotion, with the drawbacks that constraint will bring with it. Most spiders can regenerate legs in a subsequent molt, if the loss occurs early in an instar (Foelix 1996a). *Misumena* cannot recover all of that loss in a single molt, as readily ascertained by the not infrequent presence of short (though complete) legs on male individuals.

Leg loss has been studied in a wide variety of spiders and harvestmen (Phalangidae), which vary in how readily they shed them. Harvestmen differ from spiders in that they cannot regenerate their legs. No doubt due to the extreme length of their legs they, as well as pholcid spiders (daddy-long-legs spiders: Pholcidae), have generated the most attention. Various studies have reported no ill effects of leg loss in spiders and harvestmen and subscribe to the spare leg hypothesis (Brueseke et al. 2001; Dodson and Schwaab 2001; Johnson and Jakob 2001), which posits that these animals possess "extra" appendages that permit normal functioning in the likely event of such a loss.

Selection may favor autotomy, however, even if it results in significant subsequent costs. Autotomy may be highly adaptive in the short term, as it extricates individuals from life-threatening situations. However, long-legged spiders have evolved fundamentally different body forms from crab spiders and include species of such extremely different dimensions that it may be dangerous to make a general argument for the spare limb hypothesis. The loss of a leg entails significant reprogramming costs, as well as the obvious morphological loss. The nervous system of these animals is intricately programmed to coordinate leg movement, which proceeds in a precise pattern (see Barth 2002). Loss of a leg will require the reprogramming of this order of leg presentation in such a way as to enable a new pattern of coordination. They can accomplish that feat (Barth 2002), but not surprisingly, even where loss of a single leg does not result in immediately measurable impairment of function, loss of subsequent legs can present a difficult solution for efforts to maintain optimal performance.

Leg loss in *Misumena* is widespread, but appears to be by far the most common in adult males. Over a five-year period, 29.4 percent of the males in our populations ($N = 381$) had one or more missing legs upon collection, with 8.1 percent of these individuals missing two or more legs and 4.2 percent having one or more visibly shortened legs. Leg loss peaked during a very dry summer that Rebecca Lutzy and I investigated these losses (Lutzy and Morse MS). Most of the losses involved the two anterior pairs of raptorial legs, the sets with which they initially contact other objects, and the limbs used in fighting and prey capture. Several factors could account for the particularly high loss in adult males. Male raptorial legs are especially long for their size, a full 1.5 times as long as those of juvenile females of the same body mass and considerably longer than the legs of penultimate males of the same mass as well.

The marked increase in leg length at maturity speaks to the importance of these structures in adult male activities, which include both mating and contesting for females with other males. Staged male-male interactions resulted in 3.4 percent of the males losing a leg ($N = 95$), and in mating experiments with females, a further 2.1 percent lost legs ($N = 88$). In dense populations with much fighting, the loss in male fights might account for such numbers, but in these low-density populations, male-male fighting is infrequent (Chapter 9: Hu and Morse 2004), which argues against such an explanation. Similarly, if males mated, or attempted to mate, several times, as they do in our populations (Chapter 9: LeGrand and Morse 2000), these actions could account for the high frequency of lost legs. However, we found that the frequency of leg loss does not change after the beginning of the breeding season, which argues against attacks by females (or males) accounting for the majority of these losses, but rather that these individuals enter the breeding season with these disabilities. (This result differs from that of high-density *M. formosipes* populations with frequent male-male interactions, which exhibited an increasing proportion of individuals with missing legs as the season progressed [Dodson and Beck 1993].)

We also found that male *Misumena* missing and not missing legs did not differ in age; leg loss was random between the two anterior pairs (roughly similar numbers of right front, right rear, left front, and left rear legs lost). Since regeneration normally occurs, resulting in replacement with small legs, these losses do not emanate from losses in earlier instars. On the other hand, the modest number of individuals with short legs independently suggests that penultimates did not frequently suffer this level of loss.

This observation leads to the remaining possibility that these individuals lose limbs disproportionately frequently either at the last molt or when moving about shortly after that time and getting them lodged and subsequently lost in the vegetation and substrate (wedged into the petiole of a newly expanding grass shoot, etc.). In that their legs are longer and slimmer than those of similar-sized conspecifics (juvenile females and penultimate males), they should become trapped more often in the vegetation, particularly if they are unaccustomed to their extra length since the previous molt. Over the many years of this study, we have found a handful of individuals trapped in this way in the field; however, they may have been the exceptions that did not quickly autotomize when entangled. Furthermore, the long, slender legs of the adult males dry rapidly at molt, and the window of opportunity to extricate themselves from their cast skin may be much more constrained in these individuals than in other age or sex groups. We have noted that males molting in the laboratory during particularly dry summers not infrequently experienced difficulty in extricating all of their legs from their cast skin unless we took special precautions to provide them with a moist environment. Systematic work is needed to answer the questions about autotomization adequately.

We attribute the low levels of leg loss in females to the relatively shorter and more robust nature of their limbs. The increased overall size of the adults may cause them to experience lower molting losses than other members of this species, simply because the resultant increased drying time will provide them with greater opportunities to escape their molt.

A substantial part of adult male *Misumena* movement consists of crossing lines in the vegetation, either theirs or those of other individuals, including those of potential mates (see Chapter 9). Individuals missing a raptorial limb negotiated lines more slowly than intact individuals, a pattern that held independently of their mass, carapace width, or age. The difference in speed recurred in individuals that crossed more than one line in the experiment. Those missing more than one limb, as expected, moved even more slowly. A small sample of individuals with one or more short legs performed nearly as well as the completely intact ones and markedly better than those missing an entire leg.

Lutzy and I have also conducted experiments on the ability of adult males from which we removed a raptorial limb to negotiate the line-crossing task. Surprisingly, these individuals crossed lines at a speed similar to that of intact controls. Most likely, individuals in the field lose bodily condition as a result of losing a leg. The poor performances of the latter individuals may reflect their condition rather than a lowered ability to perform line crossing.

Adult males walk about on the vegetation or substrate as well as on lines. We tested males with and without missing forelimbs in 30× 30×30 cm screen cages, which gave them the opportunity to walk over the substrate but did not provide them with lines. These trials yielded results similar to those obtained from individuals moving free in the field away from any flowers, upon which they might stop and hunt (see Sullivan and Morse 2004). The cages proved especially useful for testing adult males because one loses these individuals at an unacceptably high rate in free-field trials, and the results in the two settings, cages and free in the field, are similar. Intact individuals both moved more often and moved a greater total distance than those missing legs. In contrast to the results on the lines, individuals missing more than one leg performed as well as those with one missing leg, although the sample was again small. However, in contrast to the line trials, those with small regenerated legs scored even significantly lower than those missing a leg.

Although the results from both the line and cage tests demonstrated costs to missing legs, they tested slightly different aspects of the movement patterns of the male spiders, as revealed by the differences in performances by individuals missing more than one leg and those with regenerated legs. The result from the individuals with regenerated legs was most interesting and not easy to interpret (Johnson and Jakob 1999; Barth 2002). Regenerating a leg entails major energetic costs, which might be demonstrated in the cage tests, which differed from the line tests in incorporating activity rates over a considerably longer period than did the line tests. These individuals may simply have a lower activity level in general, a condition in some ways analogous to the behavior of extremely old, and apparently increasingly senile, individuals (Morse and Hu 2004). Extremely low activity levels will decrease the ability of these individuals to compete for females through scramble competition, the apparently predominant method of securing mates in this low-density population (LeGrand and Morse 2000). Given their mediocre performance, one might wonder whether individuals with regenerating legs would not be as well off if they did not regenerate legs at this late stage. Perhaps this problem accounts for why harvestmen do not regenerate limbs following a molt.

Overall, therefore, leg loss exerts a measurable cost in *Misumena*, indicating the inadequacy of the spare-leg hypothesis for this species. The difference in performance suggests that males missing legs may cover no more than two-thirds of the terrain covered by intact males. Under these circumstances, in which scramble competition predominates and females are apparently cryptic to males (see Chapter 9), by searching only two-thirds of a normal-sized area they would enjoy a fitness of only two-thirds that of intact individuals. Whether this difference is realized, or the males missing legs find a way to accommodate for this deficiency, remains to be determined.

Some Stochastic Factors

Some conditions prevent an individual from responding to them nonrandomly. In spite of this forbidding probability, such events hypothetically could take on enormously greater significance to this individual than predicted by their frequency, presenting it with the opportunity of producing an especially large set of progeny. In a small population, that individual may have a major impact on the gene pool. In our region *Misumena* occur in many small populations; such effects may be marked locally, even though they may not spread from the sites at which they occurred.

Evidence to date has demonstrated that *Misumena* have but imperfect information about the quality of their hunting sites before encountering these sites and that the visitation rates of their insect prey are remarkably sporadic over time. Even during the middle of the day when prey visits take place most frequently, intervals between their visits to flowers or inflorescences are stochastic. Visits to high-quality umbels differ from three or more individuals per minute to periods of several minutes between visits during the peak flowering time of milkweed (Morse and Fritz 1982). These fluctuations present a complex and potentially confusing situation to sit-and-wait predators such as *Misumena*.

It would be inappropriate to consider the capture of a bumble bee by an adult *Misumena* to be a stochastic event over the period of a day. However, with capture rates of large prey such as bumble bees averaging no more than 3 to 5 percent in many instances (Morse 1979), a high level of uncertainty is built into the rate of prey capture, which is exacerbated by differences in prey visitation rates to flowers within a local area. At the same time, these events are extremely important ones for the spiders. A single kill of a bumble bee may provide a spider with enough food to double its mass, most of which will go into producing eggs for its single clutch. For spiders otherwise enjoying only modest foraging success, such a kill may mean the difference between laying a clutch and not doing so. The low success level of capturing bumble bees creates an extremely strong selective force for finding and occupying sites that attract large numbers of these insects, in this way minimizing uncertainty.

A closely related aspect interjecting further uncertainty into *Misumena*'s prey capture involves its arrival times at hunting sites relative to the visits of large insects to these sites. We have shown that these prey exhibit a random visitation pattern to these sites within the hunting time frame of the spiders (Kareiva, Morse, and Eccleston 1989). Assume that a spider occupies a site that attracts only infrequent prey, say one bumble bee per day, a rate similar to the poor-quality umbels occupied by a few spiders (Morse 1986b). The time when that bumble bee arrives, relative to when the spider takes up occupancy of that site, will make a great difference to how the spider evaluates the site. If a bumble bee arrives at 11:15, one minute after the spider arrives at a flower, and the spider at least encounters that bee, the spider's perception of that site will be far different than if the bee had arrived one minute before the spider's arrival (11:13) (Morse 1984). Since the spider's recent experience on that site will influence its giving-up times there, a minute's difference in timing will grossly affect that spider's movement patterns over the following day, one of a limited number during its adult stage (Morse 2000c).

Although the consequences of such variation may appear trivial, they involve the potential capture of an extremely rewarding prey item. By itself this item would not permit the spider to lay a minimum-sized clutch of eggs, but if the spider could procure a second such item it likely would suffice to lay a small clutch. And after attaining a minimal increment for a brood, any additional kills will significantly increase the number of eggs produced—10 to 20 percent on average. However, given the consequences of suboptimal foraging patterns on milkweed (Chapter 3), such an unlikely event as a kill on a poor site might provide a spider with information prompting it to seek out similar poor sites in subsequent moves. If so, rather than enhancing subsequent prey capture in the future, a kill on a poor site might minimize the chance of obtaining a subsequent kill required to lay even a minimal-sized clutch.

Adult female *Misumena* appear to have a fixed probability of leaving milkweed umbels that varies among individuals; time since the last visit by a prey is the critical variable (Kareiva, Morse, and Eccleston 1989). Thus, rather than all employing the same fixed giving-up time (Krebs, Ryan, and Charnov 1974), which would cause them to leave when the resources available from the site fell below the average value for the patches as a whole, a certain proportion will leave after a given period without prey visits has elapsed, another proportion after another period, and so on. This predisposition may be related to their energetic condition, but if so, it merely shifts the question back to identifying the variables responsible for generating this range of energetic conditions among the spiders. Stomach fullness, or gut distension, could provide the mechanism acting in this instance. Combined with the fixed probability of moving, it could result in individuals spending most of their time on high-quality umbels (Kareiva, Morse, and Eccleston 1989). This possibility could be tested by taking prey from the spiders immediately after the spiders capture them. Thus, this type of mechanism does not result in the same frequency of abandoning high-, middle-, and low-quality umbels as might initially be surmised, because visitation rates to these umbels are so different. Such variable fixed giving-up times could account for why a certain proportion of individuals leave a high-quality umbel at an early point, while others will remain for inordinate periods on poor-quality umbels. What remains unanswered is what programs these individuals adopt to obtain certain fixed-probability giving-up times. Do they retain constant giving-up times in subsequent situations, or do they "reprogram" following each decision? The results from both adult and penultimate spiders that had and had not been switched between flower species suggest that these individuals retained traits characteristic of their original flowers for some time (Morse 1999b, 2000c). We have yet to perform comparable tests on the early instars, although they are a high priority.

The adoption of variable fixed probabilities of leaving a site initially seems somewhat of a paradox: calculations (Kareiva, Morse, and Eccleston 1989) suggest that adoption of a threshold giving-up time would yield considerably greater resources, estimated to be roughly twofold. Kareiva, Morse, and Eccleston suggest three possibilities for this result:

1. Individuals do not optimize giving-up time because decreasing their variance in prey capture is more important than maximizing capture.
2. Inadequate information is available for making decisions.
3. Optimal behavior is not greatly superior to alternatives.

In the third alternative the optimum is modest enough that in a complex system it becomes difficult to single out any particular stereotyped optimal behavior.

Under the latter circumstances, simulations using different costs and variances of costs, in terms of prey lost as a result of moving, yielded sets of curves showing little tendency for large advantages to be gained by any single combination of giving-up time and average yield over a wide range of variables (Kareiva, Morse, and Eccleston 1989). In general, increases in mean visitation rate decreased the clarity of the optimum—that is, the extent of superiority of the optimal over alternatives. Therefore, the penalties of deviating from the optimal behavior are modest. The combinations of means and variances presented in Kareiva, Morse, and Eccleston (1989) are representative of the field situation we study. The sets of curves generated were typically either jagged or flat, with few points rising conspicuously above the others (low clarity of the optimum). Although more simulation runs would doubtless have smoothed the jagged yield profiles, we confined them to the numbers (few chances) that a spider might realistically experience in the field. Such curves suggest that time is not a particularly sensitive variable in capture success (perhaps an appropriate condition for a species that does not move quickly), has limited perceptual abilities associated with patch choice, and possesses an extremely efficient metabolic function.

We have recently begun to investigate a second highly stochastic aspect of *Misumena*'s foraging life: the capture of relatively "gigantic" prey by newly emerged second instars (Chapter 3). In the field on rare occasion, newly emerged spiderlings capture hover flies *Toxomerus marginatus* that average 4 mg in mass, over six times the spiderlings' mass (Erickson and Morse 1997). We have subsequently explored this relationship in the laboratory in controlled experiments with known sib groups (to control for a likely source of variance) and found that when hover flies are presented to naive spiderlings housed in 7-dram vials, a substantial percentage of the spiderlings will eventually capture the flies. We challenged three groups of spiderlings to prey regimes of all hover flies, an initial hover fly followed by fruit flies *Drosophila melanogaster*, and a diet consisting only of *D. melanogaster*. The individuals fed solely on hover flies grew highly significantly faster than did those of the other two groups, and the group given only the original bonus of a hover fly, followed by a standard regime of fruit flies, did no better than the all-fruit-fly group. Members of different broods varied in their frequency of capturing the hover flies. We did not explore this variable, but we need to investigate further. One would predict that broods of large individuals experience higher rates of success than broods composed of small individuals.

We do not yet know the importance of capturing outsized prey in the field, although this ability usually marks the difference between reproductive success and failure in the adults. It is highly questionable whether spiderlings capture enough of these especially large prey, or others comparable to them, to catapult their mass to that reached by a few individuals in the laboratory, especially since a single hover fly did not produce any long-term size advantage to young in the laboratory. Young we have observed in the field that gained mass with exceptional speed usually took advantage of a superabundant supply of dance flies, roughly similar in size to fruit flies. We do not know how many of these flies a spiderling will take, but when the flies are superabundant, even naive spiderlings capture them with a high rate of success—even higher than their initial success rate in the laboratory with fruit flies.

The difficulties facing these spiderlings thus fluctuate greatly. If, as newly emerged spiderlings, they had to depend on a diet of hover flies, it is questionable how many of them could capture these ubiquitous prey often enough to attain a size at which the prey became a staple (Erickson and Morse 1997), even though we have shown that those making the necessary kills under laboratory conditions grow at a rate exceeding other feeding regimes. Since large early instars appear to maintain a size advantage over other individuals (Morse 2000b), they should do very well in subsequent life. It would be of considerable interest to compare these offspring with those that coincide with a surfeit of dance flies. There, predictability of prey capture will be very high, but temporal predictability low, for in only four years out of 20 did spiderling and large dance fly emergences coincide. *Misumena* populations would not survive if they had to depend on either of these pure strategies.

Synthesis

Misumena's life cycle, which involves passing through several independent instars, but fewer for males than females, results in many potential constraints. The difference in instars should reduce inbreeding but at the same time may create situations in which adults of one sex are active in the absence of potential mates. In our *Misumena* populations, these periods occur in both the spring and late summer/fall. In the spring they occur as a result of *Misumena*'s protandrous condition. The sensitivity of differences in maturation times of the two sexes appears to be buffered by the relatively low mortality rate of adult males. A few females may also mature in the late summer at a time when few if any adult males still remain. However, such females are rare and stand little chance of rearing a brood because of severe time constraints associated with the impending end of the season. It probably is not a coincidence that penultimate females appear to curtail their prey intake at this time. This likely decreases the probability of molt into the adult stage (Morse 1995), which would doom them, since adults do not survive over the winter.

Contingencies may also result from the sensitivities of different instars to different conditions. The small end of the size range allows minimum leniency in dealing with environmental variables such as energetics and water balance. Small individuals do not have as many reserves (relative to their body size) and will probably have metabolic demands at least as high as larger individuals. Energetic problems for small individuals are primarily a concern during times of low food resources, with the period leading up to winter a particularly difficult one. The spiders can minimize these drawbacks by laying clutches early, resulting in offspring that are older, larger, and more resilient as they approach winter. Nevertheless, females must obtain considerable food before they can lay a clutch, which will limit how early they can produce it. To add to the problem, the spiderlings' own time of emergence from the nest may often place limits on when they can lay a clutch.

The issue of possible emergence times raises the question of why individuals do not simply lay an earlier, smaller clutch, followed by a second later clutch if possible, rather than waiting to lay a maximally large clutch. Since they lay only one clutch in our area, it may well be that the one-clutch strategy still maximizes output, because these individuals usually gain large amounts of mass after the very earliest females have laid. On the other hand, this very situation should select for progressively earlier egg-laying. Some individuals do abandon their clutch before the young hatch and not infrequently will begin to capture prey again and increase in size. In this way they do somewhat shorten their reproductive cycle, if at some cost to the members of their first clutch, which will suffer increased predation both in the nest and as they emerge (Morse 1988a, 1992a). Nevertheless, they never seem to have time to launch a second clutch successfully in our study area. That they seem "unaware" of this inability suggests that this strategy remains a viable option under some rare circumstances, or that it has been viable within the recent past. This condition also leads to the prediction of double-brooded females south of our study area, as noted by Fink (1986) in her green lynx spiders.

At present we do not know how many *Misumena* in our study area actually have a three-year cycle rather than a two-year one. Clearly, a three-year cycle would lower reproductive output, all else being equal. However, such penalties may be minimal in populations that do not fluctuate greatly in size and in which adults cannot survive the winter. With poorer food resources or shorter seasons, as at the northern edge of their geographical range, three-year cycles may be an important option. (Elongation of spider life cycles occurs regularly under constraining conditions such as short seasons [Leech 1966].) With fewer instars than their females, males may become adults in the year following their birth. By providing an optimal diet in the laboratory, we can raise males to a penultimate stage within five weeks, or less, after their emergence. Such a feat is not approached by their female sibs or, as far as we can tell, by males in the field.

The overwintering bottleneck may also elongate generation time among individuals that promise to be the earliest layers the following spring. It is not clear how great a price these individuals really pay, if by delaying their molt they will de facto synchronize themselves for maximum gains hunting on milkweed, the resource from which the largest clutches are produced, during the following season. Since milkweed is so patchy, however, that factor may further complicate the situation—and the very earliest clutches may even be laid before milkweed is in bloom. If adult food sources are scattered so widely as to be unpredictable, tendencies toward specialization at this stage may not be as strong as those noted for the newborns, which can specialize on the most highly predictable resource, goldenrod.

~ 6

Experience, Learning, and Innate Behavior

$\mathcal{V}$ARIATION IN BEHAVIOR COULD RESULT directly either from differences in experience and learning or from genetic differences or maternal effects. I assume that experience and learning-based differences in behavior have a genetic basis as well, but that they result in considerably more flexibility in actions than those directly controlled by genetic differences. Since one usually confronts these two factors as alternative modes of operation, comparisons are in order, while acknowledging that a continuum exists between them.

Some Initial Background

Experience and learning are related phenomena. By experience I simply refer to an animal's participation in previous events. Participation need not imply active performance; the mere witness or perception of an event might suffice. Incorporation of this event into that animal's memory may be taken to constitute learning. Through learning an animal may generate a reproducible response to a phenomenon that it has previously experienced. Experience could hypothetically even influence responses in ways other than through learning. (For instance, if earlier events have resulted in a morphological change that favors a particular type of future response, they could affect performance totally independently of learning. The fortuitous capture of a relatively huge food item by a young individual could so affect its size that it would subsequently perform differently as a result of its increased size. However, learning is likely to be the first possibility to come to mind, and probably the most important to animals with complex neural systems. The term *experience* makes no assumptions about the type of phenomenon responsible for any particular change.

Alternatively, behavioral patterns may be innate, that is, not learned. Generally, innate behavior probably has a direct genetic basis, but the term does not necessarily indicate such an origin. For instance, the response to a stimulus by a recently emerged spiderling could also be tempered by maternal effects acting in a variety of ways (producing healthy eggs, providing a nest, guarding, feeding, etc.).

Predictions

Theory predicts that conditions of moderate to relatively high environmental predictability will favor learning, but if predictability is perfect or nearly perfect, a "hard-wired" program (genetically programmed) would function most efficiently and probably at lowest "cost" (Dukas 1998c). At the other extreme, it may prove impossible to develop learning abilities under completely unpredictable conditions— experience would provide no insight for response to a future event. Thus, learning should be favored at ranges of intermediate predictability (the Goldilocks Principle: Kerr and Feldman 2003). All of these predictions rest on the ability to learn, which itself is taken to be under genetic control. Although learning has traditionally been considered primarily the province of "higher" vertebrates, it has been increasingly recognized that many other animals have a much greater ability to learn than previously recognized (e.g., Abramson 1994). Nevertheless, the actual limits to the ability to learn remain a rather open question and may depend on a certain minimal size of nervous system (see Bonner 1988; Harland and Jackson 2004), an issue of considerable interest in *Misumena*'s ontogeny.

The Alternatives in Practice

Although the extremes involve a complete reliance on learning on the one hand and innate traits on the other, animals with complex behavioral patterns will most likely feature some mix of the two. The relative importance of learning and innate traits will probably differ along both time and space gradients. The mobility of animals relative to their needs should strongly influence how they characterize patches. Birds, for example, are as highly mobile relative to their needs as any organism, simply by virtue of their ability to fly. Comparisons with mammals (with the exception of bats) along this energetics/mobility gradient are particularly enlightening. Flying insects should occupy positions comparable to birds at a very different size range, but forms such as adult female *Misumena* possess very limited mobility relative to their demands. That, in large part, is why they are routinely food-limited and why the issue of patch size is such a vital issue for them, dwarfing other variables such as prey selection.

Cognitive Skills and Limitations

Cognition is a set of neuronal processes that permit the perception, acquisition, and manipulation of information, thus depending on experience and encompassing learning and memory (Dukas 1998b). These processes do not affect fitness directly but through the behavior they elicit (Shettleworth 1998). Cognitive processes have been studied most intensively in humans and have traditionally been associated primarily with vertebrates. However, as more is learned about invertebrate behavior, it is becoming clear that they have developed these abilities to varying degrees as well. The ways by which organisms process information may differ from species to species, though doubtlessly affected by phylogenetic differences, and constraints from other aspects of their biology may prevent organisms from dealing with a problem in the most obvious way (Shettleworth 1998). Organisms can only process a finite amount of information at one time, making it necessary to discriminate among candidates for information processing. The extent to which they can accomplish this task likely differs with the circumstances experienced and is at best imperfectly understood. Much of our work with *Misumena* explores certain aspects of these capabilities and their limits within naturalistic settings.

Experience and Learning in *Misumena*

Our study on the early experience and learning of *Misumena* has focused on their role in determining adult foraging patterns, which play such a central part in dictating their lifetime fitness. Changes could, however, occur at any stage of the life cycle and thus take on interest of their own, as well as for the possible information they may convey about the adults. Opportunities for experience and learning appear to vary with the stage in question, in part dictated by the predictability of the conditions. Specifically, their effect may differ depending on the subject's size and developmental stage.

Second Instars

Background. Second instars are the first truly free-living stage of *Misumena*. The first instars are semi-embryonic and confined to the nest. Characteristically, they possess large bulbous abdomens stuffed with yolk and short, stubby limbs they periodically wave, often in unison. They molt into the second instar within the nest, so active individuals may be found inside the nest as well as external to it. In contrast to the first instars, second instars are mobile and very active, yet so small that they cannot quickly cover long distances on foot. However, they carry out most of their movement on lines that they or their sibs make in the vegetation, which increase their mobility considerably. They may also move by ballooning, which increases it greatly but apparently under only limited control.

The spiderlings' nest site typically is not a satisfactory hunting site, though it is likely to be in the immediate vicinity of such a site (Morse 1993a). Upon leaving this area, they must immediately decide whether the site onto which they secure their line will provide them with adequate foraging opportunities. Although the spiderlings still carry some reserves in their yolk sac, that resource will only sustain them for a few days (Morse 1993b). Foraging-site decisions may be among the most important ones they will ever make. Naive young that have just left their nests and have had no contact with substrates other than their nest sacs may thus provide considerable insight into the basis for choice. Spiderlings exhibit a clear preference for flowering goldenrod in our study areas (Morse 2000a, 2005). Goldenrod is ubiquitous, and it is the principal foraging site of flower-visiting insects when most spiderlings leave their nests. Numbers of insects on flowering goldenrod often greatly exceed the early instars' ability to capture them. In addition to their favored sites of flowering goldenrod, they may frequently contact flowering asters *Aster umbellatus* or wild carrot *Daucus carota* (Morse 2000a, 2005).

In much of our work on substrate choice (Morse 2005), we have placed spiderlings on one of a variety of substrates in the field (milkweed leaves, aster flowers, wild carrot flowers) that they would most likely encounter near their nest site. We then observed them carefully for 30 minutes, noting any moves or changes in behavior. Taking advantage of the large number of individuals in a single brood that permitted us to work with groups of full sibs, we exposed these groups to common substrates and in some experiments shifted the substrates of half of them on a subsequent run. This procedure allowed us to compare between-brood responses to within-brood responses on the same and different substrates. Thus, we can begin to address the question of whether the spiderlings' responses to these substrates have any direct genetic basis. We typically collected nests from the field shortly before the spiderlings began to emerge, thus ensuring that none of them had ventured outside.

Responses of Naive Individuals. We predicted that the spiderlings would favor flowers over buds and yellow buds over green buds, for goldenrod in flower attracts vastly more prey than do yellow buds, and green buds attract even fewer prey. Naive spiderlings placed on flowers and buds strongly favored flowering goldenrod over either yellow or green buds of goldenrod, and yellow buds were favored somewhat more than green ones (Morse 2000a). Of those that did leave these sites, the flower-leavers took longer to do so than the bud-leavers, with those on yellow buds taking longer than those on green ones (Figure 6.1: Morse 2000a). Spiderlings that left buds to occupy other sites usually selected higher-quality substrates than the ones they left, and the small number of individuals leaving flowers usually selected other flowers (Morse 2000a). Although we predicted this result, it is important to emphasize that the spiderlings were naive individuals that had never experienced any of these substrates. Thus, their responses were apparently innate (=unlearned). The basis for this choice takes on im-

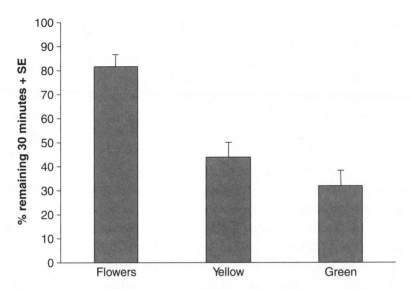

Figure 6.1. Percentages of spiderlings per brood of six (13 broods) remaining 30 minutes or more (+ SE) on assigned goldenrod hunting sites. Yellow = yellow buds, Green = green buds. Data from Morse (2000a).

mediate interest, in that we ran these experiments both in the absence of potential prey and with normal low-prey densities. Responses to substrate were similar both with low-prey densities and without prey (Morse 2000a).

We ran a control to ensure that the spiderlings responded in a way relevant to their normal behavior when leaving their natal nests. We placed spiderlings on milkweed leaves, which was the usual substrate they first contacted upon leaving their nests. Leaves also make up the substrate of their nests. The nests themselves are packed with foamy silk. Thus, the spiderlings might not contact leaves while within their nest, and if they did contact a leaf, it would have been coated with silk. Spiderlings inevitably responded to milkweed leaves as unsatisfactory hunting substrates. The rapid rate at which they either moved off the leaves or ballooned away during pilot observations prompted us to set the time limit for these experiments at 30 minutes. We have retained this time limit as the criterion for all of these experiments. Not one spiderling from 13 broods (78 spiderlings) remained on the milkweed leaves for 30 minutes, even though the average times to departure differed from brood to brood. In similar experiments we have run at other times, no more than an occasional spiderling has remained more than 30 minutes on a milkweed leaf.

Since we collected the nests from the field shortly before the spiderlings emerged, it was important to establish that we did not test them before they would normally disperse. Second-instar spiderlings used in such runs often are hesitant to move under these circumstances before naturally emerging from their nests. Thus, if offspring in an experiment did not move from the milkweed leaves within 30 minutes, we would question whether we would obtain relevant data on substrate choice from their sibs placed at the same time on flowers or buds on the assumption that they would quickly quit unfavorable sites. If the milkweed-leaf controls left within 30 minutes, we felt safe to conclude that their simultaneously run sibs had made a decision to remain on other substrates if they did not leave within that period.

What could be the precise basis for this choice of open flowers? Open flowers have a different texture than buds, and tactile perception of size might differ as well. The size difference could account for the slight preference for yellow buds over green, though we eliminated the bud category in subsequent tests, in order to simplify the design. The similarity of color cues from flowers and yellow buds also leads to the conclusion that the spiderlings use tactile cues to discriminate among substrates in this test. (The flower and yellow bud colors do not differ in the ultraviolet range.) Chemical cues might also differ in that the open flowers are producing nectar and pollen. The failure of adult males to use chemical cues (or females apparently to produce them) (Anderson and Morse 2001: Chapter 9) tends to dissuade us from a chemical explanation, but we have not yet tested this possible variable with the spiderlings.

If flowers attract the most prey, why not always concentrate on them? This question closely resembles the one presented earlier (Chapter 3): why adult females do not always select the highest-quality hunting sites, since they are in unlimited supply. In any mosaic environment such as that of goldenrod, in which flowers, yellow buds, and green buds are interspersed, a question arises regarding the spiderlings' ability to sort out these areas with complete accuracy. Since some prey species walk over buds in the process of visiting flowers, the quality of bud hunting sites may exceed initial expectations. Censuses of insects

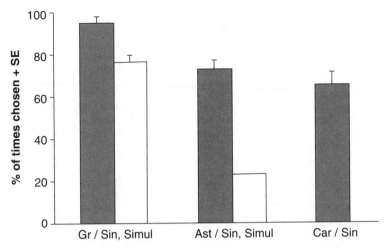

Figure 6.2. Single (Sin) and simultaneous (Simul) choice of naive spiderlings on goldenrod *Solidago canadensis* (Gr), aster *Aster umbellatus* (Ast), and wild carrot *Daucus carota* (Car) + SE. Single-choice tests (dark bars) recorded the percentage of spiderlings still present on site after 30 minutes. Simultaneous choice tests (light bars) compared goldenrod directly with aster (left set of bars) and aster with goldenrod (center set of bars), recording the percentage present on the two flowers after 30 minutes. No simultaneous tests run with wild carrot (right bar). Data from Morse (2005) and unpublished.

revealed a higher percentage of insects on these sites than predicted on the basis of their foraging value to the insects.

We have also compared the responses of naive spiderlings to gold-enrod flowers and other flowers. The spiderlings prefer goldenrod to either flat-topped white aster *Aster umbellatus* or wild carrot *Daucus carota* in both single-flower and simultaneous presentations (Figure 6.2: Morse 2005). In single-flower presentations, we placed spiderlings on one of the three flower species and tallied the number remaining at the end of the 30-minute test period. Flat-topped white aster and wild carrot are the next most common species (to goldenrod) flowering at this time in our principal study area. They frequently bloom close enough to goldenrod or to each other that flowers occasionally make contact with each other. In the simultaneous presentations, spiderlings were placed at the interface of contiguous goldenrod and aster flowers (left legs on one flower species and right legs on the other) and thus forced to decide whether to move to goldenrod or aster (or carrot). The preference for goldenrod in the single presentations was stronger than that in the simultaneous presentations, again potentially tactile in nature. Although it sounds extreme to presuppose an innate preference for a specific substrate (i.e., goldenrod), goldenrod is an overwhelmingly abundant flower throughout this geographic region when the spiderlings emerge.

Since a myriad of goldenrod species occurs within our geographic region, it would be of interest to compare the spiderlings' responses on *S. canadensis* with those of other goldenrod species at our study area, *S. juncea* (early goldenrod) *and* rough goldenrod *(S. rugosa)*. *Solidago juncea* and *S. rugosa* are two very common species whose flowering periods overlap that of *S. canadensis* (*juncea* averaging earlier and *rugosa* later than *canadensis*). We suspect that such *Solidago* comparisons would reveal differences in responses. The density of flower heads along a branch differs as much as threefold within *canadensis* clones growing on the site. This difference is probably responsible for marked short-term differences in spiderling hunting behavior (Morse 2006), in which individuals on sparse clones hunt and capture prey actively, while those on the dense clones crawl down inside the dense cover of the closely apposed flower heads. Since both the spacing and size of *S. juncea* flower heads exceed those of *S. canadensis*, we would expect to find clear differences among the spiderlings' responses to this species from those directed toward *S. canadensis*. Size and density of *S. rugosa* flower heads closely resemble those of *S. canadensis*, but its branches are much more widely spaced. This information suggests that the spiderlings exhibit a somewhat flexible response to goldenrod flower heads, since the species composition of goldenrods varies somewhat from area to area.

In short, naive individuals respond innately to the sites they first experience, with these responses serving to place them on the most favorable hunting sites available. Nevertheless, since differences occur within even the most regularly encountered substrate, these spiderlings are likely to be flexible in their choices. Since little evidence remains for maternal effects still acting (the spiderlings have left the site where their mother may still be located), these responses most likely have a direct genetic basis. Although these preferences present themselves essentially as goldenrod-specific traits, we do not yet know precisely to what aspects of these flowers the spiderlings are responding.

Individuals with Limited Experience. Many of the experiments on newly emerged spiderlings continued for up to five days, thus permitting us to observe whether the initial responses to substrates affected subsequent ones, and if so, in what way. The subsequent responses would determine whether previous experience affected subsequent behavior. In the first set of experiments (Morse 2000a), we did not change the presentation of the substrates to the spiderlings. Hence, any change in choice noted would most likely be a response to an earlier presentation of a particular substrate, since performances of naive and once-tested spiderlings of the same age did not differ. Next I outline experiments in which we changed the substrate over a set of experimental runs.

I reported on the first-day part of these experiments in the preceding section; here, I direct my comments primarily to days 2 to 5 of that experiment. These subsequent days did not reveal any striking changes from Day 1, which is not surprising in light of the absence of new stimuli. Individuals that quit buds did exhibit a progressively stronger tendency to leave these sites in subsequent runs than in the initial run. In contrast to the bud-leavers, however, individuals did not increase their frequency of leaving flowers on subsequent days. Thus, the spiderlings followed the same general pattern as in the first run, but with somewhat improved efficiency. On the basis of this information alone (but see below), the change in efficiency is most likely the consequence of experience. If it were simply an ontogenetic effect, we would expect similar changes in all groups. In either instance, this behavior would naturally concentrate spiderlings on good foraging sites.

In further experiments on goldenrod, we altered the time between exposures. In one set, we kept age and energetic condition constant, but ran an experimental group only on days 1 and 5, rather than on days 1 through 5, and compared the two groups on Day 5 (Figure 6.3). We found no significant difference in performance between individuals that were run five times and those that were run but twice. Combined with the preceding experiments, this result suggests that any learning that takes place at this time requires only one presentation.

In an analogous experiment, we kept experience constant, but varied age and energetic condition by comparing individuals run on days 1 and 5 with those run on days 1 and 2. We found no differences in performance in response to this manipulation. This finding further sup-

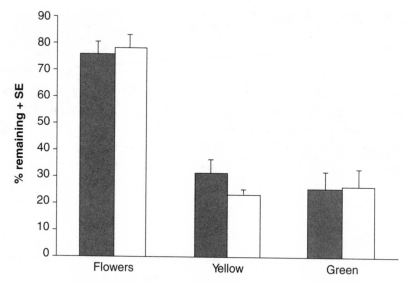

Figure 6.3. Mean percentages (+ SE) of spiderlings remaining on goldenrod inflorescences in different stages of flowering after second (dark bars) and fifth (light bars) 30-minute run on days 2 and 5, respectively, of experiment. Yellow = yellow buds, Green = green buds. Modified from Morse (2000a). With permission from Elsevier.

ported the argument that the modest differences from the naive condition resulted from a single exposure to a substrate and that the information gained was retained for several days.

Evidence from Manipulations. The spiderlings tested on goldenrod were exposed to a single stimulus—flowers, yellow buds, or green buds, which were presented at varying times or at certain times. We have also directly manipulated energetic condition and substrate, providing additional insight into the role played by experience (Morse 2005). To simplify the experimental design of these tests, given the general similarity between responses to yellow and green buds noted in the first experiments, we confined ourselves to a single bud category, yellow buds, which we contrasted with flowers.

A direct way of exploring the effect of energetic condition on substrate choice at this early stage is to provide individuals with food. We manipulated energetic condition by providing spiderlings with a single fruit fly after an initial run similar to those of the previous studies and then comparing their subsequent tendency to exploit goldenrod flowers and buds with individuals not provided with a fruit fly. (Spiderlings provided with a fruit fly every other day grow and molt [Morse

2000b].) We predicted that more of those given food would remain on their site than those without food, and that the effect would be especially marked on the buds if prey capture played an important role in substrate selection. If substrate alone is the resource in question, however, no change would be predicted. Neither those tested on flowers nor buds changed their choice of substrate in response to addition of this food item from similarly run controls without the reward. Again, access to the yolk sac may provide resources that will diminish the probability of this test revealing a difference at this time. It should be noted that this manipulation does not include the actual experience of capturing a prey item in the field, which would be the most sensitive and naturalistic test. We have not yet run that experiment, which would be difficult to perform but feasible.

In another experiment (Figure 6.4), we ran spiderlings once as before and then shifted their substrates on the second day (those on flowers were moved to buds and vice versa). On Day 3 we shifted the spiderlings back to the substrate of Day 1, and on days 4 and 5 we returned them to their substrate of Day 3 (and Day 1). We obtained striking changes in both the flower-bud-flower and bud-flower-bud manipulations, which were of the sort predicted if these individuals responded innately. On Day 2, spiderlings moved to buds (flower-bud-flower manipulations) significantly decreased their tenancy, and those moved to flowers (bud-flower-bud manipulations) increased their tenancy. The responses of these two groups on days 3–5 differed markedly. Spiderlings moved to buds on Day 2 (flower-bud-flower) subsequently remained on flowers significantly more frequently than on Day 1. In contrast, spiderlings moved to flowers on Day 2 (bud-flower-bud) showed a stronger tendency to leave buds on Day 3 than on Day 1, but by days 4 and 5 all signs of a difference from the Day 1 response had disappeared. Thus, the negative experience on Day 2 had a greater lasting effect than the positive experience. Although this experiment demonstrated over- and undercompensation, the overall results were qualitatively in the direction predicted, consistent with a primarily innate response, but with an effect consistent with the experience-related differences found in the initial set of experiments reported above (Morse 2000a).

This difference may be important because we have been looking for juvenile traits carried over into the adult stage that might help to explain the basis for unpredicted patterns of patch choice by adult fe-

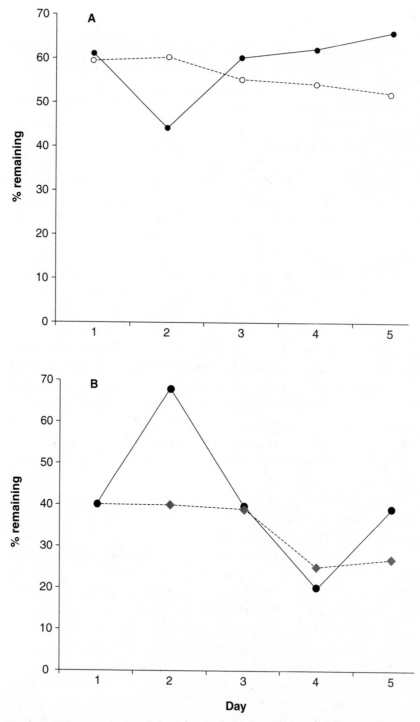

Figure 6.4. Percentage of spiderlings remaining after 30 minutes when (a) placed on goldenrod flowers on Day 1, buds on Day 2, and again on flowers on days 3–5 (solid line). Control on flowers on days 1–5 (dashed line). (b) Placed on buds on Day 1, flowers on Day 2, and again on buds on days 3–5 (solid line). Control on buds days 1–5 (dashed line).

males. Although it is asking a lot for traits from early-instar stages to be carried on to the adult stage, our results demonstrate that these individuals exhibit experience-related differences that are retained for at least a few days. The asynchronous response of the two groups, in which presumably negative experiences have a stronger long-term effect on subsequent behavior than do positive experiences, has been reported in honey bees (Richter and Waddington 1993) as well as humans (Kahneman and Tversky 1979).

We have run comparable manipulations between pairs of flower species (goldenrod, aster, wild carrot), though these experiments ran for only two days. However, when previously naive individuals were moved from their initial flower species to another flower species on Day 2, they changed their tenancy to one that closely matched those of sibs initially placed on this substrate. This result further supports the argument that early-instar *Misumena* spiderlings choose sites largely in response to innate traits. Since we ran this last set of experiments for only two days, we do not know whether over- or undercompensation occurred in a way comparable to that found on goldenrod flowers and buds during days 3–5.

Experience and Learning in Prey Capture. We have also explored experience and learning of the spiderlings as they relate to prey capture itself, testing their responses every third day to fruit flies in small petri-dish arenas (Morse 2000b). We investigated orientation time (time from initial prey movement until the spiderling oriented its body to the prey and stretched its two anterior pairs of legs horizontally) and prey capture (time from initial prey movement until contact made with prey during a successful capture) (Figure 6.5).

Both measures of spiderling time decreased initially. Orientation time stabilized after two to three runs, while capture times became highly variable at this point, with several individuals not even capturing prey. Those that did capture prey at these times often consumed very little if any of them, losing mass over this period even if they made kills. The decrease in performance apparently resulted from oncoming molt (Figure 4.8: Sullivan and Morse 2004) rather than satiation, since individuals in a second experiment, which were presented with the same prey daily, rather than every third day, initially readily captured prey at this accelerated rate. After molt, which required a few days, spiderlings at first retained part of their gain in orientation time, but none of their

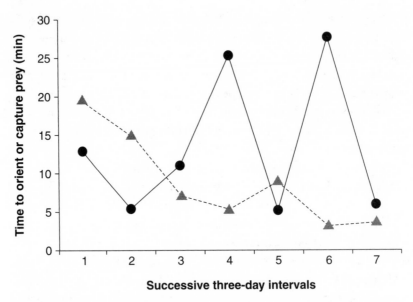

Figure 6.5. Time in minutes required by spiderlings to orient to (dashed lines) and capture (solid lines) *Drosophila melanogaster* prey in successive three-day intervals. Data from Morse (2000b).

gain in capture time, though they captured prey more rapidly than immediately before molt, when their rates were slow and variable. Thus, as a group they did not improve this aspect of their hunting ability (essentially they had to start over from scratch), but the fastest of the individuals did exhibit shortened kill times (Morse 2000b). It will be important to determine if these spiderlings turn out to be the ones with greatest eventual success.

On average, large individuals captured prey most rapidly, a trait that would serve them well under natural circumstances when prey densities are low. Although they did not grow relatively faster than smaller individuals, they retained their initial advantage. Since size and number of eggs are directly related (Fritz and Morse 1985), the successful large individuals of a brood need to be compared against the larger numbers of small spiderlings that could alternatively be produced from the same biomass. Which group will make the greater contribution to the following generation? This study was not designed to explore that variable.

As a whole, these experiments demonstrate little sustained increase in hunting prowess. What remains to be seen is whether individuals that did improve their prey-capture rates made up a disproportionately large part of the small minority surviving to form part of the next breeding population. Although the arenas used were artificial, the traits in question should enhance success under natural circumstances as well. Perhaps a combination of choosing maximally favorable hunting sites (Chapter 3), plus the gains in prowess made by some of these individuals, would provide the most highly adaptive set of traits. As noted earlier, however, patch choice assumes a particularly important role in foraging success, such that the highly accurate choosers might prosper even if they did not exhibit outstanding prey-capturing abilities.

Middle Instars

We have done considerably less work with the middle instars (fourth and fifth) than the second and third instars, but these investigations have uncovered some very interesting patterns and provide suggestions for further experiments. To date this work has consisted of comparing wild individuals collected from the flowers of several plant species with individuals reared in the laboratory up to the fourth and fifth instars under conditions devoid of any normal features of the substrate.

Wild-caught fourth- and fifth-instar spiders strongly favored daisies over buttercups when placed immediately under a bouquet containing equal areas of the two species, likely in part a response to earlier experience. This experiment took advantage of the spiders' response to move upward on vegetation, providing a period (two hours) adequate for them to choose between the two species once they reached the bouquet. However, since we did not know the past history of these wild-caught individuals, it was difficult to ensure that experience accounted solely for the responses. It would be preferable to use wild-caught individuals of known recent history to establish how previous experiences with substrates affected responses to the substrates. This method provides an appropriate comparison with the presumably naive individuals reared in the laboratory. Individuals captured from sites in which only daisies or buttercups grew would be superior subjects for this comparison.

In contrast to the wild-caught individuals, when first exposed to daisy and buttercup flowers, laboratory-reared individuals showed no preference for one species over the other. However, when run a second time one week later, these laboratory-reared individuals showed a clear preference for daisies over goldenrod, similar to that of the wild-caught individuals. This result strongly suggests that the preference for daisies has an innate nature to it, but that it may be overlain by other factors. The specific basis for this behavioral change remains unclear, though it appears that the young are not limited by a sensitive period, even though the wild-caught individuals have already developed these capabilities (see Clayton and Lee 1998).

It is difficult to compare the responses of the naive middle instars to other parts of *Misumena*'s life history. Subsequent runs of these individuals will be required to establish the limits of their ability to develop choices at this stage and whether their choices in subsequent instars are only developed in response to the presence of insect prey, as in the adults. This result will assume great interest because it will help to establish when the basis for the spiders' pattern of substrate choice changes. The presence of prey is not a necessary part of substrate choice in second instars, but the presence of insects is necessary for positive substrate responses by the adults. Neither do we know the response of naive second instars to the daisy-buttercup choice, which, though the major choice between foraging substrates for fifth instars in late spring, is not available when second instars emerge from their nests. The relationship of the flowering phenology of the plant community and the dependence of the spiders upon it remains a question to be addressed. Do individuals of a particular instar respond "appropriately" to stimuli that they would normally never experience in that instar? The daisy/buttercup versus goldenrod comparisons are the most obvious tests to be made, using second and fifth instars, respectively. To address this question one could rear spiderlings out of the normal phase to match the phenology of the flowers, and preferably the reciprocal as well.

Penultimate Females

Daisies and buttercups are also important hunting sites for penultimate female *Misumena* and are the most common flowers in bloom when most females are in this stage. Penultimate females exhibit a strong preference for daisies over buttercups, as predicted from the superior

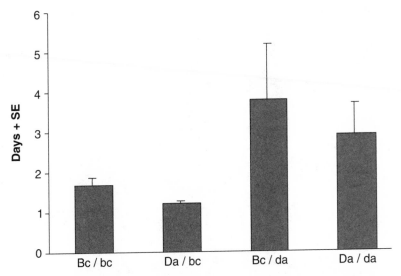

Figure 6.6. Days (+ SE) penultimate females remained on flower when shifted from buttercup (Bc) to another buttercup, daisy (Da) to buttercup, and so on. Modified from Morse (1999b). With kind permission of Springer Science and Business Media.

numbers of insects that visit daisies and the corresponding differences in the numbers of prey captured by the spiders when hunting on these two flower species (Figure 6.6: Morse 1999b). When placed on a substrate, individuals collected on daisies remained on daisies significantly longer than did daisy-collected individuals moved to buttercups. In contrast, those collected on buttercups remained significantly longer on daisies than on buttercups. Perhaps even more revealing, individuals from buttercups moved to daisies remained even longer than did those simply moved from one daisy to another, and individuals moved from daisies to buttercups remained shorter periods than those moved from one buttercup to another. Since we collected all these individuals from sites that provided no opportunity to forage on the other species, the results strongly suggest an ability to compensate in response to their current substrate. Such flexible behavior should favor them at sites containing both daisies and buttercups, since considerable variation in their flowering occurs from year to year. One year we could not run this experiment because the daisies failed to flower in the same areas as buttercups in our study area, probably because of drought the

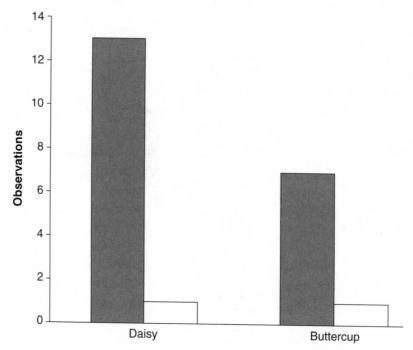

Figure 6.7. Flower preferences of newly molted adult female spiders as a function of flower species upon which they molted (dark bars—chose same flower species, light bars—changed flower species). Modified from Morse (1999b). With kind permission of Springer Science and Business Media.

preceding summer (Morse 1999b). We have not reared naive spiders to this stage to test whether they retain an innate preference for daisies over buttercups.

The vast majority of penultimate females molting on a flower selected a flower of the same species when making their next choice, whether the most often favored daisy or buttercup (Figure 6.7: Morse 1999b), thereby showing considerable conservatism in choice. Most likely, these individuals carried species-specific foraging information over from one instar to the next, which contributed to the first adult choice. Alternatively, in a phenomenon analogous to imprinting, these individuals merely responded to the first substrate they encountered after molting. In either instance, the spiders ended up on the same flower species as before, which could help to explain the suboptimal patch-choice behavior exhibited by a minority of the adult females (Chapter 3).

Adults

In some of our earlier work on *Misumena* foraging at milkweed, adult females collected from other species of flowers appeared to exhibit a lower fidelity to milkweed than did those collected from milkweed. At that time we viewed the matter as merely a potential source of error in the foraging experiments and simply ensured that we used spiders captured on milkweed for these experiments; we did not give the matter much further thought. However, as emphasized earlier, previous experience is a potentially critical source of relevant variation. Therefore, we subsequently revisited the problem to make explicit comparisons of individuals that had immediately previously foraged on milkweed and on pasture rose, the most frequent alternative hunting site to milkweed.

Pasture rose attracts fewer of the large insects required by adult females than does milkweed, with capture rates of these insects correspondingly low on pasture rose (Morse 1979). We would therefore predict a preference for milkweed when the spiders have a choice between flowers of the two species. And that is exactly what we found (Figure 6.8: Morse 2000c): the spiders strongly preferred milkweed to rose. Spiders that captured prey also were more likely to remain than those that did not capture prey, a trait that further increased the difference in residence times on the two flower species. Of the spiders that did move, however, those captured on milkweed and shifted to rose did not remain as long on these flowers as did those captured on milkweed and then returned to milkweed. Of the spiders in the experiments that moved, those captured on rose and returned to rose remained as long as did those captured on rose and moved to milkweed, however. Thus, novelty appears to play a role in the giving-up times of these spiders, regardless of the flower species (Morse 2000c), and permits them to evaluate these sites.

Detailed comparison of hunting on rose and milkweed is difficult, however, because pasture rose only blooms for a day. Moreover, the spiders cannot count on another flower to bloom on the following day immediately alongside the one they have just exploited. Thus, we could not run these experiments past one day, although a milkweed stem often guarantees hunting opportunities for two weeks or more. However, the choices as seen are consistent with those of the penultimates

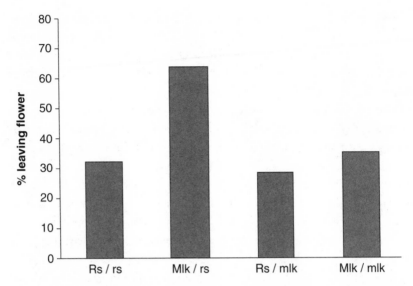

Figure 6.8. Percentage of adult females moved from rose (Rs) to another rose, milkweed (Mlk) to rose, etc., that left flowers before the end of the day. Data from Morse (2000c).

and strongly suggest that experience at the previous hunting site plays a role in the giving-up times of these individuals. Although they strongly suggest an effect of past experience, the effects are not as strong as those for longer-flowering blooms.

It would be possible, though difficult, to run this experiment through a second day if sites could be reliably located in which rose flowers bloomed in an adjacent position on the following day. In addition to correcting the obvious limitation in these experiments, the runs could address the interesting question of whether the spiders gave up these sites at higher than predicted rates on the second day, since sequentially available flowers would rarely be available. One might predict that spiders with and without previous experience on pasture rose would differ in their performances, and that performance would further differ among experienced individuals, depending on whether they had been confined to the normal one-day stands or had had the opportunity to experience sequentially available adjacent flowers.

Changes over Ontogeny

The relative roles of innate factors and experience thus appear to shift strikingly over ontogeny. Innate factors play a dominant role in early

ontogeny, with the role of experience in nonnaive second instars severely limited and largely confined to improving performance in certain basically innate tasks. In contrast, recent experience, combined
with direct responses to prey, plays a significant role among adult females, which exhibit the effects of some learning. Penultimates appear
to use experience and learning in ways comparable to those of adults.
Major changes occur between the third and penultimate instars and are
an important part of our current interest. Although we do not know
exactly when this change takes place, work to date suggests that intermediate conditions exist in the fourth- and fifth-instar period. We also
are uncertain whether the change occurs gradually over this time or
whether it is abrupt: a step-change.

It is interesting to ponder the basis for the change in behavior seen
over *Misumena*'s ontogeny. If individuals occupy variable environments, they should enjoy a strong advantage by incorporating recent
experience and responding directly to prey and other cues. Since the
young are born with only a limited nutrient supply (yolk), any simple
regime that enhances early prey capture should be selected, as in the
innate traits that facilitate the choice of favorable hunting sites. Because their mothers often place their nests near favorable hunting sites
for the young (Morse 1993a), chances are high that the young will soon
find such a site. The mothers' nesting choices will thus give the young
a fairly dependable start, and information subsequently gained could
allow them to fine-tune their responses to the environment, which will
be in continual flux because of the changing populations of flowers that
constitute their hunting sites. Not only do individual flowers senesce,
thus requiring a change of site, but most flower species have relatively
circumscribed flowering periods, thus requiring that the spiders continually readjust their responses to the environment. All of these factors should select for individuals that take advantage of information received, which would hasten their ability to find and to exploit new
flowers of a particular species efficiently. Thus, initial innate responses
should favor naive young, but environmental input should subsequently favor the use of experience, learning, and the ability to respond
directly to stimuli experienced.

Learning in arthropods may often, but not inevitably, be lost over a
period of hours or days if not continually reinforced (Abramson 1994;
Keasar et al. 1996). This reversion can be readily observed in *Misumena*'s hunting sites, with the dominant flower species changing over

the summer from daisies and buttercups to milkweed and pasture rose to goldenrod and flat-topped white asters. However, the details differ from site to site and may involve other flower species, which in their own right may constitute a yet more formidable problem. Other species likely to be used frequently as hunting sites at certain times of the year include dandelion *Taraxacum officinale* and various clovers *Trifolium* spp. early in the season; yarrow *Achillea millefolium,* black-eyed Susan *Rudbeckia hirta,* and common St. Johnswort *Hypericum perforatum* during midsummer; and various other aster species *Aster* spp. and wild carrot *Daucus carota* in the late summer.

Although we do not yet know whether the shift from innate patterns of substrate choice to direct responses to prey occurs gradually or suddenly, we do know that penultimates respond similarly to adults in reacting directly to their prey, and that second and third instars respond in a basically innate way. Naive fourth and fifth instars initially appear unable to choose between two commonly available flowers, ox-eye daisies and common buttercups, consistent with the patterns of adults in the absence of prey. However, in subsequent runs they favor daisies over buttercups, in common with the usual distribution of *Misumena* in the field, and in agreement with both the spiders' hunting success and the frequency of visitation to these sites by insect prey species (Morse 1999b). This information suggests that the fourth and fifth instars are the ones in which the critical changes occur.

As the spiders get incrementally larger, it is possible that at some stage their nervous system reaches such a size that they can exploit learning more fully than the second instars, though even fruit fly larvae far smaller than these juvenile spiders can learn (Dukas 1999). Has the nervous system at a particular molt increased in size such that it is only now capable of learning skills that were not possible at the immediately previous size of the nervous system? Other examples are known in which attainment of a certain minimally sized nervous system may permit the acquisition of more complex behavioral patterns (Bonner 1988; Barth 2002). As emphasized by Bonner (1988), fundamental changes in organismal complexity are likely to emerge as a result of simple shifts in body size; these shifts are often grounded in the species' relationships to certain environmental variables or its relationships to other species in their communities. Such a shift would be timely, given the increase in experience and opportunity for learning. However, this proposition is still largely speculative, both with regard to *Misumena*'s behavioral patterns and to neural function in general (Dyer 1998). If change in the importance of learning by *Misumena* is a stepped process, it may warrant investigation from a morphogenetic perspective.

The Problem Posed by Flowering Phenology

I have emphasized certain aspects of the seasonal cycle that impinge on *Misumena*'s success—the collective phenology of the flowers it frequents, the seasonal climatic variation, and the inability of adults to overwinter combined with the varied impact of that season on the other population members. These constraints will tend to program *Misumena* in terms of the optimal periods for prey capture, which may or may not closely match the blooming periods of the flowers they frequent as hunting sites. The blooming time of flowers, in turn, is likely to be restricted for reasons not directly related to the constraints of the spiders. The combination of constraints on the spiders and flowers should result in the co-occurrence of certain combinations of instars and flower species, and the scarcity or absence of others. These limits, and the fact that several of the instars may potentially encounter a particular plant species, should favor a fair degree of flexibility in response to the flowers. Flexibility should select for maximizing the use of experience. On the one hand; a stereotyped reaction to a particular flower species would lower the options open to an individual that could respond flexibly to the conditions it was currently experiencing. On the other hand, where the conditions are more likely to be highly predictable, as in goldenrod for second instars, innate responses are likely to be advantageous. Second instars likely possess the fewest possibilities for flexibility. They have not yet attained much experience, and their nervous system possibly limits alternatives to an innate response. Matching the phenologies of the flowering substrates and the prospective prey that visit them to *Misumena*'s schedule represents an as-yet largely unappreciated challenge.

Predators themselves may play a role in plant demography—for instance, by affecting seed set positively (Louda 1982; Romero and Vasconcellos-Neto 2004) or negatively (Suttle 2003) and hence potentially affecting the times at which plants may "schedule" their reproductive phenologies. These scenarios take plant-pollinator or plant-herbivore relationships as important factors that drive plant reproductive relationships. Actions on the pollinators by predators may thus affect seed set, thereby creating an effect that presumably acts against the directional pressure generated by plant-pollinator relationships.

For the most part, the importance of this force in natural systems remains to be seen. However, thus far we have not detected any effect of *Misumena*'s presence on pollinia removal rate or seed set in milkweed (Dukas and Morse 2005). Milkweed, which has large umbels that often produce far more fertile fruits than the plant can rear to maturity due to the energetic constraints on it, may resist this type of predator intrusion more effectively than single flowers on which the predators are conspicuous. Milkweed's clonal nature will likely further complicate demonstration of such an effect. *Misumena* hunting on milkweed often conceal themselves, further complicating the issue. The absence of a short-term effect does not, however, permit the conclusion that milkweed is unaffected, for current theory suggests that the massive oversupply of flowers produced by many species of plants may permit them to select the best sires for their fruits, with the poorer ones selectively aborted (Willson and Price 1980). Thus, if the spiders decreased the delivery of outcrossed pollinia to the flowers of a clone, they could decrease the fitness of milkweed, though it would be extremely difficult to demonstrate if it did occur.

Predators would more likely exert an effect on plants with conspicuous single flowers upon which *Misumena* could not hide, such as pasture rose, than on milkweed. Even in this instance, however, the question of whether pasture rose plants are resource limited, and would merely shift their resources to the reproductive efforts of another flower if not pollinated as a result of a spider's presence, remains to be seen. Dukas (2005) has just provided a striking example of predator impact on seed set, again involving predation of bumble bees. In heavily predated sites about beewolf colonies, the seed set of plant species dependent on bumble bee pollination declined significantly in bee-depredated areas.

Flowers may compete for their pollinators, however (Proctor and Yeo 1973), in which case flowering times of different species should minimize their overlap. This factor would lead to substantial flowering in a community over maximally long parts of the season. Consequently, the flowering phenology of a plant community would become maximally spaced, though Poole and Rathcke (1979) have pointed out that it may be difficult to separate such a phenology from a random one. A community with maximal spacing of flowers, however conceived, would provide optimal conditions for *Misumena*.

Some Genetic Implications

We have focused on observable changes in traits, both behavioral and quantitative, at the phenotypic and individual level, but it may be profitable to speculate on why *Misumena* exhibit the traits we have observed and how they maintain them. Behavior is a potentially flexible medium that permits a variable response, especially if combined with experience. It provides another dimension to phenotypic plasticity, the ability of an organism to change with the environment, but usually associated with morphological, physiological, and similar factors. Although behavior may provide a broad range of responses to selective forces, those responses are predicated on genotypes amenable to this change. Whether such phenotypic variation can be attained depends on whether it is matched by a genetic response to selection. That factor in turn depends on whether appropriate genetic variation exists to permit a heritable change; this is not a foregone conclusion owing to the heritability of these characters, not to mention the possible complications of pleiotropy and linkage disequilibrium. Critical factors for any population, they assume particular interest in our *Misumena* populations: they might help to explain why such major variation in patch choice and other behaviors remains in the face of the heavy selective pressure generated by successful patch choosers.

Differences between Broods

Working with groups of known sibs and controlling environmental variation (Morse 2000a), we could obtain an estimate of the genetic contribution to variance in the performance of spiderlings by comparing results between broods. Although members of a brood showed very similar responses to goldenrod flowers and buds, they varied considerably in the time they took to vacate their test substrates (Morse 2000a); that is, they consisted of "fast" broods and "slow" broods. This variance in rate acted as if it operated through a "speed governor."

Broods did not change relative to each other over the period tested. As an extreme, in a single brood whose members were exposed either to goldenrod or aster inflorescences, those spiderlings responded like most broods to goldenrod, but it was impossible to keep brood members on asters long enough to run the planned experiments. Thus, the differences among the broods are substantial in some instances and warrant further attention.

The similarity of responses to the substrates by the broods takes on considerable interest, given the variance in this behavior exhibited by adult females. Obviously, this comparison involves different substrates by individuals of markedly different experience, size, and physical attributes. Of possible importance, though, the naive young appear to negotiate substrate choice more efficiently than do the adults. However, it must be kept in mind that the young are occupying what is, arguably, the most predictable substrate they will encounter in their lifetime, a resource that can be relatively safely exploited by an innate response. The subsequent divergence in the behavior of adults from each other is thus most likely a consequence of the adults' different lifetime experiences, keeping in mind that differences in foraging rates of the young could, if retained, account for some variance in adult behavior. Admittedly, this conclusion is a broad extrapolation based on a modest body of information and no direct evidence.

The information on our populations is sufficient to suggest the presence of considerable within-population variation of a probable genetic basis. We have not had the opportunity to compare these variables with those from populations isolated from them, but the presence of considerable within-population variability is consistent with between-population variation.

Riechert (1982) and Hedrick and Riechert (1989) found considerable behavioral variation in both foraging and aggressive behavior in the funnel-web spider *Agelenopsis aperta* (Agelenidae), which was related to differences in habitats occupied by these spiders in the southwestern United States. Members of a population from desert grassland with scarce prey and few predators attacked prey more readily and fought frequently and aggressively over territories. In contrast, those from riparian woodland with more common prey and heavy predation were choosier about their prey and fought less than those from the desert grassland, traits consistent with their high vulnerability to predators. The spiders exhibited similar patterns in the laboratory, as did the members of a second laboratory-reared generation. However, the predator-vulnerable group did not attack a narrower range of prey, as predicted from foraging theory, apparently the consequence of gene flow from the desert grassland (Riechert 1993; Riechert and Hall 2000).

These results suggest that it would be instructive to compare *Misumena* populations from different habitat types. The striking differences in behavior between Riechert's populations in the low- and high-predator settings suggest that it would be profitable to compare our populations with those from areas with high populations of sphecid wasps, either immediately inland or to the south of our study areas.

A Simple Genetic Basis for Behavioral Traits?

It is generally assumed that complex patterns such as most behavioral acts are under the control of multiple genes. One may further anticipate that even if single-gene control occurs, pleiotropic effects (one gene affecting multiple traits) or other genes, often closely linked ones, will play significant modifying roles. One would thereby expect patch choice, as used for the choice of hunting sites, to be under complex genetic control and therefore not amenable to study using anything short of quantitative genetic analysis. Nevertheless, increasing evidence suggests that certain behaviors, including foraging behavior, may have a simpler basis than others, sometimes even controlled at a single locus.

Sokolowski and her co-workers (Sokolowski 1980, 1998) have found a simple Mendelian basis for two distinctly different foraging behaviors of fruit fly larvae on media in petri dishes, which she has referred to as "rovers" and "sitters." These behavioral types either moved around a great deal on the plates in the process of searching or consumed media within a small area. Sokolowski noted this same behavior on fly larvae occupying decaying apples in orchards. Later, her group identified the gene that controlled this behavior (Osborne et al. 1997). This gene *(for)* is associated with control of PKG (a cyclic guanosine monophosphate-dependent protein kinase) activity, which is known to play a role in learning and memory in other animals. The rover-sitter syndrome may not be confined to *Drosophila*. For example, a homologue *(Amfor)* has recently been found in honey bees (Ben-Sharar et al.

2003)—it turns on foraging through a precocious change in phototaxis. In contrast, the ant homologue *(Pbfor)* represses foraging; thus, though widely conserved across insects, the gene's regulation may evolve (Ingram, Oefner, and Gordon 2005). Several other species as well show evidence of foraging control by *for*-like processes (e.g., *Caenorhabditis elegans:* Fujiwara, Sengupta, and McIntire [2002]; review by Fitzpatrick et al. [2005]).

We have not undertaken direct study of the genetic basis of behavioral characters in *Misumena*, but some of the results emerging from the work on choice of hunting sites by adult female *Misumena* are strongly suggestive of behavior associated with *for* or its equivalents. In six independent assessments of individuals leaving and remaining at sites, the proportions of individuals remaining ranged between 69 and 80 percent ($Ns = 20$–75), with a mean of 74.0 percent (Morse and Stephens 1996). These figures fall tantalizingly close to the 3:1 ratio predicted from a simple Mendelian dominant and suggest the importance of exploring this relationship more closely.

The similarity of *Misumena*'s patch-choice behavior to the rover-sitter relationship alerts us to the possibility that the patch-choice conundrum of adult female *Misumena* is a manifestation of the very same set of alleles as those of rovers and sitters in *D. melanogaster* (leaver = rover, remainer = sitter). Leavers move considerably more frequently than remainers in our populations of *Misumena* (Morse 2000c) in a way that is very comparable to the relationship between rovers and sitters in Sokolowski's populations of *D. melanogaster.* However, if such a genetic basis to foraging behavior exists, it is of interest to resolve the possible conflict between it and the putative relationship between spiderling and adult patch choice noted in the preceding section.

Phenotypic and Genotypic Characteristics of Tradeoffs

Tradeoffs, as I have suggested, are a center of ferment in the study of life history evolution. I wish to investigate them from the perspective of understanding what the theory of phenotypic and genotypic tradeoffs can contribute to understanding the relationships between several life history variables of *Misumena.*

This issue is of particular interest to me, for much of the discussion about tradeoffs in the life history of *Misumena* involves the behavioral acts that generate these tradeoffs. Behavior is taken to be under ultimate genetic control, but to allow an individual to respond to the changeable and unpredictable situations that it experiences. Encoding behavior at a strictly genetic level, as is likely for some imprinting and certain types of reproductive behavior, commits an organism to a set response that would sometimes be highly counterproductive in environments and social conditions less predictable than that experienced under these special conditions. These circumstances result in strong selection pressure for flexible behavior (Bonner 1988)—and for a genetic milieu that will allow it. However, if one considers the cost of strictly encoding the myriad flexible behavioral patterns used by an animal with sophisticated learning capabilities, the implausibility of direct genetic programming becomes apparent.

Learning is now understood to play an extensive role in the behavior of arthropods, thereby greatly expanding our appreciation of their functioning and potential. As Stephens and Clements (1998) have noted, although an animal may experience certain events several times in its life, the conditions under which it experiences them may not always, if ever, be the same. And experiences of an individual in such a situation may temper how it responds under similar circumstances in the future. Thus, it would appear surprising if such traits were completely under immediate genetic control.

This emphasis on genetically based tradeoffs tends to lead to the assumption that any advantage is "bought at a cost," and that the participant is involved in a zero-sum game. Yet, the options within the set of variables tested should be broader than this, and the frequency with which nonconformities to this prediction are found suggests that it is an incomplete model (see Bell 1984; Scheiner, Caplan, and Lyman 1989). For instance, success at a particular stage may result in success at subsequent stages, which is partly or solely a consequence of the earlier success. In *Misumena* this advantage comes about through previous foraging success, which results in the spider producing such a large amount of biomass that it has a residual effect at subsequent stages (Morse and Stephens 1996). As a direct consequence of this initially large biomass, the spider enjoys advantages later in the life cycle. Depending on a single set of initial circumstances, this advantage progressed through several stages of the life cycle; this is a striking example of Grafen's (1988) silver spoon effect (see Chapter 4). This

difference reflects the fact that some individuals are simply more successful than others under a given set of circumstances, either by chance
or by the fit of their attributes to the conditions of the time (i.e., more
fit). These individuals may make contributions to the gene pool that
significantly alter the attributes of the following generation, although
in the absence of supporting genetic variance, no subsequent advantage
will be recognized. Attention to the silver spoon effect may help to
clarify the frequency and nature of genetically based tradeoffs.

Synthesis

Misumena vatia spiderlings start their independent life using innate
traits in their selection of hunting substrate. This pattern changes relatively little over a common substrate during a five-day period of experimentation and does not change fundamentally until they reach their
middle instars. Over this period, direct response to substrate wanes,
and they eventually begin to respond primarily to cues produced by
their prey. Changes over this early period consist largely of increased
rates of performance in some traits. These rates often differ among
broods, a variation that they retain under common conditions. In the
important traits of orientation to prey and attack, though showing declines in the second instar, only part of the individuals decreased their
strike time at the start of their third instar. It would be particularly interesting to see if they are disproportionately represented among those
reaching adulthood. We found one response to early experience of possible major interest—an asymmetric change associated with shifts of
substrate. Those moved from flowers to buds (high-to low-quality
sites) on Day 2 exhibited higher site fidelity when returned to flowers
on days 3–5 than did controls exposed to flowers throughout. In contrast, the reciprocals (buds to flowers and back to buds) exhibited no
tendency to over- or undercompensate. Older individuals showed
more flexibility, consistent with their use of direct cues and recent experience and perhaps increased cognitive skills. We do not know the
basis for the shift in response, though it represents a time at which
hunting sites are less predictable than early in ontogeny, individuals
have had considerable experience, and they may have greater neural
development.

Although we have not explicitly explored the genetic underpinning of the behavioral traits we have studied in *Misumena*, our analyses have permitted the initial screening of some traits. Certain aspects appear to have a direct genetic basis. Of particular interest, the responses of brood members to a particular variable often resemble each other more closely than those of other broods reared under similar or identical conditions, even though most or all broods responded in a generally similar fashion. For example, in studies of flower choice by spiderlings, members of all broods usually selected high-quality flowers or inflorescences over others, but members of some broods made these choices more rapidly than others. Although ordinarily these differences might result only in a slightly earlier start of profitable foraging activity, occasionally this difference could result in a fundamental change in resource exploitation. The tendency for about one-fourth of the adult females to select suboptimal hunting sites repeatedly is also remarkably consistent with the one-locus, two-allele rover-sitter relationship of foraging *Drosophila* larvae.

7

Some Sensory Aspects
of Substrate Choice

AN ANIMAL MUST CONTINUALLY MONITOR whether its site meets its current needs. For *Misumena*, at one time or another these needs will include a place to hunt, and evaluation of this point source may be termed *substrate selection*. Often animals must also select sites considerably larger than a point: this type of discrimination is referred to as *habitat selection*. Substrate selection and habitat selection may be considered aspects of patch choice, albeit usually at different spatial scales. Minimal requirements of a permanent site include a place to forage, avoid predators, shelter from unsatisfactory climatic conditions, and reproduce. Without all of these amenities, an animal must eventually move; dispersal is a consequence of these limitations.

As a small, slow-moving predator that must periodically change sites to track its food supply, *Misumena* faces important decision-making problems at both the levels of substrate selection and habitat selection. Since needs at the substrate level may differ from those at the habitat level, *Misumena* may experience formidable problems, not only in terms of finding sites, but in assessing them. In this chapter I will explore how *Misumena* evaluates the conditions experienced in its search for hunting sites, nesting sites, and mates. As a small species that moves relatively little over most of its lifetime, *Misumena* occupies sites that may be very small and not seemingly of the same spatial scale as that of the larger animals usually featured in studies of habitat selection. However, *Misumena* faces major temporal challenges at a site that may prevent it from occupying areas that appear to be satisfactory over much of a lifetime (Chapter 6).

Hunting Sites

As a small arthropod, *Misumena* experiences clear limitations in its use of available cues. It is probably more effective at detecting motion than fine details in the environment, even though arachnids, unique among arthropods, possess camera-type eyes (Land 1985). Although crab spiders are generally assumed to have poor eyesight, Foelix (1996a) and Barth (2002) placed them with other spiders that have good eyesight (but see Homann 1934), based on their sit-and-wait hunting strategy and presumed close relationship to the highly visual jumping spiders (Salticidae) (Jackson and Pollard 1996). In *Misumena*, visual capabilities may be greater than is generally recognized (see below), but their abilities could never be considered comparable to those of the jumping spiders. Although increasing evidence suggests that pheromones play a vital role in the everyday life of many spiders, their role in the life of *Misumena* is less significant (see Chapter 9). Tactile and vibratory cues probably play the biggest role in *Misumena*'s perception of its outside world. The importance of vibratory cues is commonly recognized in web-spinning spiders, where it is the key to prey perception, but its importance to other species may be great as well (Barth 2002). The relative importance of different types of cues probably varies over ontogeny.

Tactile Cues Used by Early Instars

Tactile information may provide the single most important source of cues for young instars, including naive individuals that have just emerged from their egg sacs. Naive spiderlings make decisions about substrates that appear to be based largely or entirely on tactile cues. These spiderlings select sites that attract maximum numbers of potential prey; however, they make these choices whether or not prey items are present (Morse 2000a). That these newborns make their choices either in the absence or presence of prey suggests that several other potential cues do not play a central role here, including visual, olfactory, and auditory/vibratory stimuli. These decisions include selecting among flowers of a given species that differ in phenology and between species of co-occurring flowers. The fine-level ability of totally naive spiderlings to select between goldenrod flowers (flower heads) and buds, and further to discriminate between yellow (older) and green (younger) buds, could result from a variety of factors, and it seems unlikely to be based on the characteristics of a single plant species. Perhaps the simplest set of useful cues for choices of hunting sites on goldenrod relates to the size of the alternatives. Flowers are larger than buds, and older buds are larger than young buds. Flowers, however, exert markedly stronger preferences than do buds. This difference could result from the contribution of nectar and pollen by the flowers, or it could more simply be a consequence of tactile differences between "petals" (ray flowers) and unopened flowers. We question the importance of nectar or pollen. We have seen little sign of these spiderlings consuming either nectar or pollen (see Chapter 3), though a few other studies (e.g., Pollard, Beck, and Dodson 1995) report that some crab spiders may use these substances as a resource. Although the moderate difference in response to older and younger buds could be a consequence of size, and possibly of color, they presumably do not result from differences in the availability of nectar or pollen.

The basis for these differences in preference is thus somewhat complex and probably depends on more than one variable. Furthermore, when we presented choices of substrates simultaneously, their choices, though similar, were not as strong as when we presented them with only one substrate at a time (simultaneous vs. sequential choice: Chapter 3; Morse 2005). The simultaneous presentation apparently provided more ambiguous information than did the single presentation of either a flower or a bud. Naive second instars also preferred goldenrod to flat-topped white aster and wild carrot, and they exhibited the same relationships in simultaneous and sequential tests as in the tests between goldenrod flowers and buds (see above), further supporting the argument for tactile cues. These differences in choice match the visitation rates of small insects to the three substrates.

In areas close to their natal nest, however, many young follow the lines of fellow littermates. Thus, the initial decision (or movement) of one individual may disproportionately affect the sites visited by other members of the brood. However, it is not clear how these lines affect the behavior of the followers after they arrive at a particular site. Does the presence of other individuals at the site promote aggregation or repulsion? Judging from the behavior of the spiderlings over time, we anticipate that this behavior would initially foster aggregation in those most recently emerged, but that in older individuals it would promote dispersal.

Visual Cues of Early Instars

Initially, the spiderlings' preference for certain flowers did not seem to have a visual aspect, because they did not recruit preferentially to similar-sized goldenrod and aster targets only 10 cm away (Chapter 3). This response is of considerable interest, for, as noted earlier, when spiderlings first disperse from their nest sites they routinely move on lines considerably longer than 10 cm, the average length of these lines being 40 to 50 cm. Thus, these early dispersers leave their nests in a random way relative to the substrates they contact with their lines. Only secondarily do they make decisions about the sites after they have proceeded to them on a line. Not surprisingly, when we tested their responses to the same targets (goldenrod vs. aster) at 50 cm, the spiderlings also showed no preference for one species over the other. The primary difference at 50 cm was that a considerably greater proportion of the spiderlings never reached either of the targets presented. Fewer left the dispersal site within the time frame (30 min) allotted. Of those that did leave, a much higher proportion moved off on lines to other more distant sites (over 1 m away) or simply ballooned away.

We subsequently discovered that at a coarser level, visual cues may play a role in choice at a distance. Our first recognition of this perception resulted from inadvertently failing to control for the size of the targets when running the 10-cm species-choice experiment. In the process, we presented an aster target twice the area of the goldenrod target, though of similar width (Figure 7.1). The spiderlings strongly favored the aster. Furthermore, they made this choice even though the flowering part of the aster target was only half the area of the flowers on the goldenrod target, which was composed completely of flowers. The remainder of the aster target consisted of green leaves. Thus, the spiderlings apparently responded explicitly to the gross difference in size of the target, rather than to any particular characteristic of it. We

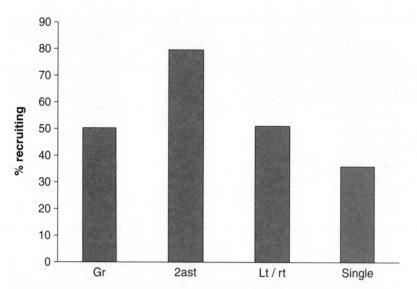

Figure 7.1. Percentage of spiderlings recruiting on lines to goldenrod or aster target from origin 10 cm away. Gr = % of times spiderlings reached goldenrod (51%), versus aster (49%); 2ast = aster twice the size of goldenrod; Lt/rt = two goldenrod inflorescences, most frequently reached is depicted (51%) (a similar aster experiment yielded the same results); Single = single goldenrod inflorescence, percentage reaching site (same results for aster). Data from Morse (2005).

have not yet determined how frequently the spiderlings are faced with this choice, how important these differences are in the field, or what their consequences may be for the site choices and exploitation patterns of the spiderlings. Neither have we yet tested whether their choices would change as differences in size of the two targets became less extreme.

A second set of experiments supported this interpretation as well. Since the first set of experiments did not include tests in which two inflorescences of one of the species were presented as a control for the choice between one of the two species, we ran a series of experiments to test this factor. They revealed no difference in frequency of "hits" from that found in the mixed (goldenrod/aster) series (Figure 7.1: Morse 2005). This finding strengthened the conclusion that the spiderlings did not distinguish between the two plant species under these conditions. We also ran companion experiments in which we presented the spiderlings with only one inflorescence (goldenrod or aster) rather than two. Without exception, whether goldenrod or aster, the spiderlings contacted the single inflorescence roughly half as frequently as the two inflorescences combined (Morse 2005). Overall, these results strongly suggest that the spiderlings do not distinguish between common potential substrates at even a short distance. Frequency of recruitment to a particular target appears to be a consequence of the size of that target, and they may respond to that variable with some precision.

Cues Used by Adults

In contrast to the spiderlings, adults probably do not depend primarily on tactile cues for selecting substrates. As noted in Chapter 3, they make these decisions to a large degree in response to the activity of large prey at flowers. Thus, substrate choice of the adult female spiders is an indirect consequence of selection by the visitors to the flowers, whose own decisions are governed by the amounts of nectar and/or pollen present (Morse and Fritz 1982). Thus, even if those resources do not attract the spiders directly (see Pollard, Beck, and Dodson 1995), they do so indirectly. However, our primary interest here is to determine whether the adults retain any of the traits on which the early instars depend so heavily. The modest, repeated (but nonsignificant) response to substrate quality by adults in the absence of prey (Morse 1988b) suggests a limited ability to evaluate substrate independent of the activity of their prey. If so, rather than exhibiting an all-or-none pattern, the contributions of innate and learned patterns shift in importance over ontogeny. However, any direct response to substrate is overridden by the presence of prey, and its overall importance is in question.

We compared the distribution of spiders from patch-choice experiments on milkweed umbels with the distribution predicted by several possible hypotheses: only the highest-quality umbel visited (the prediction from foraging theory), visits of prey to umbels of different quality on a stem, percentage of white (nectar-producing) flowers on the different umbels of a stem, flowers visited by prey on umbels of different quality on a stem, total prey caught by the spiders on umbels of different quality on a stem, and large prey (bees) caught on umbels of different quality on a stem (Figure 7.2: Morse and Fritz 1982). Only

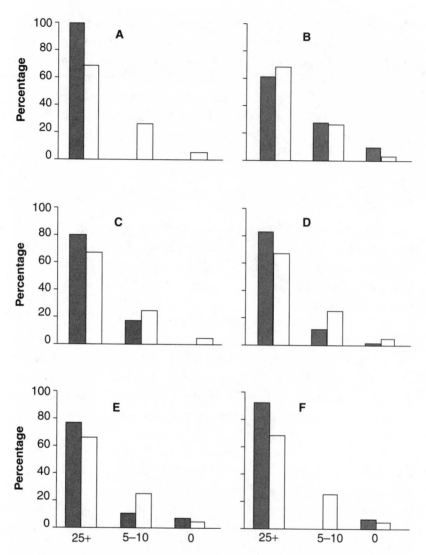

Figure 7.2. Predicted (dark bars) and observed (light bars) frequencies of adult female spiders occupying milkweed umbels with 25+ (high-quality), 5–10 (middle-quality) and 0 (low-quality) white (nectar-producing) flowers. Predicted values based on cues potentially available to the spiders. (a) = Optimal foraging theory, (b) = Number of umbels visited by prey, (c) = Percentage of white (nectar-producing) flowers, (d) = Number of flowers visited by prey, (e) = Total prey caught, (f) = Number of bees caught. All except (b) (number of visits by prey to umbels of different quality) differed significantly from the predicted result. Modified from Morse and Fritz (1982). With permission from the Ecological Society of America.

the second of these alternatives, visits of prey to umbels of different quality on a stem, did not differ significantly from the distribution of umbel choices obtained in the experiments (70 percent of visits to high-quality umbel/ 30 percent to middle and low-quality umbels). Although the alternatives exhibited a general resemblance to this distribution, they did not resemble it as closely as the umbel distribution. The simple all-or-nothing prediction from foraging theory provided the poorest fit of the lot, which emphasizes the importance of factors other than the simplest economics in the spiders' selection strategy.

Thus, the spiders do not make simple decisions to solve their patch-choice problems, but neither do they make use of some information potentially available to them. Although one might predict that total prey capture, or capture of their most important prey, bees, would provide the most accurate information about site quality, the modest number of captures made by any individual may not provide as accurate a basis of evaluation as the total number of visitors to the sites. Close approaches to the spider by the prey did not improve the fit over that obtained from touchdowns (Morse 1988b), further suggesting the importance of easily measured variables to decision making. This result held over a fourfold range of visitation frequencies—from half to twice normal visitation rate at the height of the season (Morse 1988b).

The precise nature of the cues used by the adults is not entirely clear, though it obviously relates primarily to the information obtained from the visitors. It could have a number of bases: visual, sound/vibratory, or olfactory. Although we have not investigated this question, visual and vibratory cues seem most likely, and, given the apparently limited visual capabilities of *Misumena*, vibratory cues probably are the major factor. Spiders are very sensitive to both substrate-borne and airborne vibrations (Barth 2002). Adult female *Misumena* are responsive to the actions of bumble bees that visibly shake the flowers while moving about rapidly on them. They also orient to large flying insects approaching the flowers they occupy, but are blocked from view, strongly suggesting a response to airborne vibration as well.

Whatever the nature of the cues, satiated *Misumena* do not respond to them similarly to hungry ones. At such times they select sites randomly in relation to visitation rates of prey (Morse 1988b). Adult females that have fed on a bumble bee will almost never capture a second one on the same day and usually not on the following day. The satiated spiders referred to above were tested on the second day after feeding on a bumble bee (Morse and Fritz 1982).

Individuals selecting new sites have a distinctly different problem that confronts them at a larger scale: how to determine at a distance which stem they should choose (Morse 1993c). In some instances their path is eased by leaves from different stems that touch each other, and in others they may move downwind between stems on lines. However, these options are not always available, and the spiders may have to descend to the ground to move to a new stem. At that point they confront the problem of which stem to climb (Figure 7.3). This is an important decision because they spend considerable time moving to and climbing a new stem, a factor that is multiplied if they make a wrong decision. Under these circumstances, the spiders exhibit a strong tendency to move upward. They do not choose climbing sites entirely randomly, for they mount other milkweed stems significantly more often than the far more abundant small stems of grasses and other smaller forbs (red clover, yellow-rattle, hawkweed). The spiders also selected flowering milkweeds significantly more often than nonflowering ones, though approximately half of them even rejected the flowering stems they encountered. We did not investigate the basis for this distinction, but it could have resulted from the size of the stems or the activity of the insects immediately above. Accuracy of stem choice by spiders placed at the base of flowering and nonflowering milkweed stems did exceed those of individuals released into the grass substrate underneath these stems.

In a second run of the same test (Figure 7.3), we obtained similar quantitative results, but those that made a correct choice on the second trial were a random sample of the pool. Thus, several individuals that made a favorable choice in the second run did not do so in the first (and vice versa). Because those making a correct choice remained far longer on average than those that made an incorrect choice, this action will result in a progressive accumulation of spiders at favorable sites. (Under natural circumstances, of course, those making a correct decision in the first run would not be in the pool for the second run.) The second run resulted in an increase from approximately 50 percent on flowering sites in one run to over 70 percent when adding those making a correct choice the second time to those that made it the first

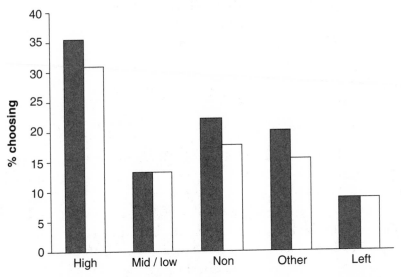

Figure 7.3. Substrate choices of adult female spiders placed on substrate-level grasses in midst of milkweed, equidistant from flowering and nonflowering stems. First run (dark bars), second run (light bars). High = milkweed with high-quality umbels (25+ nectar-producing flowers); Mid/low = milkweed with medium (5–10 nectar-producing flowers) or low-quality (0 nectar-producing flowers) umbels; Non = nonflowering milkweed stems; Other = other species of plant; Left = left area on foot or on lines. Data from Morse (1993c).

time (Morse 1993c). Interestingly, over 70 percent of the adult females make "correct" umbel choices (Morse and Fritz 1982).

Nevertheless, the failure of these individuals to obtain a higher success level in their first run suggests that they cannot make accurate decisions at this distance. Perhaps this deficiency is not surprising if they are responding to the prey, which are at least 50 to 70 cm away. Not only is the distance considerable, but the spiders must exhibit an extremely strong sense of direction (in lack of precise abilities to distinguish between slight differences in sound levels) to make correct choices from the substrate below. We have found that the spiders differ markedly in their level of response to prey over distances of 2 to 7 cm (Morse 1986b). Thus, although the spiders can call upon abilities that improve their efficiency in making stem-choice decisions, this capability is limited, probably on the basis of deficiencies in their perceptual abilities. The large ctenid spider *Cupiennius salei* could not use airborne cues for distances over 30 cm (Barth 2002). Upon reaching flowering stems, *Misumena* selected high-quality umbels as frequently as seen in earlier studies (Morse and Fritz 1982). This finding suggests that we had used individuals with traits similar to those used in the earlier work.

Do Cues Change over Ontogeny?

Thus, newly emerged and adult female *Misumena* use different cues in selecting substrates. The young innately favor substrates that attract maximum numbers of prey, but independent of the prey's presence. In contrast, the adult spiders respond directly to their prey, and substrate plays little or no direct role in their choices. The prey probably respond to olfactory and visual cues of their flowers (Barth 1985), while the spiders most likely respond to vibratory cues (and possibly visual ones) produced by the prey. As noted earlier, establishing whether adult females make any use of direct tactile cues from the substrate will provide insight into whether changes between newborn and adult are absolute. At very most, however, substrate cues play little or no role in the adult spiders' choices of hunting sites. Further information would help to establish whether experience simply overrides any remaining innate tendencies toward a direct response to cues. Individuals reared to middle-instar stage in an environment devoid of meaningful substrate cues may provide useful information on this question, though the environment itself could affect their behavior (Dukas and Mooers 2003).

In common with the patch-choice question, the issues of when changeovers in the use of cues occur, and whether they are relative or absolute, remain open and are of considerable importance. Do the two change simultaneously, or do they change at different stages of ontogeny? An answer to this question would provide considerable insight into the mechanisms of stasis and change in resource exploitation.

Nest Sites

As noted in the discussion of egg masses, *Misumena* place their nests on leaves, typically bending the distal end under the rest of the leaf. They lay their eggs in the resulting shelter, and they secure the edges with silk. The spiders almost always use milkweed leaves when available, usually on nonflowering stems at the edge of a clone, but we have found their nests on a wide variety of plants, raising the question of the basis for their preferences.

Choice of Nest Leaf

To test their preferences, we presented prepartum female *Misumena* with the leaves of several species in their habitat that the spiders either used as nest sites or that resembled leaves used as nest sites (Morse 1990). These choices included milkweed (the most common nest site in our study areas), spreading dogbane *Apocynum androsaemifolium*, chokecherry *Prunus virginiana*, and pasture rose. We presented naturally growing leaves of these species to prepartum females in every possible two-species combination, within cages that contained no other leaves or other vegetation that would provide a nest site. Under similar circumstances in the field, these females would make a nest on milkweed in one to five days. Most of the spiders used in this experiment were captured on milkweed, a reflection of their preference for this plant as a hunting substrate. However, we obtained enough females from pasture rose to run a reciprocal test against those collected from milkweed. In every instance the spiders strongly preferred the milkweed leaves to the alternatives (Figure 7.4). Dogbane and chokecherry were intermediate in preference, with pasture rose the least favored. We used leaves bent by the spiders as the criterion for nest-site choice. Earlier observations (Morse 1985b, 1990) had indicated that in almost all such instances, individuals subsequently laid their eggs at these sites.

Immediately after the spiders laid their eggs, we removed the cages from part of the nest sites in order to test for differences in success. We supplemented these results with others from spiders given only one plant species upon which to lay. Possible bases for *Misumena*'s preferences could be readily seen in the differential survival of clutches in nests on the different substrates. Nests on milkweed were more successful than those on the other plants (Figure 7.5), the result of differences in parasitism by ichneumonid wasps *Trychosis cyperia* and scuttle flies *Megaselia* sp. (Phoridae), and predation by ants. Spiders built the tightest nests on milkweed, minimizing the space between the leaf folds protected only by silk. *Trychosis* inevitably probed these spaces

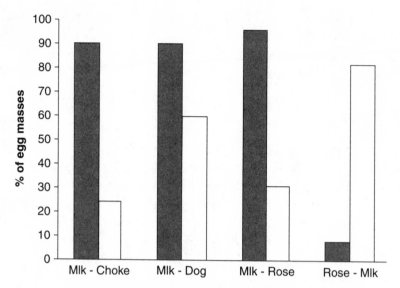

Figure 7.4. Experimental choices of nest leaves by broody spiders. Spiders found on milkweed (Mlk) were given a choice between milkweed and chokecherry (Choke), milkweed and dogbane (Dog), or milkweed and rose (right two pairs of bars). Spiders found on rose were also tested against milkweed (right pair of bars). Each bar indicates the percentage of runs in which the leaf was selected. Half of the spiders in each combination tested on the plant species they had previously occupied (milkweed or rose), and the other half tested on the alternate plant species. Data from Morse (1990).

first with its ovipositor (Morse 1988c). The tightness of the nests on milkweed leaves was a consequence of the extremely pliable and soft character of these leaves, which apparently made them easy for the spiders to work. The latex ducts in the milkweed leaves may have provided an additional deterrent, for if the wasps struck one of these ducts inadvertently with its ovipositor, it seems quite possible that the latex extruded could effectively clog the ovipositor (Dussourd and Eisner 1987; Dussourd and Denno 1994)! None of the other substrates used in the experiment permitted the spiders to make as tight and potentially invulnerable a nest as did milkweed.

Although dogbane leaflets were also pliable, they were on the small side for nests and the eggs were probably vulnerable as a result. A wasp could more easily position its egg directly on the spider's egg mass than in the larger nests built on milkweed leaves (independent of spider size). The large leaves presented the spiders with the opportunity to fit

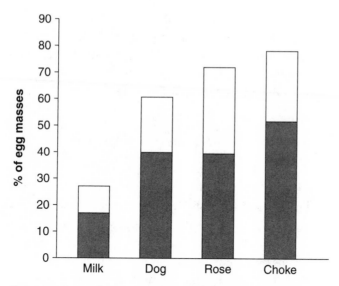

Figure 7.5. Loss of egg masses from nests on different plants.
Dark parts of bars = losses from *Trychosis cyperia*
(Ichneumonidae), light parts of bars = other losses. Data from
Morse (1990).

maximum amounts of foamy silk into the nest, which would isolate the
egg mass in the middle of the nest at an extended length away from the
probing of an ovipositor (Hieber 1992). Chokecherry leaves have a
shape and size very similar to those of milkweed leaves, but in contrast
to milkweed, they were tough and minimally pliable. Although the spiders managed to fashion nests from these leaves, they did not fold them
tightly, presumably owing to the strikingly different structural character of these leaves. As a result, nests on chokecherry routinely contained sizable areas covered only with silk; in this way they provided
much easier access for parasites than the milkweed nests. Nests on pasture rose, the least favored of the four experimental substrates, were yet
more open, a consequence of the spiders' attempts to use a compound
leaf with small leaflets as the substrate. These nests had the lowest success of all.

Although these preferences and the data on success fit closely, the
cues used to make these choices may be distinct. For instance, the ease
of handling the different surfaces could in itself dictate the spiders'
preferences. Since the spiders were placed in simple paired-leaf-choice
situations as extremely gravid individuals needing to relieve themselves

of a clutch of eggs, they were forced to make do with whatever leaves were available. They nevertheless possess a considerable ability to retain their eggs before laying them, and where poor choices of nest sites were the only ones available, they waited up to several days longer before laying than did those with access to milkweed (Morse 1990).

Cues Used about the Nest by Postpartum Misumena

Upon rare occasion we have found *Misumena* guarding formerly abandoned nests of conspecifics known not to be theirs, or guarding a *Xysticus* nest. This finding suggests that they do not inherently recognize their own nest. We subsequently investigated this question (Morse 1989), which is of particular interest in view of the insight it may provide into the cues an adult female *Misumena* uses to ensure that her nest is guarded. This information might further provide insight for the more general problem of how to choose an appropriate substrate.

We conducted these experiments on individual (within-species) recognition as reciprocal transplants. We simply shifted guarding females between nests and compared them with individuals removed from and returned to their own nests, and with undisturbed individuals. We also placed guarding females on a wide range of other substrates that resembled *Misumena* nests to varying degrees, comparing these performances with those of the controls described earlier. All our experiments used female *Misumena* that had recently laid eggs and hence would be expected to remain on a nest and guard it. This predisposition to guard a nest could result either from never leaving the surface of their nest (simplest) or from recognizing the nest (or some simpler analog of the nest). We already knew that the guarding mothers did not require food in order to maintain long-term guarding (Morse 1987). Therefore, they did not need to hunt in order to maintain a vigil, even though they captured insects that inadvertently wandered onto their nest. Hence, the simple procedure of continually remaining at the nest provided a viable option for successful guarding.

The spiders accomplished somewhat of a compromise, laying down lines between the nest and surrounding leaves or the stem. In some instances they bound other leaves about the nest, thereby providing further protection and cover. Such a network of lines would easily give a guarding spider all of the information needed to retain contact with the nest and obviate the necessity of possessing any special capacities for identifying her nest.

Since these spiders often abandoned their nests before their young emerged, either on their own volition or accidentally, these lines may not provide a fail-safe means of nest identification. One can test the intent of these individuals by returning them to their nest to see if they will remain. Those that do, particularly those doing so early in their guarding tenure, have most likely wandered accidentally off their nests. Those that persist in leaving after being reintroduced to their nests most likely have left their nests intentionally. Since some females do appear to wander off and become dissociated from their nests, they appear to have little sense of the location of their nests or of cues associated with the nests. (In contrast, some spiders that do not build webs have a well-developed ability to home to a particular site when displaced from it [Barth 2002].) Since they normally leave a dragline everywhere they go, the spiders' problem of return would usually be simple. However, should that line be broken, this possibility disappears.

Although adult female *Misumena* seldom go anywhere without laying a line, they may do so if suddenly confronted with a potential predator, whereupon they drop from their site, eschewing the use of a line. When confronted by ants, guarding females often descend on lines below the nest until the ants leave. At times these lines may break, particularly if the spider remains on them for an extended period. Under any of these circumstances, if the mother spider has retreated on a line to another (usually adjoining) plant and a strong wind ensues, the line will often break (Anderson and Morse 2001), thus isolating the spider from her nest.

The experiments showed that these spiders do not distinguish their own nests from those of conspecifics (Figure 7.6). As in other experiments, cross-fostering worked perfectly, with the percentage of female spiders accepting foster nests as high as controls (taken from and returned to their own nests), and retention as great as that of unmanipulated nests. If nests are scattered widely, as we have found them to be, the probability of adopting someone else's nest is very low. Our two records of this behavior (Morse 1989) came about as a result of intensively censusing more than 2000 nests with marked individuals over several years. The sole difference between responses of the experi-

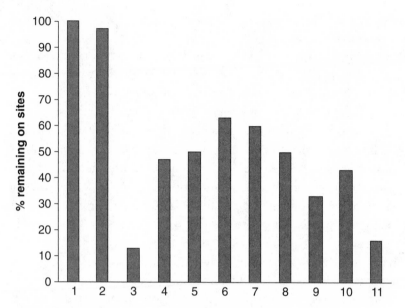

Figure 7.6. Percentage of adult female spiders remaining one hour or more on substrates bearing varying resemblance to their nests. 1 = spider's own nest, 2 = different *Misumena* nest, 3 = unaltered leaf, 4 = leaf turned under by spider, 5 = nest silk on leaf, 6 = artificial nest made by turning under leaf and adding nest silk to it, 7 = *Xysticus emertoni* nest, 8 = *Pelegrina insignis* nest, 9 = *Enoplognatha ovata* nest, 10 = *Misumena* nest with broody female, 11 = *Misumena* nest with actively hunting female. Data from Morse (1989).

mental animals to their own and to other *Misumena* nests was that they initially moved about more frequently on their newly assigned nests than they did on their own nests, as if they perhaps detected some difference. However, this initial heightened level of activity did not affect the acceptance of their newly assigned nests, for their frequency of retaining adopted nests did not differ from that of remaining on their own nests.

In subsequent experiments, we placed brooding spiders on various substrates that resembled their own nests to varying degrees and monitored their responses over one hour. They revealed that a brooding spider's tendency to accept a nest was roughly based on the similarity of the site to a natural nest (Figure 7.6). The spiders exhibited little tendency to remain on an unaltered milkweed leaf, but when the tip of the leaf was turned under, as a leaf-bearing nest would be, half of the test spiders remained on that site. A similar increase occurred when we placed sheet silk, normally used to bind a nest tightly after egg-laying, on a normal leaf. When we combined the two stimuli, the rate of acceptance rose to over two-thirds, similar to that for artificial nests in which we sewed the leaves into a form resembling that of a nest. The spiders also accepted two-thirds of the superficially similar *Xysticus emertoni* nests. *Xysticus* nests differ from *Misumena*'s nests largely in lacking substantial amounts of silk woven over their extremity, probably a consequence of the parent *Xysticus* guarding from inside the nest, rather than outside. Spiders accepted only half of the similar but smaller nests of the jumping spider *Pelegrina insignis*, and the similar-sized but much more loosely constructed nests of the candy-stripe spider *Enoplognatha ovata* were accepted only one-third of the time.

The tendency to accept nests is closely associated with the spider's reproductive stage. Fewer than half of the individuals about to lay (they had turned under a milkweed leaf, a reliable sign that they would lay within 1 to 3 days) accepted a completed *Misumena* nest presented to them. Large females that had not yet started a nest seldom accepted a completed *Misumena* nest: only one-sixth of them remained on the nests assigned to them, although we presented them with nests similar to those used in other parts of the experiments. A spider's tendency to accept a genuine *Misumena* nest decreased over the time subsequent to when she had laid her own clutch and was a function of this factor rather than simply the time an individual had been away from a nest.

Acceptance of nests thus appears to be based on the reproductive status of the individual and to be a quantitative phenomenon based on the overall similarity of the object in question, particularly its shape and tactile characteristics, to a *Misumena* nest. Shape and texture of the substrate played relatively equal roles in determining the acceptability of a site. These cues are thus quite simple but would normally suffice to enable the mother to retain her nest. However, the highly exceptional instances of *Misumena* guarding at other sites (e.g., *X. emertoni* nest) emphasize the limit of their discriminatory powers. Retaining contact with a site is thus important, since guarded nests are more successful than unguarded ones (Morse 1988a).

Once a spider is displaced from its nest plant, its problems multiply considerably. In our study areas, this problem appeared to occur most frequently when guarding spiders were confronted with ants. The spiders do not typically nest where ants are present in numbers, yet the possibility of aphids recruiting to a nest plant subsequent to the spider laying presents a problem that the spiders cannot control. In the areas in which we work, the ants *Lasius neoniger* and *Tapinoma sessile* commonly recruit to both nectar and aphids on milkweed (Fritz and Morse 1981). Aphid infestations (*Aphis asclepiadii* and *A. nerii*) frequently occur on milkweed in late summer. *Lasius neoniger* workers attending aphids are aggressive toward the spiders and will readily attack them, often *en masse*. Usually the spiders simply descend on a line and hang suspended if harassed by ants, returning after the ants leave. However, if frequently confronted, they may become separated from their nests and are then unable to find them again (assuming that the frequency of ant incursions is manageable). We have found (Morse 1991) that individuals placed under the leaves of a nest plant are no more likely to find their way up the appropriate stem than predicted by chance (comparable to their ability to find good sites as foragers at ground level [Morse 1993c]). In contrast, most individuals placed at the base of their nest stem found their way back to their nests. This was probably a simple consequence of their tendency to move upward in the vegetation, plus contacting lines they had made about the nest earlier (Morse 1991). The results from these experiments give us insight into the cues used by *Misumena* in their activities.

Cues Used by Males

We know that the cues used by newly emerging males and females are similar, and we assume that their responses remain similar through instars prior to the antepenultimate stage, at which point they can be visually distinguished from the females. As adults, their activities become so different from those of the females that they need to be considered separately. Many of the differences relate to mate search and finding, and I will consider them in that context in Chapter 9.

Synthesis

The importance of choosing hunting sites prompts consideration of the cues the spiderlings use in making these choices. The matter is not a simple one, since the flowers used for hunting sites change throughout the season, and any viable strategy will have to accommodate for the differences in these flowers and the plants that produce them. Furthermore, the spiderlings regularly encounter new flower species, and as first-time choosers just out of their nest sacs they are completely naive. They use a combination of innate and learned behavior that capitalizes on flowering phenology as well as the cues produced by the prey themselves. It may be coincidental that the young emerge from their egg sacs at the very time that flower availability is highest and most predictable—on goldenrod in the last summer and early fall. Thus, at the time that the spiderlings have no experience, the most ubiquitous flowers in the region, goldenrods, peak in bloom. The spiderlings innately prefer goldenrod to other available substrates. Although arguably the young might just as well be innately programmed to respond to prey as are older individuals, their prey might not be as reliably sensed as the goldenrod substrate, which they contact directly. And since obtaining satisfactory hunting sites assumes prime importance, these spiderlings do not tarry on unsatisfactory sites, such that contact with prey would provide a less predictable cue than substrate. Furthermore, the large prey taken by older spiders provide more easily detected cues (vibration, sound, etc.) than those from the spiderlings' extremely small, relatively slow-moving prey.

Although we do not yet know what characteristics prompt naive spiderlings to select goldenrod over other substrates, it seems questionable whether their responses could be specific to goldenrod, notwithstanding its local and regional abundance. Identifying that critical factor(s) should be a prime consideration, for it may provide insight into how older individuals select sites.

The basis for choice changes as the spiders grow, such that eventually they, as adults, depend on cues produced by their insect prey rather than directly from substrate quality. We do not yet know precisely how this change takes place, although the spiders' condition appears to shift in the fourth and fifth instars, and penultimate females, at least, respond to stimuli similarly to the adults. We also do not know why the older spiders shift from using substrate cues, but they may only be able to respond to goldenrod-type cues innately, which would not be surprising because of its ubiquity when in season. At the same time, the large prey they must capture produce signals (sound, vibrations) that they detect from some distance, providing them with apparently unambiguous information. Still, the ability to evaluate substrate in the absence of prey could increase their time on prime hunting sites, and, as generally food-limited individuals, efficient use of time should be advantageous. Several experiments have suggested that the adults may retain a limited ability to evaluate substrates independent of the prey visiting them. Even if they do, however, this ability assumes minimal importance relative to their ability to detect prey directly.

Choices of nest substrates (leaves) bear a close relationship to the expertise with which the spiders can work the leaves, and the favored leaves are those upon which they have the highest success. Losses occur largely through parasitism by the ichneumonid *Trychosis cyperia*, which focuses on parts of nests that are not tightly knit. Guarding mothers usually manage to retain contact with their nests, but their methods of retaining contact are simple and do not involve recognition of their own nests. This simple system is not often a problem because they seldom nest close to each other and also are usually attached to their nests with draglines. If they do lose contact, either via a broken line or by dropping to avoid a predator, they have relatively low success in finding their nest. They recruit to other nestlike sites roughly in proportion to the similarity of those structures to their own nests. Thus, a quite similar *Xysticus* nest is much more likely to be recruited than an unmodified leaf or even a very loosely turned leaf containing the nest of a candy-stripe spider.

8

Morphological Variation

$\mathcal{T}$HE NATURE AND EXTENT OF morphological differences between males and females vary drastically among species and groups, from monomorphy to situations in which one sex has become a tiny parasite of the other (Ghiselin 1974). Among spiders, sexual size dimorphism differs strikingly, extending from monomorphy to dwarf males (or giant females) in which adults of the two sexes may differ in mass by a hundredfold or more (Vollrath 1998; LeGrand and Morse 2000). In a relatively small number of species, males modestly exceed females in size; usually, however, females greatly exceed males in size. Although striking sexual dimorphism may be a consequence of sexual selection, the conditions facilitating these differences have a strongly ecological character. Such variables as the size, spacing, availability over time, and predictability of resource bases contribute markedly to the pattern of sexual size dimorphism that develops (Andersson 1994).

How These Differences Are Accomplished

Since males and females arise from similar-sized eggs, subsequent developmental patterns and differential growth account for their size-related sexual dimorphism. If one sex is larger than the other, the difference must entail either markedly faster growth or a longer period of growth. For individuals growing as rapidly as possible, a longer period of growth offers the only alternative, which should simultaneously permit the development of protandry or protogyny (male or female reaching reproductive condition first). These conditions lower the potential for inbreeding, a consequence with strong implications for their breeding systems, especially since some spiders show a tendency for inbreeding. These traits may be advantageous in their own right and may become driving forces where inbreeding depression is likely to be a serious consideration. Tradeoffs may occur in the process because time and size are unlikely to be optimized at the same point.

Modest differences in size dimorphism result from protandry (or protogyny), a widespread trait especially well studied in butterflies (see Chapter 9). Here, males emerge earlier than females from otherwise similar caterpillars, but a cost of early emergence is often a modest deficit in size relative to that of females.

In some instances dimorphism, also associated with protandry, may be facilitated by eliminating instars. In the simplest case, differences in size between males and females may result largely or totally from instar numbers. Thus, the sixth instar of male and female *Misumena* might be relatively close in size, but males mature in the sixth instar, while females do not reach adulthood until the eighth instar. In some ways, instar elimination offers an extremely simple way to ensure protandry, though the difference between the sexes may have a more complex basis to it than in simply separating the reproductive period of male and female sibs. Curtailing the number of instars is analogous to neoteny in that sexual maturity ensues at an earlier developmental stage than in the presumed ancestral form. This type of dimorphism occurs routinely among spiders.

External morphological differences between males and females are sometimes minimal, perhaps as a result of environmental demands constraining their ability to diverge. In some taxonomic groups monomorphy is correlated with biparental care of the young (birds, mammals). This factor is unlikely to play a role in monomorphy among spiders, however, since wherever parental care is seen in this group, it is confined to the females (Foelix 1996a). Foraging differences between males and females, which would minimize competition in finding food to feed their young (Selander 1966; Morse 1968), could lead to dimorphism. However, the food supply does not limit many spider populations in a density-dependent way (Wise 1993); this factor is also unlikely to account for the extreme dimorphism seen in *Misumena* and certain other spiders. In contrast, selection for high fecundity could favor large females (Morse and Fritz 1982) but have little effect on males. In the male spiders larger than females, this size difference results from male-male aggressive competition for access to females (e.g., Watson 1990), similar to that seen in many mammals (Andersson 1994). Attainment of superior size by either sex will probably confer social dominance, thereby displacing the subordinate sex when in close contact with it (Morse 1977a).

Dimorphism in *Misumena* and Other Species

Misumena exhibits extreme sexual dimorphism. The gravid adult females, for example, sometimes weigh up to 100 times as much as adult males, and even just-molted virgin females more than 10 times as much. They share this extreme dimorphism with a few other closely related crab spiders (LeGrand and Morse 2000), a ratio only exceeded by certain of the largest orb-weavers (Araneidae and Tetragnathidae) (Vollrath 1998). Head (1995), Prenter, Elwood, and Montgomery (1999), and others have attributed the relatively large size of female spiders to fecundity selection. Ghiselin (1974) proposed that this extreme degree of dimorphism was a consequence of females sequestering resources to reproduce in areas of very low resource and population density. Only mobile males, also rare under these circumstances, could regularly find the females. Being uncommon, they would experience little competition for the females, and under these conditions small size and mobility would be favored. Thus, selection among males would occur for efficient movement capabilities that required a minimum of resources, thereby favoring the individuals most competent at finding females within the energetic regime experienced, which normally would be at least partly a consequence of their mobility.

Ghiselin found this type of strategy more common in the deep sea than on land, though he focused most heavily on marine systems. Significantly, however, he recognized spiders as the terrestrial animal group containing a spread of dimorphism in the same range as that of some deep-sea groups. He suggested that this condition might be a consequence of spiders also being forced to strain their milieu for "plankton." In the case of the spiders, plankton was represented by the flying insects captured in the webs. That explanation would not account for the dimorphism found in crab spiders, sit-and-wait predators; the problem would therefore seem to be a bit more complex than Ghiselin suggested. However, Ghiselin pointed out that in this regard sit-and-wait predators such as *Misumena* bear possible resemblance to the wide-mouthed, deep-sea angler fishes (Ceratiidae).

Since *Misumena* females seldom capture enough food to lay a maximum-sized clutch (Morse and Fritz 1982; Fritz and Morse 1985), they effectively live in a food-sparse environment as well, notwithstanding the amounts of prey some of them encounter. They are instead limited by their difficult mode of foraging. These observations suggest that it should be profitable to investigate common traits of orb-weavers and crab spiders.

Some adaptations of deep-sea organisms even exceed those of spiders. At the extreme, the male is reduced to a tiny parasite of the female connected to her circulatory system, as seen in some of the deep-sea anglerfishes, and perhaps even more dramatically, in some barnacles. Of particular interest among the barnacles, both sexual parasites and fully developed, relatively large males may occur within a single taxon (Darwin 1854); these striking differences may therefore have some primarily environmental basis. These conditions further argue for investigating the environmental factors related to extreme dimorphism in the crab spiders and orb-weavers. Why haven't any of them reached this extent of dimorphism, or is the difference merely a consequence of the larger number of marine taxa taking up a highly dimorphic existence?

Most of the crab spiders (Figure 8.1: Dondale and Redner 1978) and many of the orb-weavers do not exhibit dimorphism comparable to that of *Misumena* and the largest orb-weavers, though dimorphism is nevertheless marked in most of them. *Xysticus* crab spiders, for instance, exhibit marked dimorphism, with the largest males about one-fourth the mass of gravid conspecific females. Very likely these *Xysticus* males experience one more molt than do *Misumena* males. If we extrapolate to obtain the expected carapace width of the two large outlier *Misumena* males mentioned earlier (17.8 mg and 13.3 mg), they fall right along the main regression line for most North American crab spiders for which data exist (Figure 8.1). We might be able to test this proposition if we could induce the males to proceed through an extra instar by the use of juvenile hormone, or the synthetic equivalent,

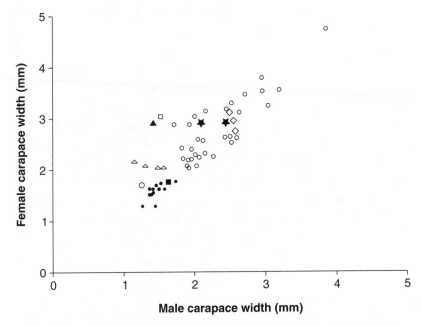

Figure 8.1. Size ratios (male × female) of North American crab spiders, including extrapolations (stars) that substitute two "giant" male *Misumena* values for the much smaller normal mean male size. Each other symbol denotes a genus. Dark triangle = *Misumena*, light square = *Misumenoides*, light triangle = *Misumenops*, light circle = *Xysticus*, dark circle = *Ozyptila*, light diamond = *Coriarachne*, dark square = *Tmarus*. Multiples of a symbol denote different species in a genus. Modified from LeGrand and Morse (2000) from measurements in Dondale and Redner (1978). With kind permission of Blackwell Publishing.

methoprene. Such an endeavor would not be guaranteed, however, since spiders are not currently known to possess juvenile hormone-like substances (Trabalon, Pourié, and Hartmann 1998; Pourié and Trabalon 2003). Juvenile hormone has, however, been found in a tick (Pound and Oliver 1979), thereby demonstrating its presence within the chelicerate arthropod line.

Discussions of extreme dimorphism in spiders have often been couched in terms of dwarf males (Vollrath and Parker 1992; Vollrath 1998), implying that males were the sex that changed the most from some presumably sexually monomorphic condition. However, Coddington, Hormiga, and Scharff (1997), using phylogenetic analysis, argued that a more plausible general explanation has females changing more than males. It would therefore be more appropriate to think in terms of giant females rather than dwarf males. They noted that, although their interpretation might not account for all instances of extreme size dimorphism in spiders, deviations from that model might best be viewed as variations on a more general theme of female change.

Change in *Misumena*, as well as closely related *Misumenoides* and *Misumenops* species, does most appropriately fit the dwarf male syndrome rather than giant female syndrome. The plot of male carapace size against female carapace size for North American crab spiders (Figure 8.1) reveals that females of these genera fall along the line of average size among crab spiders as a whole, but that the *Misumena*-like males fall well off this line as small outliers. The only other species with males this small are nearly monomorphic. That the larger *Misumena* and *Misumenoides* show more extreme deviations than *Misumenops* may be a consequence of their greater size, suggesting a pattern of allometric growth.

Allometric Patterns

Changes in size, both in ontogeny and phylogeny, are routinely accompanied by changes in relative dimensions—allometric growth. Allometric growth occurs routinely over successive molts in *Misumena*. However, by their antepenultimate instar, males have diverged noticeably from females, presumably because of sexual selection. Growth in carapace width slows, while leg length continues to increase. As a result, males become progressively longer legged, and, given the relatively small mass they are carrying, potentially more cursorial, which should improve their ability to search for females as well as to escape from predators. Overall mass does not differ between penultimate and adult stages of the males. However, forelimb length increases more markedly than at any other time in ontogeny, in synchrony with a decrease in abdomen size. Over this period, leg length has grown such that the ratio of leg length to carapace width reaches 1.5 times that of females, juvenile or adult (LeGrand and Morse 2000). Long leg length should be favored in scramble competition (LeGrand and Morse 2000).

These modifications in *Misumena*'s growth pattern differ from those of *X. emertoni*, probably because of *X. emertoni*'s litter-dwelling existence. Although *X. emertoni* retains the large, raptorial forelimbs characteristic of the thomisids, its hind legs considerably exceed those of *Misumena*, but its abdomen is smaller. Also, male and female growth trajectories of *X. emertoni* resemble each other more closely than those of male and female *Misumena*. This finding suggests that those of *Misumena* are a response to unusual demands put upon their males, a possible consequence of their low densities, their habitat, and their failure to use pheromones in mate-finding (see Chapter 9). In our study areas, *X. emertoni*'s population densities are considerably higher than those of *Misumena*. Furthermore, in contrast to *Misumena* (Anderson and Morse 2001), *X. emertoni* probably uses pheromones in mate location, a capability likely enhanced by its litter-dwelling habitat, in contrast to the herbaceous canopy frequented by *Misumena*.

These observations emphasize the importance of thinking about allometry in a much broader sense than with regard to a single parameter (Gould and Lewontin 1979; Uhl et al. 2004). Of special interest in the above comparison, the growth patterns of two related species with different ecological traits differ markedly in character: *Misumena*'s growth pattern presumably deviates considerably farther from an ancestral line than that of *X. emertoni*. Could *Misumena* (or its ancestors) also have secondarily lost the ability to use olfactory cues in mate finding as a result of secondarily derived dwarfism of the males rather than the females losing the ability to produce these substances? Life history traits, such as numbers of instars, are, in common with other traits, variables subject to selection and change.

Why Do Males Differ in Size?

Although I have focused on the differences in size between male and female *Misumena*, and on the small size of the males, males within a population differ in adult mass by more than threefold (2.3–8.1 mg, excluding the two large outliers: Figure 8.2), accompanied by a range in carapace width of between 1.1 and 1.9 mm. If males experience so much pressure to be small, why do they differ so greatly in size? Although all males are very small, relative to the females, the over threefold difference seems strikingly large, and particularly so if considerable directional selection is still in force. On the one hand, we have shown that large males usually prevail in interactions with smaller males (Holdsworth and Morse 2000); on the other hand, they do not routinely experience competition with other males in obtaining matings with females (Hu and Morse 2004). To date we have not succeeded in establishing a clear difference in movement rates between

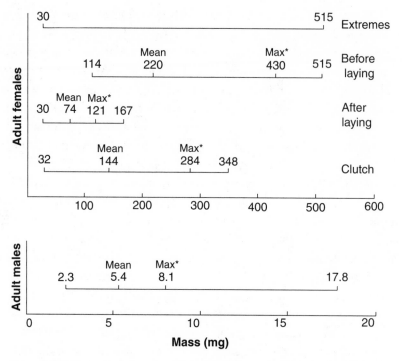

Figure 8.2. Masses of adult male and female *Misumena* and their clutches of eggs. * = maximum with exception of rare outliers. Data from Fritz and Morse (1985); LeGrand and Morse (2000); D. H. Morse (unpublished data).

large and small males (Sullivan and Morse 2004). If rates of movement do not differ, large males should be selected on the basis of their precedence when they do have to compete overtly with smaller males over mating with virgin females. Certainly this factor seems to hold where many males often contest for a virgin female, as in populations of the closely related *Misumenoides formosipes* studied by Dodson and Beck (1993). Those *M. formosipes* should be under strong selection for enhanced male size, but they do not seem to produce significantly larger males than do the male *Misumena* with which we have worked (Dondale and Redner 1978).

Thus, all male *Misumena* (and *M. formosipes* as well) may be small and light-bodied enough that they can search for virgin females equally well, while variation in size may reflect different strategies between maximizing efficiency of resource use versus fighting ability. If small males molt earlier than the large ones, they may enjoy an advantage through protandry, while the large males prevail once they reach adulthood. So far, however, size does not appear to differ with molt time. Since the data on male mass differ continuously, the possibility of any intermediate size exists, though various means of male mass we have measured (3.9–4.1 mg) fall well below the 5.4 mg median, suggesting an advantage for small size. We would sooner expect to find selection for larger males if male competition were high than if it were low, and the prevalence of small males would be consistent with low interaction rates. Alternatively, this variation may be environmental; if so, it would be extremely interesting to measure the current size distribution against the optimal size distribution within a population, given the intensity of male competition and the difficulty of finding females.

Lastly, we do not yet know whether small males are constrained in their reproductive ability (ability to fertilize eggs). We do know that after mating, males do not remate quickly, even if presented with a newly emerged virgin female. Male *Misumenops celer* do not recharge their palps for at least 12 hours (Muniappan and Chada 1970). We do not know, however, whether "recharge" time of *Misumena* differs with size. Even if males do not have the ability to fertilize as many eggs as large individuals, if their sperm supply suffices to fertilize as many eggs as they usually encounter, this potential limitation would not normally be a constraint. Because the adult sex ratios dictate that substantial numbers of females are available for the average male to mate, we are dealing with a situation in which the smallest males might experience a constraint. We also do not know whether the smallest males survive as long as do the larger ones. However, extended survival would assume decreasing importance as the season drew on to a point in which virgin females became few and far between. As noted in Chapter 5, small males are subject to greater desiccation stress in the laboratory, but we do not know the importance of this factor in the field, especially since extreme drought usually occurs after most reproduction has taken place.

Gabritschevsky (1927) reported variation in the number of male instars, with some males within a brood reaching sexual maturity in one less instar than the others. This variation would facilitate the development of size differences, with diminished size potentially enhancing protandry. Several individuals that we reared through early instars also molted into recognizable male form one instar ahead of the others. We have not found any unequivocal evidence of this dimorphism in the field, though it would be extremely difficult to detect unequivocal evidence except in confined individuals. However, we had fed these individuals hover flies *Toxomerus marginatus* ad lib, a diet that few if any young could match in the field due to the difficulty of initially capturing these flies, which often weigh six times as much as a just-emerged spiderling (Chapter 5). Individuals from the same broods that were fed on a *Drosophila* diet did not exhibit this accelerated growth. Gabritschevsky also fed his spiderlings a rich diet, which may have been the basis for obtaining this result.

Our two extremely large outlier males, which have likely undergone an additional molt, may resemble the ancestral condition from which *Misumena* and probably other closely related species have arisen. Since most other crab spiders have relatively larger males (Gertsch 1939; LeGrand and Morse 2000), and their degree of dimorphism more closely resembles that of the outlier male *Misumena* than that of the "normal"-sized males, if strong selection occurred for larger size, this trait should spread within populations. *Misumenops tricuspidatus* exhibits considerable flexibility in the number of both male and female molts, which vary with environmental conditions; the molt number increasing as the food supply and/or temperature declines (Hukusima and Miyafuji 1970).

Why Do Females Differ in Size?

Male size differences have commanded considerable attention because they may help determine the form of breeding systems and the advantages that size bestows on different individuals. Potential advantages include synchronizing their breeding cycle to maximize reproductive opportunities (protandry), finding females, fighting for females, and physical ability to inseminate eggs. As we have seen, male *Misumena* exhibit some, but not all, of these traits, and their importance in some instances is open to question. Sizes of females within these populations may also assume major importance, even though from a male perspective they may tend to be viewed primarily as differentially valuable targets for the males. The size of females also differs widely, and this variation is not simply due to the number of eggs produced by adults.

Size Range of Females

First, given the considerable variation in size of adult females, we should ask whether these differences correlate with those of their male sibs. If so, it would suggest that male size ranges could be controlled in part by factors important to females (or vice versa). However, the wide dichotomy in size between the two sexes, at the extremes of free-living terrestrial animals, suggests that one sex does not seriously constrain the other in terms of the size that it eventually achieves. In fact, in grossly dimorphic systems such as this one, the constraints are arguably as low (or nonexistent) as we are likely to find among free-living organisms. The tactic of varying instar number provides an opportunity by which the two sexes deviate widely in size, though we assume that even this convenient mechanism required a shift in hormonal signaling or some comparable effect.

Adult female *Misumena* range from 30 to 515 mg (over seventeen-fold) (Figure 8.2), and their carapace measurements vary from 2.8 to 4.1 mm (nearly twofold). When compared with those of the males, these figures are somewhat misleading because, in contrast to the males, much of the variation in female mass results from the production of their single clutch of eggs, which averages about 65 percent of the prelaying body mass. Furthermore, we have never found a female in the field immediately prior to egg-laying that weighed less than 114 mg. Since we have found several that approached that mass, these individuals probably approached the minimum at which they will produce a clutch under field conditions. The range of 114 to 515 mg (four- or fivefold) somewhat exceeds the threefold range of variation in mass of the males (excluding the two large outliers). However, the two ranges do not differ drastically, especially since the 515-mg individual was itself an extreme outlier that exceeded the next largest individual (430 mg) by 85 mg, or 17 percent. Using the range of 114–430 mg, they vary by almost fourfold. However, the range in mass at molt into the adult stage may be a more appropriate figure to compare the females with the males, for it does not include the increase in mass subsequent to molt. That range varies from 30 to 80 mg (nearly threefold), still a wide range, and comparable to the threefold variation in male mass. The range in carapace width is also the same (about one and one-half fold for both sexes). Thus, any initial impression that male and female patterns of size dimorphism vary largely independently of each other requires further attention.

To explore this variable further, we should compare sibs to determine whether their size ranges fit this model. That is, we need to test the proposition that both males and females of any given brood will be relatively large or small, or that the sizes of the two sexes within a brood will differ randomly among themselves. For the moment we cannot state with confidence that males have become as small as required to maximize their fitness. To the contrary, considerable selective pressure seems to favor large size in the females, at least as adults, although we have not adequately explored this issue. Even though large females are successful by virtue of the large clutches they produce, they approach maximum size so seldom that little selection probably exists for larger body size.

These comments relate to the genetic basis for size, and any more explicit investigations would have to be performed under carefully controlled conditions, making sure to match the field conditions as closely as possible. Size in spiders and other arthropods has a strongly environmental aspect, with the result that feeding regimes can dictate to some degree the eventual size attained. Thus, good feeding conditions during the early instars may result in large individuals (Beck and Connor 1992), which, as adult females, produce larger clutches than do smaller ones. However, it remains to be seen how often spiders realize this advantage in the field, since most individuals never have the opportunity to reach their maximum possible biomass, with their flower-hunting sites senescing about them and with considerable costs incurred by attempting to find fresh sources. Thus, most such females elect to lay their single egg mass at the time that a major flower source senesces (Chapter 3).

We might expect the members of small populations to differ in size from others, independent of their success at capturing prey. Size is a difficult variable to investigate, since it possesses such a major environmental component; success at capturing prey, both as early instars and as adults, grossly affects skeletal dimensions as well as body mass. However, by far the largest individuals (as a group) that we have yet discovered came from a small roadside patch that did not differ understandably from individuals on other sites. It would have been profitable to investigate the number of instars of these females, but in recent years trees have grown over the site and now completely shade it. The result is such a severe decrease in flower numbers as to cause the extinction of this small population. If the large size of these spiders had a genetic component to it, they could well have been minimally equipped to deal with an increasing shortage of food resulting from the decrease in flowers. However, the overgrowth of the trees completely suppressed the forbs, so that no population of insectivorous predator could now survive at the site. This demise likely occurs frequently to *Misumena* populations in their typical successional habitats.

Selection for Change in Size

Since females can seldom match their reproductive capabilities, only occasionally reaching a possible maximum mass, they must experience little selective pressure to develop larger body size (LeGrand and Morse 2000). Although they must attain adult size to kill their favored prey, bumble bees, all adult females appear able to do so and depend on this resource or other large insects to reach a large mass. Thus, whether additional size would pay off on this basis alone is open to question in that it would bring increased maintenance costs along with it. Since the extreme importance of high-quality hunting sites eclipses capture success as the most important factor determining success in our system, the argument for increasing size is further weakened. Because adult females are already somewhat ponderous, increased size might, if anything, constrain their ability to find these premium hunting sites, a simple consequence of their decreased mobility. Blanckenhorn (2000, 2005) has noted the scarcity of information countering selection for large size in females (fecundity selection). However, the problems confronted by female *Misumena* in finding both patchy hunting sites and satisfactory nest sites may counter fecundity selection.

Given their limited mobility at this time, gravid females appear loath to operate in the risk-averse way required to approach that maximum mass on a more regular basis, which ought to select for an early, and likely smaller, brood and then an attempt to produce a second brood. However, work to date suggests strongly that at least in our study areas the season is too short, under any natural circumstances, to produce a second brood (Morse 1994). Although under these circumstances they might have time for, say, 1.8 broods, one brood is the maximum available to them, since brood number is a discrete measure.

Some Constraints on Female Size

Our spiders exhibit certain constraints in feeding that may be related to the consequences of gains in mass—as if they were watching their diet! We have observed at least two instances that may be related to such concerns.

1. Although slow growth normally is counterproductive, late in the season penultimate females may decrease their rate of capturing prey, thereby failing to molt into the adult stage before winter (Morse 1995). We interpret the penultimates' decline in intake as a response to their highly vulnerable situation, in which high prey intake might result in molting into adults with little or no opportunity to produce young before the end of the season (Chapter 5).

2. The behavior of some adult females prior to egg-laying may also reflect constraints on unlimited food intake. A few females do succeed in approaching a likely maximum reproductive mass. When feeding these individuals in the laboratory, we have observed that they usually become very reluctant to take additional prey when approaching what we calculate to be their maximum possible mass. This reluctance occurs even though they might seem good candidates to "top off" their mass to a level matching that achieved by similar-sized individuals that reached a higher mass by imbibing heavily on a very large prey item (e.g., bumble bee) when starting from a lower mass. As a result, those individuals that have approached, though not reached, their estimated limit sacrifice the opportunity to reach a truly maximum mass. One might attribute this failure to their making a conservative decision to enter reproductive condition immediately and lay their eggs. Thus, they can get a jump on the season, with possible advantages for the survival and success of these early offspring. After all, if they do perceive an extreme limit for any particular body dimensions, merely taking a part-meal may not justify the loss in time required to imbibe it and to mature the eggs associated with it. However, if such individuals cannot readily stop at the "allowable" part-meal and continue on to imbibe a full meal, they may be at heightened danger of mortality, as suggested from Higgins and Rankin's (2001) results. We have as yet no direct in-

formation to support this proposition. However, we should note that large adult females feeding primarily on bumble bees almost never take more than one such prey per day, and only exceptionally do they take them with that frequency (to the point that in the laboratory we only offer large prey every other day to individuals reared to maximize growth rates). We have recorded a single exception to the one-per-day "rule," a mature female that completely consumed two bumble bees during a day and a third bumble bee on the next day. She died the following day (Chapter 4). Although a single example scarcely suffices to draw a conclusion, the result does not seem surprising and may provide another manifestation of the possible adverse consequences of unusually large intakes of food.

Synthesis

One of *Misumena*'s most characteristic features is its striking sexual dimorphism. Indeed *Misumena*, along with close relatives, ranks among the most dimorphic of land animals. This pattern, which reaches its extremes among certain crab spiders and orb-weavers, is likely facilitated by manipulating their molt cycles. In fact, molting animals (Ecdysozoa: Aguinaldo et al. 1997) appear to have the basis for controlling their size over an extremely wide range. That being so, extremes of sexual-size dimorphism might be predicted within this group. Nevertheless, it is of interest that certain lines of spiders have used this strategy to an extreme. Our current understanding of spider taxonomy suggests that development of extreme dimorphism in crab spiders has an evolutionary basis separate from change in orb-weavers.

Adoption of extreme dimorphism relegates male and female adults to totally different lifestyles, although a majority of their lifetime will be spent as sexually monomorphic forms. There is little if any difference in lifestyle of juveniles as far as we can yet ascertain, though much work remains to be done on this point. The males are in a sense neotenic, and it would be of extreme interest to determine whether certain traits apparently missing in these individuals (explicitly pheromone use) are the consequence of a morphologically juvenile condition. If so, it is important to note that highly sexually dimorphic male orb-weavers do retain this ability.

Considering the extreme degree of sexual dimorphism in *Misumena*, the high within-sex variation assumes considerable interest. Much of this variation is a consequence of environmental factors, but the variation appears to exceed these levels. If so, it suggests that mutations favoring change in instar number might be subject to positive selection.

Although changing numbers of instars may provide these spiders with unusual potential for sexual dimorphism and morphological change, it is important to note that these attributes may have negative aspects as well. Instar change would only allow stepped change, as opposed to graduated change seen in nonmolting forms, which nevertheless achieved marked sexual dimorphism in some instances. However, it is probably significant that the most extremely dimorphic species, certain spiders, including *Misumena*, depend on stepped change. Forms with stepped change (ecdysozoans) obviously are very successful, making up the great majority of known eukaryote diversity (insects, crustaceans, spiders, etc.).

9

Male-Female Interactions

$\mathcal{F}$EMALES ARE GENERALLY LIMITED BY the resources they can obtain for producing offspring, and males are limited by the number of eggs they can fertilize. Thus, natural selection controls the limits of success to females, and sexual selection controls the limits of success to males. Considerable information supports the argument that natural selection controls female success in resource acquisition (Andersson 1994). However, quantitative data on male reproductive success have traditionally been much more limited (but see Clutton-Brock, Guinness, and Albon 1982; Clutton-Brock 1988), owing to the extreme difficulty of assigning paternity. Modern techniques of genetic fingerprinting have now largely closed this gap. Traditionally, male success has been taken to be much more variable than that of females, in large part the consequence of male–male competition for mates, a relationship that becomes extreme in some polygynous systems.

In this section, after providing some appropriate background, I will briefly discuss *Misumena*'s reproductive biology and then address factors that regulate the reproductive success of male and female *Misumena*, as well as how they respond to these factors. Where appropriate, I compare *Misumena*'s reproductive relationship with those of other species.

Theoretical Approaches

I have thus far focused on situations appropriately addressed by the application of simple optimality principles. However, simple optimality procedures are not strictly appropriate for conditional decision making, even though in both instances they represent an organism's best efforts to maximize its lifetime fitness under a particular set of circumstances. Put somewhat differently, optimization methods analyze evolutionary equilibria and thus cannot be used to study the dynamics of adaptation (Gomulkiewicz 1998), including learning.

In many of the situations I consider in this chapter, the optimal response of an individual depends on the response of other individuals, and these responses are frequency-dependent. Frequency-dependent phenomena have no fixed optimum, and the most appropriate response to a challenge will be to act in a way that resists invasion by some other phenotype—in this instance, resistance to an individual performing in a different way that would allow it to take advantage of you. This tactic is referred to as an evolutionarily stable strategy (Maynard Smith and Price 1973). Behavioral interactions normally take such a form, having a fundamentally selfish or cooperative basis to them. In addition, other traits such as sex ratios fit a frequency-dependent pattern at a population level. This set of conditions is referred to as (evolutionary) game theory, based on the fundamental principle that the actions of an individual affect the fitness of others and must be considered when examining the evolution of a trait (Maynard Smith 1982; Dugatkin and Reeve 1998; Nowak and Sigmund 2004). The range of behavioral games is vast: bluffing and escalating over a resource, including a mate; cooperating in food acquisition; and selecting the intensity of territorial defense (Stephens and Clements 1998) only serve as examples of the breadth of such possible interactions. These subject areas include both primarily antagonistic and cooperative behaviors. All are dynamic phenomena readily subject to change, in some instances, within the lifetime of an individual. For that reason, they seldom fall under tight genetic control (Stephens and Clements 1998). Indeed, these are the conditions that strongly favor learning, though they do not require it.

Such individual-to-individual level interactions may have significant population-level consequences as well. Successful parents maximize fitness by investing most heavily in the sex that will most effectively represent their genes in the subsequent generation. The decision of how to accomplish that goal will differ with the existing sex ratio. Although an individual's investment decision is about maximizing its own fitness, it should be made in response to the characteristics of the population members with which it interacts.

Many aspects of both male-male and male-female interactions in *Misumena* should thus have a conditional basis. Confrontations of males vying for females should be affected by the relative sizes of these males, and interactions between males and females by the female's reproductive status and earlier experience.

Reproductive Biology of *Misumena*

In contrast to many spiders, *Misumena* exhibits an extremely simple mating routine. Upon detecting a female, a male moves directly toward her, and in a small minority of instances (< 10 percent) even meets her face to face. He then climbs onto the dorsum of her abdomen, most often from her posterior or laterally, but in some face-to-face encounters directly over her head. Once on the abdomen (Figure 9.1), he may briefly move rapidly about it before attempting to move under her abdomen into a ventrolateral position, his head pointed in the same direction as the female. We have not yet ascertained whether these rapid movements on the female's abdomen facilitate her response, but, if they do, any effect is subtle. Some males do not indulge in this behavior, although Bristowe (1958) reported that it is widespread among crab spiders. In contrast to some *Xysticus* crab spiders, males show no sign of binding the female to the substrate with silk prior to copulation (e.g., Bristowe 1958). If the female is receptive, as most recently molted virgins are, she will soon raise her body, thereby providing the male with convenient access to her venter and paired external genital apertures. At this point he inserts one of his pedipalps into the adjacent genital aperture of the female. If, however, the female does not raise her body at this time, he may experience considerable difficulty wedging himself far enough under her to achieve insertion. In a typical bout, after accessing the first of her genital apertures, he will move to the dorsal side of her abdomen and then shortly repeat the procedure on the other side of the female with his other pedipalp, contacting the female's other genital aperture. The entire routine (left and right sides)

Figure 9.1. Male on abdomen of female *Misumena* prior to mating. Illustration by
Elizabeth Farnsworth.

requires only a short period, about four minutes on average (Morse
1994). Some individuals insert a pedipalp more than once on a side.
When multiple insertions take place, the male may remain for considerable periods on the dorsal abdomen of the female between bouts.

After completing the insertions, the male typically jumps off the female rapidly. Making distance from her quickly, the male seems to suggest that she is a dangerous object at this time. In fact, under some circumstances the females are dangerous to the males, occasionally
cannibalizing them (Morse 2004; Morse and Hu 2004). In contrast to
many spiders (Elgar 1995; Sasaki and Iwahashi 1995), male *Misumena*
typically service both genital openings of a female with one mount of
her abdomen. In this way they significantly lower their exposure to
cannibalism by the female.

Males also routinely mount or attempt to mount already-mated females, which often are considerably less accommodating to the males, and frequently aggressive. Therefore additional matings are frequently prevented (LeGrand and Morse 2000), although the level of reception varies markedly and may depend in part on how recently she has mated. Males appear unable to determine whether a female has already mated without mounting and inspecting her (Holdsworth and Morse 2000), in common with some species but not others (Suter 1990; Simmons et al. 1994; Gaskett et al. 2004). Mounts with nonvirgin females are highly significantly shorter than those with virgins (Morse 1994) and so are unlikely to achieve intromission. If we remove males mating with virgin females at the same time that these males usually abandon already-mated females, we have not as yet obtained a fertile clutch of eggs from one of these females (Morse 1994). Males will also regularly mount penultimate females, which vigorously resist these actions (Holdsworth and Morse 2000). Virtually all male guarding takes place at this time, part of it in the mounted position. Presumably such behavior will enhance the guarder's opportunity to mate with this female immediately after she molts (Elgar 1998). Most males, however, abandon these penultimate females well before they molt into adults (Holdsworth and Morse 2000).

Although *Misumena* produce only a single brood in our study areas, one can often obtain a full-sized second brood by a supplementary feeding of the females (Chapter 4). Thus, they are physiologically capable of iteroparity but normally constrained from it by the short season. We therefore anticipate that they would sometimes lay second broods in areas south of our study area. However, in 30 seasons we have never obtained a natural second brood in our study areas, though by now we have reproductive data for well over 2000 females. Thus, although we may some day find such a second brood, it is unlikely to have more than token significance to population numbers under current conditions. This is not a trivial matter, for having a second brood has important implications for the paternity opportunities of late-emerging males (LeGrand and Morse 2000).

Normally semelparous species exhibit varying abilities to produce a second brood under abnormal circumstances. In general, it appears that if initial efforts frequently end in failure or near failure, some plasticity may be retained (Tallamy and Brown 1999). The desert eresid spider *Stegodyphus lineatus* usually produces a single brood that it feeds and that eventually cannibalizes her. However, if her brood fails or only a few young survive, she will lay a second clutch (Schneider and Lubin 1997; Schneider, Salomon, and Lubin 2003).

Protandry

Males commonly molt into reproductive condition somewhat before females, a trait referred to as protandry. In this way they may maximize their opportunity for matings and subsequent reproductive success. Some males can reproduce upon emergence as adults, but others do not become sexually mature until a few days after their last molt (Andersson 1994). Protandry often results in some cost to ultimate size (Wiklund and Fagerström 1977), with the result that males of protandrous species often are smaller than females. Vollrath and Parker (1992) have even advanced protandry as a possible basis for sexual size dimorphism. Females that can control the sex of their offspring may vary both the timing and provisioning of young in ways that facilitate protandry. For example, the ground-nesting sphecid wasp *Crabro monticola* lays its male eggs in shallower cells and provisions them less lavishly than its female eggs, which facilitates protandry (Kurczewski 2003). Not surprisingly, the trait is widespread among haplodiploid animals, for they can readily control the sex of the offspring. I am not aware of any comparable tactics by spiders, however.

Male *Misumena* achieve protandry in a very different way, undergoing at least two fewer molts than females and thus not surprisingly becoming adults before their female sibs. Most males molt into the adult stage before the females, though the difference between final male and female molts is not as great as expected from the difference in instars between the two sexes. Noncorrespondence between instar number and time of final molt has been described in other spiders as well (Higgins 1992; Higgins and Rankin 1996). Most male *Misumena* molt into adults during late May and early June and subsequently outnumber adult females for a short period. Numbers of reproductive males and females in these populations fluctuate widely over the breeding season. They range from the strongly male-biased condition characterizing the start of the season to a relatively short midseason, during which numbers of virgin females outnumber adult males, to the latter part of the season, during which the numbers of males again greatly exceed those of the virgin females (LeGrand and Morse 2000). Most females molt into the adult stage in the midseason (late June and early July). Males should therefore experience strong selective pressure to reach prime breeding condition by that time.

A tradeoff occurs between male size and emergence time if males compete overtly for females and if protandry is accomplished at the cost of size. Small male *Misumena* in dense populations may experience difficulty mating in the presence of larger males, judging from our experimental manipulations (Holdsworth and Morse 2000). Large *Misumena* males may preempt smaller males from access to virgin females during the uncommon situations in which more than one male is present. Large males predominated under these circumstances in dense populations of the closely related *Misumenoides formosipes* (Dodson and Beck 1993). Large males typically enjoy superior physical access to females (Andersson 1994), and size should be an important factor affecting the degree to which protandry will progress. However, it should not benefit males to extend protandry beyond a certain point.

Males gain nothing by emerging far earlier than the females, and advantages of protandry should decline as numbers of males build up over the season. In many species, if protandrous males become immediately active, they may become highly vulnerable to predators (Magnhagen 1991; Andersson 1994), thereby lowering their probability of ultimate success in reproduction. Moreover, because males usually do not live as long as their females (LeGrand and Morse 2000), an extremely premature emergence might even shorten the part of their adult life span during which they would have access to reproductive females. However, male *Misumena* appear largely unaffected by these potential constraints. They enjoy low mortality as adults (Chapter 5), as well as a fairly long adult life for male spiders (Morse 2004), which may somewhat mediate the cost of any excess in protandry. Furthermore, a given male may mate through the majority of the breeding season (LeGrand and Morse 2000). A set of four individuals in the laboratory mated multiple times (9 to 17 times, mean = 12.5), until nearly the end of their lives. On average, they mated successfully until four days before death, with one individual accomplishing this feat within two days of death (LeGrand and Morse 2000). By anyone's standards these were old individuals at the time of their deaths, and at this time very few penultimate or unmated females remained in the field.

In general, however, the differences in size between males and females that can be directly attributed solely to protandry appear modest. From an evolutionary perspective, they could nevertheless create initial differences in size between males and females, upon which selection might subsequently act.

Naturally, these size differences do not exist in a world independent of competing interests, which may present strong interacting variables. For instance, small males are most vulnerable to severe environmental constraints at times of drought because of their surface area/volume relationship. In the laboratory at ambient temperatures during hot weather and drought, small males will die especially quickly from excessive desiccation if not given water (Chapter 3). How often they would reach this condition in the field remains to be seen. Large individuals will require more food resources to maintain themselves than small ones. These are some of the factors that could prevent male size from reaching a point that optimizes the tradeoff between body size and protandry.

Some males in the population may be incapable of molting into the adult stage early in the season. One reason may be that they belong to late broods unable to progress through the juvenile stages soon enough to match the emergence of the "early risers" discussed earlier. Others may have had the misfortune to live in areas with relatively poor resources.

Male *Misumena* molt into the adult stage over a period of several weeks (LeGrand and Morse 2000), as might be predicted solely as a consequence of the relatively wide range in time that second instars emerge from their nests in late summer. Not all male *Misumena* reach adulthood in near synchrony with the peak emergence of virgin adult females, and I would not expect the success of late entrants to compare favorably with that of the earlier emerging males. However, the late-emerging males may contest more successfully for virgin females that do appear at this time than the older males, since younger males generally succeed over similar-sized older ones in encounters associated with procuring matings (Hu and Morse 2004), thereby "making the best of a bad situation." Late- and early-emerging males do not differ in size.

Whether this advantage is important in these low-density populations, remains to be seen, though it could be important in the dense *Misumenoides formosipes* populations studied by Dodson and Beck (1993). We do not know the ultimate reproductive capabilities of small and large males at this time. Nor do we know whether any male *Misumena* normally mate enough times to approach the maximum possible. However, since the vast majority of broods have extremely high hatching success (Morse and Stephens 1996), it is questionable whether many matings suffer from sperm depletion, especially since a single copulation often suffices to fertilize two complete broods and always more than one complete brood under experimental conditions (Morse 1994). It is also questionable whether any males in the field would become permanently exhausted of sperm or energy, since the number of copulations accomplished by the four above-mentioned males considerably exceeded the mean number of expected matings in the field, 12.5 to 5.1 (LeGrand and Morse 2000). Neither did the experimental males show any decline in success over time.

Sexual size dimorphism in *Misumena* thus greatly exceeds what one would predict solely on the basis of the perceived advantages of protandry. This conclusion leads to the proposition that the extreme dimorphism observed in *Misumena* has a more complex basis.

Sex Ratios

One can generally expect the number of males and females in a population to be roughly equal. That assumption does not hold for *Misumena*, however, at least for considerable periods of the year and parts of the life cycle. Sex ratios not only fluctuate considerably, but may also have important implications for members of the populations.

Primary and Secondary Sex Ratios

Natural selection, in the simplest sense, acts to maintain a 50–50 percent sex ratio, or more accurately, equal resource allocation to both sexes (Fisher (1930). These ratios may shift strikingly over the life cycle, to the point that adult (secondary) ratios of spiders may be highly skewed in favor of one sex or the other (Foelix 1996a). For these purposes, I define a primary sex ratio as the one existing at emergence from the natal nest, though one might argue that embryonic loss should also be included, and a secondary sex ratio as that found later in life, as at maturity. If one sex is more expensive to produce than the other, deviations from that ratio are to be expected, since Fisher's (1930) allocation variable relates to the overall costs of investment in a sex rather than to the sex ratio per se. Thus, if a male is twice as expensive to produce as a female, allocation of resources per individual should favor males but result in more females (33.3 percent males, 66.7 percent females). According to this argument, variation in environmental conditions may impinge on these allocation rules and produce noticeable deviations from a 50–50 ratio. Thus, the more expensive sex to produce should be the one most likely produced under good conditions, while the other sex will be disproportionately represented under poorer conditions (Sutherland 2002).

In this regard, *Misumena*'s sex-allocation system takes on considerable interest. Given the similar sizes of newly emerged male and female *Misumena* spiderlings within a brood found by Edgar Leighton (pers. comm.), allocation costs for males and females should be similar, in spite of the different growth trajectories they take in later instars. This prediction obtains because the later needs are not "start-up" costs, and thus are unlikely to be charged to the sex-allocation costs described in Fisher's model. Producing similar-sized males and females with different growth trajectories provides one interesting way to balance allocation expenses for the two sexes. The sexes have such different future needs, however, that they may encounter strikingly disparate mortality rates resulting in changing sex ratios, and perhaps selection for deviation from the 50–50 ratio.

Evidence to date suggests that the majority of spider species maintain a 50–50 primary sex ratio (e.g., Gunnarsson and Andersson 1992; Vollrath 1998). Thus far, only some social spiders (Avilés 1986; Evans 1995; Vollrath 1998), the sheetweb weaver *Pityohyphantes phrygianus* (Gunnarsson and Andersson 1992) and *Misumena* (LeGrand and Morse 2000) are known to have a female-biased primary sex ratio. Adult sex ratios also take on great interest, for they may indicate the type of male-female relationships to expect, as well as give some insight into differences between the sexes, such as relative size and other aspects of their morphology.

We found a primary sex ratio of 1.5 females/male (20 randomly chosen individuals from each of 21 broods) in our *Misumena* populations. This ratio, for as yet unclear reasons, varied markedly from brood to brood (Figure 9.2: LeGrand and Morse 2000). The extremes ranged from two broods composed entirely of females to two strongly male-biased broods of 66.7 and 69.4 percent males (LeGrand and Morse 2000). To make this measure, we used the criterion of rearing individuals from emergence in the second instar until they could be externally sexed (when males reach their antepenultimate stage, their legs become striped). Since virtually all of the eggs hatched and reached the second instar, the ratios obtained should hold for the egg stage as well, so that the skew was not a consequence of differential egg or first-instar loss. (Since estimates of single broods were based on samples of 20 young randomly removed from their broods, it probably is safer to say that these broods exhibited a very strong sex bias than to claim that they consisted exclusively of one sex or the other.)

It would be worthwhile to investigate whether the sex ratios of broods vary with the sites they occupy, what environmental conditions exist at these sites, and to what degree these conditions affect the sex ratio. The mean sex ratio subsequently diverged over ontogeny from 1.5 females/male until two exhaustively monitored, marked field populations of 96 and 123 individuals contained 2.6 and 5.1 times as many adult females as adult males (Holdsworth and Morse 2000). This finding confirms our overall impression of the sex ratios in the study areas. In these adult ratios, we either systematically marked all individuals until we ceased to find additional unmarked individuals in the population, or we systematically removed all individuals until no more were to be found. Another intensive study yielded similar results (Figure 9.3).

These ratios have strong implications for the individuals involved. First, if all members of a population are to reproduce, each male, on average, will have to fertilize several females. Becoming mated could prove a difficulty for the females because of the relative scarcity of males. All these factors should affect the sexual behavior of the two sexes. They suggest behavioral patterns that are dynamic, shifting with changes in the sex ratio in accordance with game theory predictions.

The ratios observed, however, are consistent with an evolutionary response to a system in which some males obtain a disproportionate

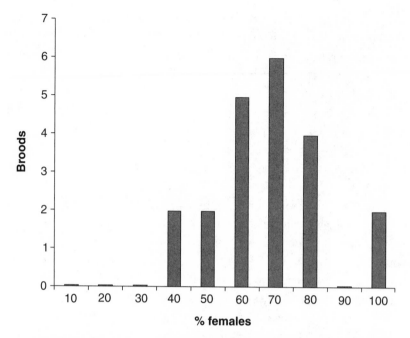

Figure 9.2. Primary sex ratio (at emergence from nests) of *Misumena* broods. Modified from LeGrand and Morse (2000). With kind permission of Blackwell Publishing.

part of the matings. Unfortunately, we do not know the variation in success of different males in inseminating virgin females, which would provide an estimate of the contributions of individuals with different characteristics, as well as how many, if any, of these males are superfluous to the reproductive process. If the model of scramble competition holds, we would expect to see fast-moving or cognitively superior males obtaining a majority of the copulations. If small numbers of males accomplish most of the matings, mothers would do best to produce few males, as the rest might accomplish little. However, in contrast to haplodiploid animals, we do not know whether the females can determine the sex of their offspring, or, if they can, how they do so, or to what extent. Nevertheless, the continued appearance of female-biased primary sex ratios warrants exploration of the basis for sex determination in this species. Gunnarsson, Uhl, and Wallin's (2004) recent description of the probable basis for variable sex ratios in *Pityohyphantes phrygianus* suggests a way by which diploid females might manipulate their sex ratios.

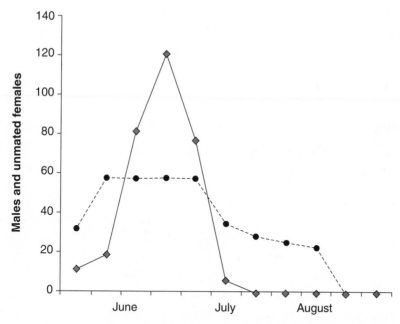

Figure 9.3. Numbers of adult males (dashed line) and unmated adult females (solid line) in population over the season. Modified from LeGrand and Morse (2000). With kind permission of Blackwell Publishing.

Operational Sex Ratio

Not all individuals are reproductively available at the same time. Some are still immature, while others may have already mated and left the reproductive pool. That factor introduces the concept of the operational sex ratio, the ratio of reproductive males to reproductive females in a population (Emlen and Oring 1977), which accommodates for variation in reproductive condition. Although the secondary sex ratio may define the outer boundaries of reproductive opportunities open to the members of a population, the operational sex ratio quantifies the actual reproductive opportunities open to the population members at any given time.

The operational sex ratio is an important measure, for it affects the opportunity of both sexes to achieve a mating. Not surprisingly, it influences or reflects the type of breeding system encountered (lek, territory-based monogamy, etc.: Andersson 1994). In principle, the operational sex ratio is a simple concept, but a number of practical problems arise. With regard to *Misumena*, it remains questionable whether to eliminate females from the tally once they have mated. We have chosen to do so (LeGrand and Morse 2000) because it remains unlikely that many do mate a second time, in common with many other spiders (Elgar 1998). Those females that appear to mate a second time exhibit significantly higher levels of aggression at these times than during initial matings (Morse and Hu 2004). It is doubtful whether intromission occurs, since male-female contact is so brief (Holdsworth and Morse 2000).

Operational sex ratios may change strikingly over a season, as predicted by protandry. In our *Misumena* populations, the operational sex ratio shifts from an initially male-biased one to a brief female-biased period, and then reverts to a male-biased condition (Figure 9.3). Thus, although a male-biased system exists over most of the summer, the majority of matings take place during the female-biased period. In fact, the fit between the population size of adult males and the time at which females become reproductively active is remarkably tight, as is the fit between the availability of reproductive females and the time at which males apparently die in the field (Figure 9.3).

Rapid seasonal variation in itself should increase uncertainty and minimize the tendency to be choosy, even if operational sex ratios do not vary spatially within a population. In the simplest situation, the operational sex ratio will not vary over the range of a population, rendering the problem of spatial variation moot. Yet the variation in sex ratios we obtained among broods (LeGrand and Morse 2000) is great enough to suggest a somewhat patchy local variation in sex ratios (within an individual's cruising range). This variation should affect choosiness among both males and females. Shuster and Wade (2003) provide a useful discussion of the spatial problems encountered by males experiencing varying distributions and densities of females.

A potential difficulty of calculating operational sex ratios, probably common to a majority of studies of this sort, concerns the relevant unit for study. The standard definition of a population does not strictly hold for our purposes, for although a population relates to a group whose members freely intermix, that mixing takes place over a longer time scale than the reproductive period of any given individual. Operational sex ratios incorporate only current information that will determine how the sexes of a group of individuals should ideally respond to each other. If individuals are not in contact with each other over their mating period, they are not relevant to the issue of operational sex ratio, *sensu stricto*. Since they may communicate in ways we do not routinely monitor, an additional difficulty relates to establishing their information-gathering abilities.

Probably the question of our determining who is in such close contact would be simplest in large, conspicuous species living in the open, which could easily identify and "tabulate" each other. Even such species as these, however, would likely move about so much as to make the problem difficult. For individuals that cannot be easily and continually monitored (probably most species), it may not be possible to establish this measure directly or with perfect accuracy, and the individual (and the investigator) must proceed using indirect methods. For instance, Foellmer and Fairbairn (2005a) used numbers of males on a female web as a measure of operational sex ratio, which assumes that any males not on the web are irrelevant to any interactions that will occur. Their method is rather more restrictive than the one we have taken here.

Misumena presents a prime example of a species with locally restricted knowledge resulting from limited sensory capabilities. What boundaries should be used in such a calculation? Presumably the answer will differ from that of a species that makes substantial use of pheromones to attract mates, as in other spiders studied to date (Tietjen and Rovner 1982; Watson 1990). If *Misumena* obtain significant information from the physical presence of lines (Leonard and Morse 2006), that factor should help the individual to estimate the relevant operational sex ratio. We have addressed the issue by intensively monitoring the members of marked groups in discrete patches within close range of each other and are thus confident that our estimates of operational sex ratio (LeGrand and Morse 2000), which fluctuate widely over the season, are realistic. However, although we could make these calculations, we cannot be certain that we precisely matched the spiders' abilities. These studies involved bounded areas of roughly 0.25 ha; it remains unclear whether ranges of these sizes would hold in effectively unbounded situations. And since the operational sex ratio shifts so rapidly over short periods of time, thoroughly up-to-date information, though potentially highly useful for the participants, will generally be inaccessible. We thus have little confidence that individuals can make highly accurate assessments under these circumstances. Encounter rate will thus probably have to suffice as a surrogate measure.

Mate Choice

Mate choice is often an important aspect of fitness, with the resulting quality and/or number of young serving as the criteria on which fitness will ultimately rest. Thus, an individual's efforts to obtain the best possible mate may overshadow other parts of its life history if it mates but once. Choosing an infertile mate could prove catastrophic if an individual mates but once, a likely event for many female *Misumena*. The costs of selectivity in mate choice are often considerably higher for females than males because of the basic asymmetry of male and female expenditures to any given offspring or brood. However, some highly modified conditions occur among spiders (not *Misumena*), in which males can only mate once because physical constraints such as damage to pedipalps resulting from insertions (e.g., *Argiope aemula*: Sasaki and Iwahashi 1995) or apparently programmed death as in *Argiope aurantia* (Foellmer and Fairbairn 2003). Even if an individual mates more than once, choosing a diseased or parasite-laden partner could be highly damaging as well (e.g., Freeland 1976; Hamilton and Zuk 1982), as could choice of a partner whose mating routine could generate physiological damage sufficient to compromise subsequent matings (e.g., Rice 1996, 2000). Relatively little information exists about the importance of these factors in the wild (Arnqvist and Rowe 2005). However, even reaching the point at which an individual can invest in mate choice is of fundamental importance as well, providing the ground base by which the game of mate choice can be played.

General Background

Mate choice is a game between the sexes. Traditionally, students of reproductive behavior assumed that the best interests of both sexes are played out (reviewed by Chapman et al. 2003; Arnqvist and Rowe 2005). Arguments in favor of this mutualistic relationship often had an implicit if unwitting group-selective aspect. More recent theory and growing databases make it clear, however, that such harmony is not the norm (Parker 1979; Holland and Rice 1998; Gavrilets, Arnqvist, and Friberg 2001). Both sexes have their own interests in "mind," and what works best for one sex may not work best for the other. This approach contains far more selfish implications than a mutual best-interests argument, and the apparent best interests of one's mate may really be no more than an incidental consequence of this perpetrator's acts. Since neither sex presumably plays a passive role in this interaction, the result will likely end in a compromise, in some instances to the detriment of one of the parties. The extent to which the two sexes' reproductive strategies match is imperfectly understood at this time, but most outcomes are probably unstable and well described as coevolutionary arms races (Arnqvist and Rowe 2005). Both sexes should continually strive to improve their own fitness, at each other's expense if expedient.

Strictly monogamous pairings provide the simplest relationships, though the initial choice of a mate will here take on primary significance. However, even supposedly monogamous relationships contain far more promiscuity than meets the eye. Which member of a pair should seek matings outside the pair bond, and under what circumstances that sex will vary, can provide considerable insight into male-female relationships.

Overtly polygamous relationships may enhance the opportunity of improving the quality or size of one's investment. A male may enhance its fitness either by maximizing its fertilizations or by ensuring that females mate with it alone, as may be surmised by the frequency of mating plugs and postcopulatory guarding (Andersson 1994). Females may profit from subsequent matings (e.g., Double and Cockburn 2003), especially if they can thereby either improve the quality of their offspring or increase their genetic variance when living under variable circumstances. Countering the advantages of multiple matings are the enhanced dangers of resulting disease or predation. Heightened aggressiveness of mated females in response to attempted subsequent matings could result from these dangers. It would be of interest to determine if female aggressiveness is higher where acquisition of disease through mating is an important concern.

The ability to display effectively may be associated with identifying one's self as a good mate. That advantage in part permits us to account for the conspicuous, and sometimes gaudy, display structures characterizing one or the other sex, often the male. These modifications vary in their degrees of exaggeration in size, shape, color, and the like, as well as differences in other sensory variables, including olfactory cues, which human observers usually cannot recognize. The primary difficulty of investigators lies in linking these traits to those that provide genuine measures of fitness. For instance, will mating with the male peacock that exhibits the most elaborate train result in a female's ability to produce young that will enjoy the highest fitness in the population? Petrie (1994) has in fact confirmed this point. It has frequently been argued that only males in good physiological condition can afford to carry such elaborate displays (Zahavi 1975), which obviously exact high production and maintenance costs (e.g., Clutton-Brock 1988). The frequent high correlation between that character's expression and its parasite or disease load may be an important reflection of its physiological condition.

Theory of Mate Choice

Mate choice may take a variety of forms, which Darwin (1871) separated into male contest and female choice. In classic contest competition, males fight among themselves for access to females or prevent access by other males, and in the simplest situation, females do not exert any input into this decision. Traditionally, this type of mate choice has been associated with species that experience regular combat for access to females, and standard examples include species exhibiting striking sexual dimorphism, with the large size of the males typically attributed to obtaining access to females through fighting. Alternatively, males compete for females via displays through female choice. This mechanism traditionally received much less attention but has reached center stage over the past 25 years (Shuster and Wade 2003). These examples have usually focused on mammals or birds, though in fact in most taxa females are the larger sex.

We now know, however, that alternative male strategies occur in some systems seemingly dominated by male contests of one type or another. These males, generally referred to as satellites or sneakers, may partially parasitize the above-described relationship by obtaining some of the copulations, in spite of the best efforts of the socially dominant males (reviewed by Shuster and Wade 2003). Females may employ a variety of inconspicuous mechanisms that to varying degrees counter the dominance of males and make competition more likely in others (reviewed in Birkhead and Møller 1998; Simmons 2001). If multiple matings occur, sperm competition may take the question of mating preferences to the level of a female's reproductive system, where variables of sperm precedence, mobility, and viability may influence or decide the extent of a male's paternity (Parker 1970; Waage 1979). Females may even be able to manipulate these sperm in ways that give them considerable control over the paternity of their eggs (Thornhill 1983; Eberhard 1996). A female may be able to expel a male's sperm en masse (Eberhard 1996; Burger, Nentwig, and Kropf 2003). Even more impressively, females may have varying abilities to select among sperm accumulated from various males (Eberhard 1996). The ubiquity and extent of this possibility is unclear, owing in part to the difficulty of investigating and evaluating it.

These sometimes subtle manipulations, which are hard to measure but potentially of utmost importance, present a genuine challenge for analysis. Male contest for access to copulation may thus be significantly modified by female reproductive anatomy and physiology, thereby generating additional variance in male success. The ability to determine paternity may therefore take place in ways and to degrees not previously appreciated. Thornhill (1983) termed this phenomenon *cryptic female choice*. Although cryptic female choice is extremely difficult to demonstrate conclusively, Eberhard's (1996) treatise has provided fresh insight into the wealth of ways a female might control her reproductive output. Thus, from the standpoint of the ultimate reproductive success of females, male contest and female choice are not clear-cut opposites. Furthermore, females may refuse to copulate, and even if the male succeeds in forcing copulation, it is not ordained that he will achieve paternity.

Scramble competition (Ghiselin 1974), a potentially important, but much understudied, phenomenon provides an alternative to overt male contests. Scramble competition, rather than contesting for females via display, favors the males most proficient at finding females. This strategy should select for a different set of attributes, which favor mobility and the ability to respond to cues that facilitate locating females, rather than the ability to prevail in an overt contest or display competition. Conditions facilitating scramble competition include low densities, low resource availability, and habitats not favoring long-distance perception. Under these circumstances, males might experience the same female strategies as those associated with male contest competition, though with implicitly lower numbers of males and fewer opportunities for multiple copulations.

Mate Choice in Misumena

Most of the literature on mate choice in spiders focuses on situations where both sexes are common and contact each other frequently by chance alone. This condition provides both sexes with opportunities and constraints. Potential choices of mates may exist, but they may not be available owing to intrasexual or intersexual interactions.

Our experiences with *Misumena* differ fundamentally from that type of relationship. Our populations also have far lower densities than those studied in one of *Misumena's* close relatives, *Misumenoides formosipes* (Dodson and Beck 1993), and male-female interactions in our *Misumena* populations (Hu and Morse 2004) correspondingly differ greatly from theirs.

In our low-density populations with female-biased sex ratios, the norm where we have studied *Misumena*, overt contests between males are the exception, and scramble competition usually prevails (LeGrand and Morse 2000). The first male that finds a virgin female stands a high probability of mating with her in the absence of other males, and competition largely consists of finding a female before another male does. Such "winning" males usually proceed to mate with these females without incident, though the females are always potentially dangerous counterparts capable of fatal attack (Morse and Hu 2004). In this way sexual selection may nevertheless proceed, but selection favors the first, the fast, and the "bright" (i.e., any other tactic that enhances early contact). From a theoretical viewpoint, this search bears many similarities to foraging behavior—a male could be said to be foraging for females!

Under these circumstances males encounter females sequentially, and we have not encountered situations in which a single male had simultaneous access to more than one female, a likely consequence both of the modest density of females and their typical aggressive response to each other when they do come in contact. Under experimental conditions, males will not mate with more than one female in rapid succession (D. H. Morse, unpublished data), but since they seldom experience this circumstance in the field, it seems unlikely that this constraint significantly impacts their fitness. We do not know, however, whether a male will reduce its searching behavior after mating. We could seldom get these males to mate on the following day in our experiments, so we adopted a timetable of mating them with a different virgin female every second day, under which circumstances they regularly mated.

Matings usually take place on flowers exploited by the females as hunting sites (Holdsworth and Morse 2000). This is not surprising in as much as they spend most of their time on these sites and the males concentrate there as well (Chien and Morse 1998; Holdsworth and Morse 2000). Matings, except for those occurring at molt, take place within the context of the female carrying on foraging behavior as usual. This activity is a vital and time-consuming one since few females manage to capture enough prey to maximize their reproductive effort (Morse and Stephens 1996).

In a few instances, males obtain access to a female through precopulatory guarding. To do so, they attend a penultimate female that will soon be molting, at which point she becomes sexually competent. Males not infrequently guard such females, but they apparently have no effective way of determining in advance when such a female will molt (see below). The majority of these males give up and leave the female before she molts (Holdsworth and Morse 2000). Penultimate females are unreceptive to males and continue to be so up to their molt. Males not infrequently attempt to mount them, and when they do, the females typically resist their advances strenuously, often lunging at the males if approached. If a male does manage to access the penultimate's abdomen, she will usually vigorously attempt to extricate the male by scraping him off with her long raptorial forelimbs. As a result, such males seldom remain on a female for more than a minute. Since experimental males mounted 10 percent of the overtly resisting penultimate females and 27 percent of the nonresisting females in the experiments (Figure 9.3), guarding males must regularly attempt these mounts. The mounts may accomplish two ends, likely accounting for the males' persistence. First, the males apparently cannot identify whether or not a female is an adult or has previously mated without mounting her (Holdsworth and Morse 2000). Guarding a female from this location may also maximize priority of access if another male should appear, even though guarders seldom remain long enough for the female to molt (Holdsworth and Morse 2000). The operational sex ratio favors male guarding early and late in the season. In low-density populations, however, the males might not easily ascertain the current ratios, and if they seldom contact females, guarding might nevertheless be favored.

Extended precopulatory guarding is thus the exception. In experiments testing this phenomenon, only 10 percent of the males guarded from 2 to 4 days, and only 18 percent remained until the penultimate molted (Figure 9.4), each of which soon mated with her (Holdsworth and Morse 2000). These figures probably exceed natural conditions, since we initially placed all of these males in a highly favorable location for finding the females, several of which were nearly ready to molt into adults. By failing to provide clear information about the imminence of their molt, the penultimate females provide the males with a difficult-to-solve problem: when they should guard and, if so, when they should quit (Parker 1974). Such ambiguity may favor females, judging from their frequent extremely energetic resistance to mounted males. The males may compromise the females' hunting success at this critical period. The females' continued effort to extricate the males supports this interpretation.

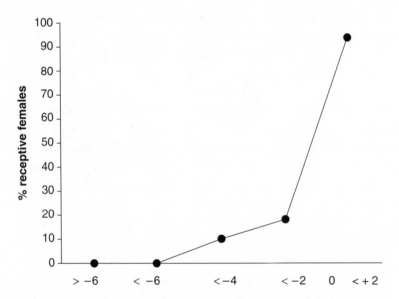

Figure 9.4. Receptivity of female spiders to males during days before (–) and after (+) molting into adult stage. Receptive females allowed males onto their abdomens, as in Figure 9.1. Data from Holdsworth and Morse (2000).

Only occasionally do we see more than one male guarding a penultimate female. However, if priority of access is likely to be determined in advance (see below), and probability of success is low for any male at this time, not surprisingly males seldom remain for long. Hence some momentary multiple guardings are likely to go unnoticed. Nevertheless, their frequency is likely to be very low, given the small number of penultimate females guarded at all (7.3 percent of 41 intensively observed penultimate females at one time or another; 1.4 percent of the total observations on these spiders). In fact, we noted no simultaneous multiple guardings in 109 spider-days of observations (Holdsworth and Morse 2000). Nevertheless, continued attention to penultimates exceeds that accorded to adult females, many of which may have already mated, but are usually left immediately after mating or rebuff. We saw only 2.3 percent of 85 intensively observed adult females with an attendant male at one time or another, and we only noted such guarding during 0.5 percent of the total observations (Holdsworth and Morse 2000).

Where overt contest competition does occur between males for females, large males usually win, at least in staged encounters (Holdsworth and Morse 2000). Although this trait should in itself favor increased male size, we have seen that this advantage does not frequently determine a mating in the *Misumena* populations we study, and other factors may counter it. Over the period from 1976 to 2005, we have observed only two matings in the field where more than one male was present. The larger male prevailed in both instances. Similarly, missing forelimbs, a frequent problem for the males, decreased male success in staged encounters (Holdsworth and Morse 2000). A further variable, age, also affected access to females during encounters, with younger males prevailing over older ones of similar size (Hu and Morse 2004). Although size differences may partly counter age differences, over half of the randomly selected pairs of males in a population were of similar size (within the 10 percent carapace width determined by Dodson and Beck [1993] to qualify as equal-sized). Age is thus not a trivial factor in male-male interactions, determining the outcome of nearly one-quarter of the male-male encounters (Hu and Morse 2004).

This difference is accompanied by a higher level of aggression among the younger individuals, which counters the theoretical prediction (Parker 1974) that older individuals, which have "less to lose" than younger ones because of their lower reproductive value, should mount more aggressive, and hence more dangerous, offenses than younger ones. However, most of the encounters won by the older spider were initiated by that individual (Hu and Morse 2004). Although priority of access to the female resulted from these encounters, in a minority of instances the loser subsequently mounted the female. Given the apparent sperm precedence in this species (Holdsworth and Morse 2000), however, it is questionable whether this second mating accomplished much, as first-mating male *Misumena* did not indulge in postcopulatory guarding and never attacked these secondary males when they subsequently mounted the female. Furthermore, the low frequency of contact between males contesting these females suggests that aggressive male fighting does not produce major benefits.

In general, interactions between *Misumena* males were not highly aggressive. They consisted of momentary contact and retreat, probably because of their normal low density and female-biased sex ratio (Holdsworth and Morse 2000). In about half of these encounters, however, the two males grappled with their forelimbs. Three such interactions (out of 90 encounters) resulted in the loss of a forelimb (Figure 9.5: Hu and Morse 2004). We also recorded a single death during similar encounters in other projects (Holdsworth and Morse 2000), so that, even though infrequent, these male-male encounters are not entirely symbolic and are potentially dangerous to the participants. All such encounters between males recorded in field experiments or casual observations took place either on flowers in the presence of females, or on the backs of the females themselves (Holdsworth and Morse 2000).

The patterns discussed here differ markedly from those observed in the similar *Misumenoides formosipes* by Dodson and Beck (1993). Their *M. formosipes* population was much denser than our *Misumena* populations and had an adult sex ratio of unity. Several (up to 10) males might gather about a penultimate female, and such variables as male size took on a far more important role in determining mating priority than they did in our *Misumena* populations under natural situations. Furthermore, the outcome of past encounters with other males affected the males' standing in the *Misumenoides* groups (Hoefler 2002) and thus helped to dictate which individual had access to the female when she molted. Past encounters are unlikely to play a major role in the mating success of males in our low-density *Misumena* populations. *Misumenoides formosipes* males also engaged in more overt encounters than

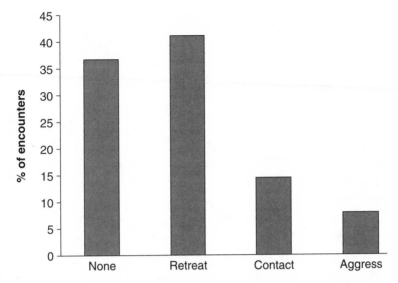

Figure 9.5. Results of staged encounters between male spiders. $N = 90$.
None = no interaction; Retreat = one individual moves away with no contact;
Contact = contact initiated, but one leaves immediately; Aggress = individuals
spar with one another. Data from Hu and Morse (2004).

those between male *Misumena*, with frequent loss of limbs and several
deaths recorded. Nevertheless, the behaviors observed in the staged
Misumena encounters bore considerable similarity to those of *M. formosipes*, even though the *Misumena* did not escalate these encounters
with the frequency reported for *M. formosipes* by Dodson and Beck
(1993).

Thus, male-male encounters probably play only a modest role in determining the mating agenda of individuals in our populations. Under
experimental conditions, however, the males exhibit traits similar
enough to those of the highly aggressive *M. formosipes* population
studied by Dodson and Beck (1993) to suggest that if found in high
density and/or even sex ratios, male-male aggression might play an important role in determining mating priorities.

Cues Used in Mate Finding by Male *Misumena*

The fundamental challenge faced by an adult male *Misumena* is to find
virgin females. Tactile and visual cues involved in substrate choice
(Chapter 7) may play key roles in finding mates as well. Although the

cues spiders use in finding mates may differ quite broadly among species or groups, the issue has been investigated in detail for only a few species, none of which closely duplicate *Misumena's* lifestyle. Mate-finding cues available to *Misumena* differ from those of other spiders studied to date.

Pheromones

Pheromones play an important role in the mate finding of spider species investigated to date (Tietjen 1977, 1979; Tietjen and Rovner 1980), including both airborne pheromones and pheromones placed on lines. However, we have yet to find evidence for pheromone use in mate finding by *Misumena*, either in airborne form or on lines (Figure 9.6), though we have employed a wide variety of experiments and observations to test for the use of either airborne or line-borne pheromones in this context. We have been encouraged to test the spiders in a Y-tube apparatus in the laboratory, in which an individual is presented with odors from two sources—the other two arms of the Y-tube. However, even if the individual responded to an odor associated with one arm of the tube, that reaction would not establish that the spider would respond to this odor in the field, our criterion, even though such a reaction would assume considerable interest.

When placed in a small vial (7 drams: 5 cm long, 3 cm diameter) with a fresh adult molt, only one of 10 male *Misumena* moved in the direction of the molt, and no individuals in a small nylon tricot bag ($25 \times 15 \times 10$ cm filled with grass) found a just-molted adult female. However, most other possible distance cues have provided no more leads for finding females than have the pheromones (Figure 9.6). Only one in 10 adult males placed in a $30 \times 30 \times 30$ cm enclosure with a recently molted virgin female found the female within 1 h, and it made that discovery by chance alone. Even when placed on adjacent ox-eye daisy inflorescences, only 40 percent of the males found a recently molted virgin female. That figure rose to 60 percent for males placed on the opposite side of the daisy (2 to 3 cm away) from a recently molted virgin female. However, if we gently pushed a male into contact with a female using a sable-hair brush, he almost instantly mated with her (Holdsworth and Morse 2000). These results strongly suggest that the males depend primarily on their activity and on contact cues to find

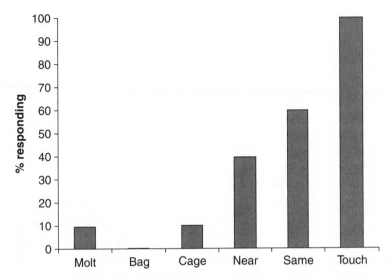

Figure 9.6. Response (percentage) of adult male spiders to various potential cues associated with newly molted virgin females in a variety of contexts. Molt = fresh adult female molt, Bag = in small bag with female, Cage = in small cage with female, Near = on opposite side of same daisy flower, Same = same side of daisy flower, Touch = male and female pushed into contact.

females. If so, they enhance the advantage of protandry, superior speed of movement, and ability to choose hunting sites similar to those of prospective mates.

Failure to use line-borne pheromones could result from the habitat in which *Misumena* conduct their activities, the upper stratum of herbaceous cover. This site differs strikingly from the litter layer frequented by male *Xysticus* spp., crab spider relatives that respond strongly to apparent pheromone analogs (Aldrich and Barros 1995; D. H. Morse, unpublished data). In a parallel relationship among wolf spiders, *Rabidosa* (formerly *Lycosa*) *punctulata*, a species that concentrates its activity in the lower part of the herbaceous stratum, depends more heavily upon line-borne pheromones than does *R.* (formerly *L.*) *rabida*, a species that inhabits the upper parts of that stratum (Tietjen 1977; Tietjen and Rovner 1980). Female *Misumena* are also likely to be extremely sedentary when they find superior hunting sites. On average they remain four days on such a site (Morse and Fritz 1982), while pheromones on lines retain chemical activity for a day or less in species using this strategy (Pollard, MacNab, and Jackson 1987). Thus, even if *Misumena* used pheromones in finding mates, cues would be severely limited at times.

Contact Cues

The males' use of contact cues resembles that of the just-emerged second instars, which depend primarily on contact cues for selecting substrates. However, the adult males appear to move randomly in relation to the females, and only after arrival do they make decisions about the quality of the sites contacted. The case of the adult males differs from that of the spiderlings in that the spiderlings apparently use gross features of the vegetation to recruit (factors common to species and condition of plant substrate) to a site. The females themselves are apparently too small to provide a usable visual target for the males. Otherwise, the males should have responded to nearby females in a more directed way than the seemingly random discovery pattern observed when we placed the two sexes on opposite sides of a daisy.

Adult males use already-laid lines, even to poor sites, in apparent preference to making a new line; these lines likely serve as a means of resource-saving movement (Anderson and Morse 2001). Subsequent experiments (Leonard and Morse 2006) further suggest that adult males obtain some information from lines that help them to find females, although they apparently cannot determine the direction that the linemaker has traveled (Anderson and Morse 2001). Draglines of spiders differ markedly with body size (Osaki 1996, 1999), so that size or other structural cues might account for the preferences of male *Misumena* (Anderson and Morse 2001). (Tietjen and Rovner [1982] argued that *R. rabida*, an upper-stratum species, regularly used structural cues in line-following.) Draglines of some species may differ in the number of threads used (Barth 2002).

Adult male *Misumena* preferred lines of adult females to their own lines or those of other adult male *Misumena* when provided with alternatives. However, they crossed lines of virgin adult females, penultimate females, and other adult males with equal frequency when these were provided one at a time (Leonard and Morse 2006). Given the usual low densities of virgin females, and the mates' less than perfect ability to choose between alternatives, it is unclear how frequently this discriminatory ability will improve their opportunities of finding females. Adult male *Misumena* did not follow the lines of another crab spider *Xysticus emertoni* when given a choice between them and female *Misumena* lines. Only reluctantly and occasionally did adult male *Misumena* follow these *Xysticus* lines when not provided with an alternative. They did not favor lines of male conspecifics over those of lines from similar-sized juvenile female conspecifics. They made these choices, even though the *Xysticus* and adult female *Misumena* lines, and the adult male and juvenile female *Misumena* lines, likely were comparable in size (Leonard and Morse 2006). The result suggests that *Misumena* exhibit some, though extremely limited, discriminatory abilities.

Two other factors increase the difficulty for searching male *Misumena*. Since females are likely to be so sedentary, the number of lines they produce will often be severely limited. Furthermore, lines in the canopy break during high winds (ca. 20 m/s) (Anderson and Morse 2001). Such a velocity occurs every few days, most frequently with the arrival of a new weather front. However, since the wind will wipe away old, likely irrelevant information, lines available after high winds should provide more useful cues for males than those present at other times. Nevertheless, lines usually are unreliable indicators for tactile cues.

Penultimate Males

In contrast to adult males, penultimate male *Misumena* did not prefer the lines of females to those of other penultimate males when provided with a choice. This finding indirectly suggests that the traits exhibited exclusively by the adult males could play a role in helping them to find females. In addition to the differences in choices of adult and penultimate male *Misumena*, penultimates seldom made multiple crossings of the lines. Adult males made many such crossings, often secondarily to the less favored sites. Because penultimates do not have the possible benefit of a mating at the end of the line, and a very realistic possibility of a dangerous predator lurking there, this difference in overall activity on the lines is quite understandable (Leonard and Morse 2006). This difference is also consistent with other contrasts in activity observed between penultimate and adult *Misumena* (Anderson and Morse 2001). However, we do not know what characteristics of these lines the males are selecting, though the experiments on adult and penultimate males indicate some degree of discrimination.

Since penultimate males did not distinguish between the lines of virgin or penultimate females and lines of other males, in contrast to the discrimination exhibited by adult males, the adult males' response to these cues has changed from their penultimate condition. We know that the setae of male *Misumena* palps increase strikingly in size and number between the penultimate and adult male stages. This is consistent with these structures providing the extra degree of tactile sensitivity exhibited by the adults (Leonard and Morse 2006). The modest development of penultimate setae serves the penultimates well in light of the minimal advantage and high liability associated with these lines (Anderson and Morse 2001).

Comments

In short, we have not yet satisfactorily resolved how male *Misumena* find females. To date the procedure does not appear to be a particularly efficient one. Nevertheless, the males manage to fertilize virtually all of the females in the population, though at a modest rate (see below). Size of lines alone does not provide a definitive cue, for adult males distinguish between similar-sized adult female lines of *Misumena* and *Xysticus*, but do not discriminate between similar-sized lines of juvenile female and adult male lines of *Misumena*. The males' modest ability to locate females contrasts strikingly with the effectiveness of wolf spiders (Tietjen and Rovner 1982) and orb-weavers (Herberstein, Schneider, and Elgar 2002). The failure of the simplest and most obvious male strategies suggests that female strategies run counter to the males' best interests.

Female Crypticity

Female *Misumena*'s apparent failure to provide their males with cues to their location and reproductive status, in contrast to those reported in a number of other spiders, suggests that the females are "deliberately" cryptic to the males. If so, it is unclear whether this condition is adaptive, or the consequence of their apparent inability to produce appropriate pheromones. Being cryptic to prospective mates might seem a counterproductive strategy for the females. However, all or virtually all of the females in our populations eventually mate, but, on average, only after 2 to 3 days, even though they will readily mate immediately after molting into the adult stage (LeGrand and Morse 2000). By mating before her carapace has completely hardened, a female spider possibly subjects herself to a heightened probability of damage, though one might question whether the tiny males would be capable of inflicting it.

Since the females can depend on being mated, the males most capable of finding these females should enjoy a reproductive advantage. If these males are the same ones that provide the traits most satisfactory for maximizing female fitness, then the basis for sexual selection is established through scramble competition. Although our males fight with each other upon occasion and those males with certain traits (large, young) win (Holdsworth and Morse 2000; Hu and Morse 2004), these encounters are infrequent in the populations we have studied. Thus, interactive competition for these females is sporadic at the moment they molt. Selection for speed, or for the ability to search areas both systematically and efficiently, may thus prevail over fighting ability. Small males should prosper under these conditions because they move rapidly and can maintain themselves at little cost. Either of these traits could select for the evolution of small males. Ghiselin (1974) has also remarked on the advantages of mobility sometimes enjoyed by small males. More specifically, Moya-Laraño, Halaj, and Wise (2002) have proposed that small male spiders should be favored biomechanically in climbing to find females or in eluding predators on vertical surfaces. This argument is consistent with the sites male *Misumena* must search to find females. However, the presence of small males in situations where considerable male fighting takes place, as in at least some populations of the similar *Misumenoides formosipes* (Dodson and Beck 1993), may complicate this explanation. In fact, the considerable variation in male size in *Misumena* (LeGrand and Morse 2000) could be a result of countering selection for small, fast, energy-efficient individuals and large, formidable fighters, as proposed for *Argiope aurantia* (Foellmer and Fairbairn 2005b).

A virgin female *Misumena* will generally accept any male that finds her, though retaining a low but variable tendency of aggression toward the males. Since virgin females exhibit little tendency to choose even when they should be able to do so, the males that find them may not be a random lot. Their fathers may have been selected for their ability to find females, even though their selective abilities appear meager on first inspection (a refurbishment of the often-criticized sexy son hypothesis: Fisher 1930; Weatherhead and Robertson 1979). Assuming that differential ability to find females is a heritable trait (selection could act on any one of the traits noted above), we can conclude that this trait should favor the success of her male offspring. Since finding patchy resources is such an important task for these spiders in general, their ability to find resources with a minimum number of cues may serve a more general advantage as well, if selection acts on a broader swath of resource finding than discovering virgin females. I have already emphasized the importance of the ability to locate favorable hunting sites for adult females, as well as for newly emerged second instars. Thus, a trait that may have profound importance to the adult males could be the very one that benefits the opposite sex at the very same time, as well as the young of both sexes. Candidate genes for similar behavior have been reported from a few other species (Sokolowski 1980, 1998; Fitzpatrick et al. 2005), as noted in Chapter 6.

Behaving cryptically to males may appear counterintuitive, since it would not select for early matings, even though the females can mate immediately after emerging from their last molt (Holdsworth and Morse 2000). However, since the males experience very low mortality during the peak of the mating period (LeGrand and Morse 2000) and since they retain their fertility even after several matings (LeGrand and Morse 2000), obtaining the fastest mating may not be critical for female *Misumena*. Extremely late matings, however, may be counterproductive because of questionable male availability and poor overwinter survival of any resulting late young. We know nothing about the prevalence or importance of venereal diseases in these populations, but the evidence to date does not suggest that diseases play a large role. If the females mate only a limited number of times, as we now believe, the disease factor is not likely to be as serious as if they mated indiscriminately.

If females do not produce pheromones that attract males, what factors were responsible for female *Misumena* (or their ancestors) giving up pheromonal signaling in this context? (I assume here that the absence of such signaling is a derived condition, since other crab spiders *Xysticus*, for example, do appear to use pheromones in reproductive contexts.) Adult *Misumena*, as well as members of the genera *Misumenoides* and *Misumenops*, are mostly species of the herbaceous canopy, and they do not construct elaborate webs like those of the orb-weavers. Considerable motion of the air occurs through these canopies, which often may be tens of cm or more above the ground. It thus seems unlikely that pheromones under these conditions would remain effective as long as those of, say, *Xysticus* species, which spend the majority of their activities in the litter layer. There, such pheromones can be placed on the available substrate, which, owing to the more modest circulation patterns of air under these circumstances, seem far more likely to generate dependable odor trails and also remain longer than if placed in the relatively breezy canopy of the vegetation. (I previously noted that *Rabidosa* species differ in their use of pheromones in relation to their position in the plant strata [Tietjen and Rovner 1982].) Male *Xysticus* virtually never venture into the canopy, and when females infrequently do so, they hunt on flowers, which, however, they exploit far less efficiently than *Misumena* (Morse 1983). Pheromones of the crab spiders probably differ from those of the orb-weavers. It would be of interest to know how effectively those compounds available to the crab spiders would function in the canopy, as opposed to those of the orb-weavers. The orb-weavers of course use them in the canopy layer, though in the context of their webs.

Sexual Cannibalism

Cannibalism, or feeding upon conspecifics, occurs sporadically among animal groups (Elgar and Crespi 1992). Often it is inadvertent, in the context of adult or late-stage individuals feeding upon the young. (In Chapter 4 I briefly discussed cannibalism among the early instars.) However, I will focus these comments on sexual cannibalism, which involves courting individuals or members of mated pairs as prey. Benefits usually are confined to the cannibal, but the cannibalized individual may sometimes benefit as well. Sexual cannibalism is especially common in spiders, a likely result of the voracious nature of these often food-limited predators (Wise 1993; Elgar and Schneider 2004) and no doubt facilitated by the marked dimorphism of most species. The precise basis for sexual cannibalism is the subject of some controversy, but it probably results from a heterogeneous group of causal factors (Elgar and Schneider 2004; Morse 2004).

Theory and Background

For many predators, including spiders, any animal that falls within a rather wide size range is a potential prey item. Furthermore, predators, including *Misumena* (Morse and Fritz 1982), spend much of their time in a semi-starved condition, and it is the exception rather than the rule for them to be satiated. Thus, prospective mates, usually falling within the size range of prey, may be vulnerable, which requires that they be able to communicate effectively their identity to the opposite party. An individual may have the important option of feeding on such a prospective mate, for that prospective mate may have greater value as a prey item than as a sexual partner (Newman and Elgar 1991). Alternatively, a female may mate with that individual and then feed upon it. Although a male could accomplish the same feat, the consequence would differ little from a standard predation event. The point of interest here is that the male might assess the female as a poor-quality mother on the basis of copulating with her and elect to eat her. I do not know of any examples that fit this criterion, and if males are very small, such an outcome is extremely unlikely.

Not surprisingly, males of many species have developed elaborate courtship repertoires that likely lower the probability of being cannibalized (Elgar and Schneider 2004). With small males and large females, the males may be especially vulnerable to cannibalism, although the very smallest males may avoid large females more readily than larger, but still relatively small, males. Newman and Elgar (1991) have advanced this argument as a factor selecting for small male size.

Sexual cannibalism has recently attracted considerable attention; it serves as an extreme example of male-female conflict that incorporates problems of foraging, reproduction, and predator avoidance (Elgar and Schneider 2004), potentially in a developmental context (Johnson 2005). Several studies have been published concerning its adaptive significance; some of them focus on spiders and others on efforts to model the phenomenon. Two such models address precopulatory cannibalism, and two address postcopulatory cannibalism. Pre- and postcopulatory cannibalism differ fundamentally in terms of whether or not gametes have been exchanged. I will assume throughout that the female is the prospective cannibal.

A precopulatory model advanced by Newman and Elgar (1991) posits that a female should evaluate a prospective mate in terms of whether he would enhance her fitness more by serving as a meal or as the parent of her offspring. This model, based on the orb-weaving spider *Araneus diadematus* (Elgar and Nash 1988), and particularly focused on virgin females, proposes that the outcome of precopulatory cannibalism under these circumstances would potentially benefit the cannibal. The victim would obtain no benefits as a result of this outcome. Major factors entering into the female's decision of whether to cannibalize would include whether she had yet mated, her energetic state, the characteristics of the male, and the probability that she would encounter another prospective mate (Elgar and Nash 1988). Large males would obviously provide more resources than small males, resulting in the prediction that the females would single them out for preferential attack, which would favor small males. Small size might thus balance the advantage secured by large males in male-male interactions to obtain preferential access to females.

A model by Arnqvist and Hendriksson (1997) suggests, in contrast, that such precopulatory cannibalism might not in itself be adaptive, although it would be part of a larger syndrome that was, overall, adaptive. More specifically, if high selective value accrued to particularly aggressive predators during their juvenile stage, they would benefit from retaining that trait throughout their lifetime if they could not shift their behavior at this time. This model differs from the Newman and Elgar (1991) model in that predation on males would not be selective. That is, the female would not assess the value of the male, although she might be affected by her energetic and reproductive state. Arnqvist and Henriksson termed this model the *aggressive spillover hypothesis* and proposed it as a consequence of their experiences with the mating behavior of fishing spiders *Dolomedes fimbriatus* (Pisauridae). These voracious predators as adults also directed considerable aggression toward prospective mates. Arnqvist and Henriksson hypothesized that this combination of traits resulted from a negative genetic constraint, though they had no direct evidence for such a mechanism.

A postcopulatory model by Buskirk, Frohlich, and Ross (1984) concentrates on situations in which mating (with gamete exchange) has taken place and focuses on whether the male's fitness will be enhanced from being eaten after mating. If a male is unlikely to achieve a subsequent mating, it may behoove him to contribute his bodily resources to the female after mating. In this way he enhances his probability of eventual successful paternity and the extent of that paternity (extra offspring, high-quality offspring). If food-limited, as females are likely to be, it probably is seldom in her interest to eschew the male after mating with him. From her perspective this aspect of cannibalism would not basically differ from simple predation, unless she would otherwise mate with him in the future. The female should attempt to capture the male if it will enhance her energetic or resource state. However, if the male is especially small, as in some spider species, it may not benefit the female to take the effort and time to feed upon him, even if she can capture him. Adult female *Misumena* routinely lost mass when feeding on prey of similar size to their mates (Morse 1979), and we have not identified any special advantages of feeding on conspecifics such as those proposed by Elgar and Nash (1988) (critical nutrients, etc.). Nevertheless, the female crab spiders regularly captured tiny prey, suggesting that, as unsatiated predators, their basic strategy was to strike first.

Elgar (1992) alternatively proposed that cannibalized males might enhance their paternity if they copulated longer in the process of being cannibalized than those that escaped cannibalization. Andrade (1996) reported such a relationship in redback spiders *Latrodectus hasselti*, notwithstanding the questionable contribution made by nutrients from the male. Although Andrade (1996) notes that males unlikely to mate a second time will maximize their fitness in this way, females could be driving this system as a consequence of their generalized aggressiveness, a major feature of the precopulatory model of Arnqvist and Henriksson (1997).

Cannibalism of Male Misumena

Male *Misumena* are vulnerable to cannibalism by their females, especially when approaching the female prior to mating, though the numbers involved are low (Morse and Hu 2004), especially when compared to a few species in which sexual cannibalism is routine (Sasaki and Iwahashi 1995; Andrade 1996, 1998). Small and agile, male *Misumena* are well equipped to avoid attacks by the females, but females nevertheless occasionally do capture and kill males. Moving onto the female in advance of copulation is a dangerous act. Some females are quite aggressive to males at this time, particularly if they have already mated. Although many females probably do not mate more than once, the males' apparent inability to ascertain female reproductive condition without mounting necessitates this exposure.

A paucity or total lack of display by males approaching females probably enhances the females' capture rates. However, by almost always achieving two palpal insertions per mount of a female (Morse 2004), males must considerably lower their vulnerability relative to many spiders. As a result, they do not have to dismount and then subsequently remount as many other species do, a period of major danger to them. Thus, rates of precopulatory cannibalism are low in our populations, falling around an estimated 4 percent in a breeding season. In the populations we study, such mortality amounts to one or two males per year. Thus, this level does not make a major impact on the overall standing population of males, but is potentially high enough to act as a significant evolutionary force. At times these males exhibit avoidance behavior about adult females that should make them less vulnerabile to cannibalism. We have not yet evaluated the cost of this behavior to their fitness.

We have yet to record an instance of postcopulatory cannibalism of male *Misumena*, though the level of aggression exhibited by some females suggests that they would capture the males at this time if they could. Although we did not observe postcopulatory attacks during experiments testing male-female interactions, we have upon rare occasion observed them in other studies (Morse and Hu 2004). Males dismount the females extremely rapidly; often the males literally jump off the female after mating, most frequently off the posterior end of the female's abdomen. This behavior strongly suggests that the female remains a dangerous entity. This diversionary tactic emphasizes the potential importance of cannibalism avoidance by the males. The scarcity or absence of postcopulatory attacks is thus probably a consequence of male, not female, behavior. Of particular interest, this agility probably simultaneously minimizes conventional predation on male *Misumena* (LeGrand and Morse 2000). Since males mate several times on average (LeGrand and Morse 2000), a high premium rests on a successful escape.

We found no evidence of size-selective cannibalism, even though the mass of males varies threefold (D. H. Morse, unpublished data). Given the diminutive size of all of these males, however, they may not have been perceived as differing in size by the females, given the apparent limited visual capability of this species (Morse 1993c, 2005). Most of the cannibalism on male *Misumena* involved the oldest individuals, which may have become slower and less active than younger ones and hence more vulnerable to attack. Old males were attacked most frequently, and a higher percentage of those attacked were killed (Figure 9.7: Morse and Hu 2004). This timing coincided closely with an apparent drastic increase in mortality in the field after the last penultimate females had molted into the adult stage, suggestive of age-related mortality (LeGrand and Morse 2000). Behavioral patterns of the old individuals did not change noticeably at this time. Hence, as a result of their low reproductive value, they were seen as exhibiting riskier behavior (see Parker 1974). However, their slowdown in activity should function analogously to risky behavior (Hu and Morse 2004).

As noted earlier, these males performed successful copulations up to as few as two days before they died. Furthermore, all of the cannibalistic acts recorded on them were precopulatory rather than postcopulatory. If male *Misumena* competed overtly for females, these old males would routinely experience limited access to females, as younger males usually dominated them, thereby achieving preferred access to the females (Hu and Morse 2004). However, in these low-density communities, this stricture would seldom hold, although if their activity levels declined, these old individuals would be unlikely to encounter females as frequently as previously. At the time that most males reach this age, few if any virgin females remain anyway (LeGrand and Morse 2000). All in all, these results provide no support for either size or age playing a role in direct mate choice (see Prenter, MacNeil, and Elwood 2006).

Alternatively, cannibalism may result from the females maintaining a high level of aggression appropriate for hunting prey earlier in their life cycle, but that cannot be controlled adequately to avoid killing males when they can capture the males—the aggressive spillover hypothesis of Arnqvist and Henriksson (1997). Such an explanation should favor both precopulatory and postcopulatory cannibalism. Working with the North American fishing spider *D. triton*, Johnson (2005) suggested on the basis of experiments run under confined spa-

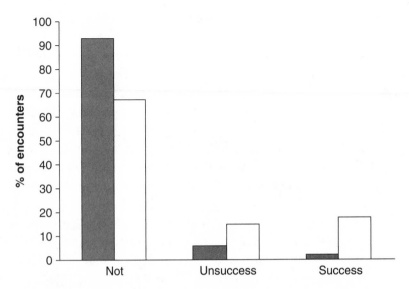

Figure 9.7. Proportions of young (dark bars) and old (light bars) adult male spiders not attacked (Not), unsuccessfully attacked (Unsuccess), and cannibalized (Success) by females in staged encounters. Data from Morse and Hu (2004).

tial conditions that the predisposition to cannibalize is strongly and positively affected by whether she has been guarded as a penultimate. Female *Misumena* are highly voracious predators, both as adults and as earlier instars, and tend to be starved much of the time (Morse and Fritz 1982), which recommends them as good candidates to fit this model. However, responses by virgin females to males are usually not overtly predatory, and those of already-mated females are not inevitably aggressive, although the latter are much more resistant of mating efforts than the virgins. Even if female *Misumena* differ in propensity toward cannibalism as found by Johnson (2005), the low frequency of guarding by these females suggests that guarding will not accurately predict cannibalistic behavior by adult females. We do not have comparable information on severely food-deprived females, which show particularly high levels of attack toward normal prey at such times (Morse and Fritz 1982).

Female *Misumena* are unlikely to realize any energetic advantages from cannibalizing males (see Buskirk, Frohlich, and Ross 1984; Elgar and Schneider 2004). In many instances males are so small that cannibalism is unlikely to account for this purported advantage; large female

Misumena actually lose weight when they feed on prey as small as male *Misumena* (Morse 1979). Although feeding on even tiny conspecific males might provide females with critical resources not readily available from other prey (Newman and Elgar 1991), females reared and mated such that they could not have fed on males had broods indistinguishable in size and survival success from broods taken at random from the field (Morse 2004). These females are unlikely to obtain vital resources from the males, and general information on nutrition in spiders (Uetz, Bischoff, and Raver 1992) suggests that individuals gathering a varied diet in the wild are unlikely to face this potential constraint. Thus, the argument for special nutrient sources does not appear to hold either for *Misumena* or for most other species for which it has been tested (Elgar and Schneider 2004). However, Johnson (2005) reported such an apparent effect in which cannibalistic *D. triton* laid heavier clutches than noncannibalistic individuals, producing apparently heavier eggs as a result. (Johnson apparently did not weigh individual eggs.) Augmenting the diet with crickets to three to four times the mass of a cannibalized male had no effect.

Comparison with Sexual Cannibalism in Other Species

Cannibalism does not appear to be as important a mortality factor for adult male *Misumena* as for the group most exhaustively investigated, the large orb-weavers (Araneidae and Tetragnathidae). In both *Misumena* and the orb-weavers, dimorphism is apt to be extreme, but behaviors of the male orb-weavers that apparently serve to maximize a single reproductive effort and that often end in cannibalism do not occur in *Misumena*, which is fully capable of multiple inseminations. This system of one mating with a single male, characteristic of some orb-weavers, would function imperfectly in the *Misumena* populations we have studied. The reason for this imperfect functioning is the heavily female-skewed adult sex ratio, which demands that, on average, each male fertilize several females, if all females are to be fertilized (and nearly all are). It is of extreme interest to ask what the evolutionary basis for this striking dichotomy in behavioral patterns between *Misumena* and the orb weavers might be. If each male *Misumena* were confined to a single mating, the presence of large numbers of resultant unmated females would produce a massive selective advantage to any individual that developed the ability to mate more than once. It is more likely, however, that ancestral spiders routinely possessed the ability to mate with multiple females and that mating with a single female is derived, as the extremely intricate nature of some of them might suggest (e.g., Sasaki and Iwahashi 1995; Foellmer and Fairbairn 2003). Male-biased operational sex ratios should favor the males' "drastic" behavior to enhance their probability of paternity. Nevertheless, when we discount the exceptions of some *Argiope* orb-weavers (Elgar and Schneider 2004) and redback spiders (Theridiidae: Andrade 1996), we find that most male spiders do appear to minimize risky behavior, in spite of high mortality in some instances (Elgar 1992).

The precopulatory models were explicitly generated with certain species in mind. It is thus of interest to ask to what degree they can be generalized to other species and groups. The results of such an analysis (Morse 2004), in which I tested the degree to which the assumptions and predictions of these models matched conditions seen in *Misumena*, suggested that none of them fit closely. The same inexact fit obtained when I compared them against data from several other spiders, for which information was available. They included the focal species that inspired the models, here run against models other than the one for which they were designed. No close fits emerged, and they revealed no apparent phylogenetic pattern. For instance, two closely related fishing spiders (*Dolomedes* sp.: Pisauridae) did not match each other any more closely than did members of less closely related pairs of species. The results further suggest that sexual cannibalism has developed in response to a heterogeneous collection of lifestyles and phylogenetic constraints (Elgar and Schneider 2004; Morse 2004).

Conflict and Coordination of Male and Female Interests

These points lead to the conclusion that male and female interests often differ strikingly and that "the battle between the sexes" is far more than a worn metaphor about human relations. The frequently antagonistic reproductive interactions between male and female *Misumena*, as well as the paucity of location cues produced by reproductive females, suggest that male and female advantages do not closely match each other in this species. Since natural selection dictates that individuals are selected to maximize their own fitness, that factor itself would, in all but the special case of genetic monogamy (mating exclusively with one partner) (Partridge and Hurst 1998; Rice 2000), frequently result in a "selfish" response that enhanced an individual's fitness, very likely to the detriment of its mate (Rice 2000; Chapman et al. 2003). Detailed investigation of social monogamy (pair-bonded with an individual of the opposite sex), which is now possible using molecular markers, has revealed previously unanticipated levels of promiscuity (Thornhill and Alcock 1983; Westneat, Sherman, and Morton 1990).

General Background of Sexual Conflict

If one sex can manipulate the other at no overall cost to itself, or gain more than it loses in the process, it establishes the motive for selfish behavior. Some of the most striking examples involve situations in which male parents do not participate in rearing their young. For instance, sexual cannibalism occurs under these conditions. Where parental care exists, the male is able to escape from sharing these responsibilities through his proficiency in manipulating his mate. In some instances the male's presence may even be a liability to the female's ability to raise young (e.g., elephant seals *Mirounga angustirostris*: Mesnick and LeBoeuf 1991). Females' tendency to mate with more than one male would also, under many conditions, enhance their fitness, but male counter-tactics (mate guarding, mating plugs, etc.) used by a wide variety of species, including spiders (Masumoto 1993; Elgar and Schneider 2004), may counter that potential advantage to varying degrees.

Between-sex tactics may take extremely negative forms, as demonstrated by a number of elegant laboratory experiments conducted on *Drosophila melanogaster*. For instance, a male's reproductive actions, mediated through seminal fluid proteins, might cause long-term harm to a female by increasing her present fecundity at the expense of future remating (Chapman et al. 1995). This initially counterintuitive effect should benefit the initial male, since any such subsequent broods produced by the female would probably be sired by other (competitor) males, whose offspring might compete with those of the initial male. Rice (1996) has shown that in only 30 generations males evolved changes in seminal fluid proteins that increased their fitness nearly 25 percent at the expense of females whose traits were held constant over that time. The less severe effect of these seminal fluid proteins on females permitted to coevolve with the males indicates that females in standard interactions with males over this period had been developing counter to the male tactics. Thus, these relationships are dynamic, changing over time and likely to differ among populations.

This body of work suggests that sexual conflict will routinely drive negative competitive relationships between the sexes, inflicting a heavy load on them. The importance of sexual conflict under natural conditions and the way the important resultant variables play out in the wild remain largely undetermined, however. Balanced against this basically selfish aspect is the vested interest of the partners in the quality (or quantity) of offspring, a fundamental enhancer of fitness that will limit the extent to which one party can exploit the other. Therefore, particularly strong negative interactions, randomly directed, stand a high probability of being self-defeating even if they initially benefit one partner.

The wide range in quality of offspring within a population could result in part from conflicts of interest between males and females in which males profit most from the number of offspring resulting from a mating and females, whose fecundity may be limited, with the quality of these young. The resulting attributes will be a consequence of the concessions wrung from each other within the confines of maximizing present and future fitness. Tactics of the two sexes may even play out in such a way that these individuals favor the genes they contribute to their offspring, even if this result occurs at the potential expense of those offspring (Haig 2000, 2004). For instance, Hagar and Johnstone (2003) have demonstrated that crosses of two inbred mouse *Mus musculus* lines concentrate their efforts at resolving family conflict in their favor at different life history stages, with the male's strategy to push the female's litter size upward beyond her optimal level, and the female's to allocate milk resources favorably toward offspring of her own line. Although the male's action likely compromises the female's future fitness, she does have the advantage of being the last player in this contest. Although Hagar and Johnstone carried out this experiment in the laboratory with inbred lines, it may be asked whether such conflicts occur routinely in the field.

How frequently do mutualistic relationships dominate male-female interactions, and how closely do they approximate genetic monogamy?

Some individuals that are socially monogamous are probably genetically monogamous by chance alone; certainly the density and spacing
of such population members will play an important role in determining
this outcome. Whether genetically monogamous individuals reach
high enough numbers to impact a population significantly is unclear,
but sparse populations should have the highest proportions of such individuals. It then remains to ask whether those conditions could produce phenotypes, and eventually genotypes, that favored genetic
monogamy. The potential advantages of exploitation to an individual
are such as to raise questions about whether genetic monogamy can be
routinely sustained as a stable strategy.

   If such strategies were to prosper, however, where would we expect
to find them, and what would they be like? It would seem most appropriate to look within the realm of socially monogamous species, especially those for which the two sexes depended strongly on each other
for services that could not easily be substituted. Severe environmental
situations should facilitate this condition, such as where survival depends on cooperative efforts.

   The survival of some permanently pair-bonded species depends on
the strong cooperation of the pair members. Carolina wrens
Thryothorus ludovicianus form pairs to survive the winter in the
northern parts of their range (Morton and Shalter 1977). Pairs jointly
defend territories at this time. Loss of one individual appears to doom
the other, for it cannot defend a territory by itself—a tight relationship
is paramount during this nonbreeding period. Territory holders are in
effect simultaneously defending prime sites for rearing young the following spring. It would be of considerable interest to know their level
of cooperation at that period, particularly since Morton and Shalter
believe that territory size is based on winter, rather than nesting, needs.
If periods other than the winter are not critical, cooperation might decline at those times. However, these birds may fall closer to the special
case of the "ideal cooperative pair" discussed earlier than do most other
species, certainly any of the spiders discussed thus far. Further attention to them and to other forms sharing some of their traits would
richly justify study. In particular, does the frequency of extra-pair copulations differ from that of other territorial species that do not attempt
to maintain year-round territories under sometimes highly inclement

conditions? Does the importance of an ideal winter partner tend to drive these individuals toward a more conservative reproductive style?

Conflicts between Male and Female Misumena

For the most part, male and female *Misumena* attempt to manipulate their prospective partners during the precopulatory period, beginning with the only moderately successful efforts of males to exploit late-stage penultimate females by actively guarding them, even mounting them if possible. This behavior suggests a strong divergence of male and female interests. Similarly, the females' apparent failure to provide males with cues to their status complicates the efficiency of the males, in addition to posing an additional source of danger from mated females, which are not remated for the most part and are more aggressive than virgins. Since males do not appear to mate routinely with already-mated females, one might anticipate that these females would in some way signal their condition and thereby avoid harassment by males at this time. However, if remaining cryptic to males enhances the quality of a female's copulations, it may not be possible to develop a signal for other times.

The copulatory and postcopulatory aspects of the *Misumena* system may be somewhat simpler than the scenarios noted earlier. Virgin females exhibit few signs of mate discrimination, but do exhibit increased aggression toward males in subsequent mating attempts by the males. The males in turn, even if succeeding in a mating effort, usually abandon her, apparently having to mount her in order to ascertain her mating status. In these low-density populations in which males and females do not frequently come in contact, sperm priority will seldom be in question, and the simple conduit-type reproductive ducts of female *Misumena* (Kaston 1948) suggests that subsequent manipulations of sperm in the female reproductive ducts are unlikely to affect paternity in a significant way. If so, the first male to mate with a female will sire the majority of her offspring in the pattern initially suggested by Austad (1982, 1984). Females of even closely related spider species may exhibit remarkably varied female reproductive ducts, as can be readily ascertained from a general manual like Kaston's (1948). This suggests that the result shown here may not hold for even related species.

Do any aspects of *Misumena*'s life history provide evidence for positive interactions between the sexes as they relate to traits exhibited? The most likely possibility we have noted to date relates to the similarities of their foraging, in which adults of both sexes favor flowers that attract large numbers of insects (Chien and Morse 1998). Although these sites are cornucopias that allow the females to maximize their rate of growth, the majority of insects visiting these sites are far larger than the males can capture, and the males might conceivably find more energetically profitable hunting sites to serve their relatively modest needs. Nevertheless, being attracted to the flowers brings the males to places where late-instar and adult females concentrate, in this way enhancing their possibility of finding mates. Given the apparent difficulty that males experience in finding females, their flower-seeking ability may play an important role and be subject to sexual selection. If males differ in their ability to identify and exploit these sites in the way that females do, this ability could simultaneously enhance the female's opportunity of obtaining a high-quality mating, further assuming a correlation between this trait and the quality and quantity of resulting offspring. This accomplishment could help to guarantee that her daughters will show a strong disposition to choose the all-important high-quality hunting sites, and that her sons will exhibit a strong tendency to select such sites as well, thereby enhancing the number of matings accomplished. It is more difficult to account for the development of a flower-seeking ability through sexual selection in the females. Differential capabilities in flower seeking strongly affect growth and fecundity (Morse and Stephens 1996), which appear more likely acted on through natural selection.

Are such mutually beneficial relationships common among animals? One could imagine that where common behavioral patterns were controlled at a single locus, such traits could be similarly selected in males and females. I have pointed out the presence of genes now known to exist in a broad range of animals that might allow such possibilities. Foraging traits of *Misumena* might be similarly or analogously controlled. Their frequency is as yet unclear, but current genomic techniques enhance the possibility of finding candidate genes for such traits (Fitzpatrick et al. 2005). However, if it is possible to lower one's costs at the expense of another individual, an antagonistic system will probably arise. Strongly positive relationships may therefore only take place under severely constrained circumstances that can only exist in the absence of negative interactions, as suggested for Morton and Shalter's wrens.

Synthesis

The extreme sexual dimorphism of *Misumena* (Chapter 8) has many consequences for male-female interactions, which result in strikingly different foraging, predator-avoidance, and life history consequences, these latter factors secondarily impinging on male-female interactions as well. The extreme dimorphism has been associated with an unusual if not unique foraging repertoire, sit-and-wait predation of hard-to-capture prey. Interestingly, the other most dimorphic free-living terrestrial forms, large orb-weaver spiders, also use a virtually unique hunting mode—webs. The two rather different sit-and-wait hunting modes usually share the problem of prey shortages, notwithstanding the frequent abundance of would-be prey. The sit-and-wait routine of crab spiders is matched by a number of sit-and-wait predatory insects, mostly true bugs and mantids. However, the insects differ by virtue of their much greater mobility, all being flighted.

The striking difference in size has major effects on male-female interactions. Given the much larger size of the females, males are at considerable risk of cannibalism in their interactions with females, for which their high mobility and small size serve them well. These small, agile males are also highly capable of avoiding predators as a consequence of these traits, plus the areas they frequent. It is unclear whether this ability derives from their small size or, conversely, whether their small size derives from their high mobility. Small size is considered an economic strategy for a male if it is not forced to fight.

Misumena's sex ratio is unusual, at least to our understanding, in that the primary ratio is not 50–50 percent, notwithstanding the similar original expenditure on males and females. By making equal investments into males and females and then programming them differently so that they eventually differ greatly in size, largely as a consequence of going through fewer instars, the spiders have devised a way to escape the standard stricture of 50–50 percent. Furthermore, since the primary sex ratio is 1.5 females per male, one would expect the males to be larger (or more expensive) than that predicted by the Fisher ratio.

The basis for this skew is unclear, although the superior survival rates of the adult males (vs. most adult male spiders and many other newly adult males) may make such a ratio possible. (However, by the time the males become mature, the ratio has become yet more female-skewed, with the only likely explanation for the obvious shift being a higher overwintering mortality of males than females. Perhaps this is a simple consequence of small size, but also possibly a male constraint, since a disproportionately high overwintering mortality has been reported in other spiders as well.)

The potential interaction of a number of variables, with their causal relationships as yet unclear, is of particular interest. Males do not appear to use pheromones in mate-finding, which runs contrary to other reports from spiders thus far, though it is known that their relative importance varies in the direction we have observed. Whether this difference is related to the small size of the males is not clear but is worthy of attention. To what degree is this difference habitat-related, and to what degree is it size-related?

The operational sex ratio should affect the strategy of the sexes in relation to each other—essentially how choosy they should be in selecting a mate, given the danger associated with mating, the probability that a higher-quality mate will appear, and so on. The overall density of populations relative to their members being able to find each other is a critical issue as well—how does one measure the operational sex ratio? Given the limited use of cues available to the males in finding females, it may be difficult for them (or at least the male) to measure this figure accurately. This appears to be part of the females' broader strategy to become as cryptic to the males as possible. One would expect this pattern to change with the density of the population. In fact, one would not expect to see female crypticity in situations where there was a problem in obtaining a mate.

10

Misumena *as Part* *of the Community*

*T*HUS FAR I HAVE LARGELY focused on *Misumena* as an entity unto itself, or in one-to-one interactions with other members of its community, such as parasitoids or predators. Although that approach typifies the major thrust of this book, it is instructive to extend these issues to a broader context as well. With a few exceptions I have not focused on *Misumena*'s role in the community up to this point. Nevertheless, its interactions with organisms at various trophic levels in its community are important to it and perhaps to other community members as well. The more we learn about this system, the greater the numbers of these interactions we identify, both direct and indirect, and regular and intermittent in nature. In this chapter I comment on some of the more apparent and potentially most important of these interactions. In many instances we have not conducted studies specifically designed to test these apparent relationships, but certain of them warrant attention. In combination they greatly enhance the view of *Misumena* in its natural world. Here I speculate somewhat beyond what I would normally do, and I seek to place *Misumena* within the community it occupies.

Some of these comments relate to indirect effects that may impinge on *Misumena*'s fitness, as well as its direct effect on other members of the community. They include the so-called top-down effects (trophic cascades—Paine 1980; Schmitz, Krivan, and Ovadia 2004) in which

Misumena directly affect members of the herbivore and nectivore
(=nectar feeder) trophic level below them, and indirectly through the
modification of the herbivore–plant interaction resulting from the effect of the predator on the herbivore. Since *Misumena*, as small predators, typically have a trophic level above them—a predator or parasitoid—the latter interaction may affect *Misumena*'s direct impact on its
herbivore and nectivore prey and the resultant indirect impact on the
plants. Alternatively, plants and/or herbivores/nectivores may exert a
bottom-up impact on *Misumena*, which in turn may affect its predators
and parasitoids.

Interactions within *Misumena*'s Trophic Level

Competitive interactions have often been assumed to dominate within-
trophic-level relationships of animals living above the herbivore level,
controlling their population size as well (MacArthur 1955; Diamond
1975). Only subsequently has this unabashed enthusiasm for competition declined somewhat as a result of careful scholarship that tested for
the phenomenon, rather than starting with it as an assumption. I will
initially focus on evidence for competitive relationships in our
system—that is, interactions with other small predators that hunt insects on flowers. My major interest relates to whether competition may
occur, and if so, what form it may take and its importance in the dynamics of both *Misumena* and the community as a whole. It will permit
a test of Wise's (1993) conclusion that interspecific competition is generally lacking, even though spiders are often food-limited. Since newer
work suggests that hunting or cursorial spiders may indeed compete
upon occasion for resources (e.g., Balfour et al. 2003), such efforts are
warranted.

Trophic chains, as well as food pyramids and food webs, provide
convenient abstractions to model the energetic relationships of community members, but the limitations of that simplification become obvious in the case of *Misumena* and other small arthropod predators.
The hazy and indistinct line between predator and prey in these forms
presents an important complication, since they take prey that are large
relative to their own body size. Under these circumstances, the distinction between predator and prey may be as arbitrary as identifying who
grabbed whom first! Two factors may further complicate the relation-

ship. (1) If different predators have different phenological schedules, major segments of two or more species may differ in their relative sizes at different times of the year. The matter of the large eating the small could be quite straightforward, but who is large and who is small may differ with the season. (2) Furthermore, since arthropods must molt in order to grow, they experience vulnerable periods during which they are soft and virtually helpless, and subject to predation by even innocuous forms that normally would be their own prey. These constraints might significantly limit the range of opportunities an individual could undertake.

Parenthetically, I should note that there is no fundamental reason why different species at the same trophic level cannot assume a mutualistic or commensal relationship. To date, however, we have not observed interactions that allow such an interpretation for *Misumena*. For that reason I will focus on putatively competitive relationships among these species.

Other Crab Spiders

Misumena is the only common crab spider frequenting flowers at our study sites; it is difficult to perceive of any other crab spiders making a significant impact on its pattern of resource exploitation. The similar *Misumenoides formosipes* and *Misumenops asperatus*, its most likely potential competitors, are uncommon in our study areas and in the surrounding vicinity. Since adult and late-instar female *Misumena* seldom make an inroad on the putatively satisfactory hunting sites in these areas and capture only a very small proportion of the local prey species (Morse and Fritz 1982; Fritz and Morse 1985), it is highly questionable whether these other two species ever depress prey populations to the point of making competition an issue. These factors render any argument for interspecific competition tenuous.

One might argue that *Misumena* itself accounts for the rarity of *M. formosipes* and *M. asperatus*, and that it has displaced them from these sites. But this argument is not convincing, even if consistent with the distribution observed. *Misumenoides formosipes* and *M. asperatus* largely take the place of *Misumena* in regions south of our study area. The apologist for competition might argue that the competitive balance shifts under different climatic conditions, favoring *Misumena* in the cooler areas and *M. formosipes and M. asperatus* in the warmer. However, the unlikely possibility of resource limitation again argues against this relationship.

If competition did occur among these species, the strongest argument might be made for the early instars, given their occasional high local densities. We have not adequately evaluated the young from that viewpoint. However, even in our study areas, the spiderlings seldom occupy a substantial percentage of the available high-quality hunting sites, except in the immediate vicinity of their natal nest. Even then, they move quickly away from these sites, in the process greatly diluting their densities. In such situations, even if such competition did occur, it would most likely be intraspecific rather than interspecific. For that reason, the argument for interspecific competition even at this age group is not convincing.

By far the most common crab spiders in the study areas are members of the genus *Xysticus*. *Xysticus emertoni* (Chapter 11) resembles *Misumena* most closely in size, phenology, and habitat use. However, several other species of *Xysticus* occupy this habitat as well, differing from *X. emertoni* in seldom or never venturing into the canopy of the old-field habitats. Since *Misumena* seldom if ever frequents the litter layer, *X. emertoni* is the only species of *Xysticus* with which it often overlaps.

Although pitfall traps baited with *d*-2-octenal suggest that *X. emertoni* densities typically exceed those of *Misumena* (D. H. Morse, unpublished data), *X. emertoni* only intermittently occupy flowers, usually for far shorter periods than do *Misumena*. They capture few large prey at such times (Morse 1983), seldom taking bumble bees, *Misumena*'s most important prey (Chapter 11).

Other Spiders

A wide variety of other spiders occupy the flowers frequented by *Misumena*. By far the most common are jumping spiders, which vary widely in size, ranging from adult *Phidippus clarus* to early-instar *Pelegrina insignis*, *Eris militaris*, and *Evarcha hoyi*. As adults early in the summer *P. clarus* reach sizes comparable to those of adult female *Misumena*, while early-instar *P. insignis* do not even match the mass of just-emerged *Misumena* spiderlings. Given the wide range in size and intensely predatory habits of both the jumping spiders and *Misumena*, their relationships are, not surprisingly, complex. They range between the potentially competitive to the strictly predatory, the competitive unlikely due to their modest densities relative to the number of available hunting sites and the latter primarily dependent on the their relative sizes, which shift over the season. When crab spiderlings occupy goldenrod inflorescences, they usually are the prey rather than the predator; this is a consequence of the jumping spiders emerging earlier in the season than the *Misumena* spiderlings. Although taking many of the same food items at flowers as *Misumena*, numbers of jumping spiders, adult or young, seldom occupy more than a tiny percentage of the potential hunting sites at flowers. Thus, the arguments for competition over food, or hunting sites, are again not persuasive. Furthermore, these jumping spiders, which are actively moving predators much of the time, almost certainly capture substantial amounts of their food on the surrounding vegetation rather than on the flowers.

The comb-footed, candy-stripe spider *Enoplognatha ovata* (Theridiidae) frequently occupies the immediate vicinity of flowers, and though it spins a web it, too, captures prey similar to those caught by adult female *Misumena*. Given its relatively modest body size, *E. ovata* captures surprisingly large prey, including the bumble bees exploited by *Misumena*. It will also prey upon *Misumena*, including the adult females, when it has the opportunity. We have not observed *Misumena* preying upon *E. ovata*, though we would expect it if *E. ovata* venture onto the flowers. However, it, too, occupies but a very small percentage of available hunting sites.

Few other spiders occupy *Misumena*'s flowers more than occasionally. We regularly find the early instars of a small sac spider (Clubionidae) on the goldenrod inflorescences occupied by early-instar *Misumena* and that probably feed on the same prey as the *Misumena* spiderlings. Adult nursery-web spiders (*Pisaurina mira*: Pisauridae) sometimes occupy flowers in early summer, but more often than not they have already laid a clutch and are guarding their young there.

Orb-weavers, primarily *Argiope aurantia* and *A. trifasciata*, make webs in the midst of the flowering goldenrod, thereby capturing many of the same prey that adult female *Misumena* capture on the adjacent flowers. The orb-weavers, especially *A. aurantia*, may attain fairly high densities in some summers and likely capture considerably more bumble bees than do *Misumena*. However, for the most part they only reach their final instar (when their webs are large and in the canopy with the flowers where they capture bumble bees) by mid-August in our study sites, after most *Misumena* have already laid their clutch. Thus, the species most likely to exert an effect on *Misumena*'s capture success is largely separated from it in time. If the orb-weaver exerts any effect, it would likely be to decrease the resource flow to the colony and thus decrease the eventual number of bumble bee reproductives, with possible consequences for future bumble bee populations. Since *A. trifasciata* matures roughly two weeks later than *A. aurantia* in the study area, its separation from *Misumena* is virtually complete. (It should be noted that although *A. aurantia* may be the most likely competitor with *Misumena*, numerous studies on it, and on *A. trifasciata*, have failed to provide convincing evidence for competition among these orb-weavers [see Wise 1993].)

In sum, other spiders are unlikely to exert a major effect on the hunting success of *Misumena*. Most are uncommon on flowers, and even where they do exploit the same prey as *Misumena*, all of these species combined capture far too few insects to suggest that they limit *Misumena*'s food supply, or that they prevent access to it.

Predatory Insects

Sit-and-Wait Predators. Spiders are not the only arthropod predators at flowers. Several predatory insects occupy similar sites as sit-and-wait predators and probably capture many of the same prey as do *Misumena*. In our study areas, these species consist primarily of true bugs (Heteroptera): damsel bugs (Nabidae), assassin bugs (Reduviidae), and ambush bugs *Phymata americana* (Phymatidae). Both the most common damsel and assassin bugs are too small to take the largest and most rewarding prey of adult female *Misumena*, but they capture the prey most rewarding to mid-/late-instar *Misumena*, such as flies, especially hover flies. No doubt they also capture early-instar *Misumena* on occasion, and the reciprocal probably occurs as well. The ambush bug, on the other hand, though of modest size, regularly captures bumble bees and probably obtains much of its sustenance from them. A specialist on goldenrod (Balduf 1939, 1941; Yong 2005), it is extremely local in distribution about our study areas. Since most *Misumena* have already laid their clutch by the time that most ambush bugs have recruited to goldenrod inflorescences, the spiders do not broadly overlap with them in season. Although not frequenting our study areas, mantids (Mantidae), including some introduced species, are characteristic inhabitants of late-summer goldenrod communities, where they feed primarily on large prey, including bumble bees.

Wasps. Although both mud-dauber and spider wasps are among the most important spider predators (e.g., Gauld and Bolton 1988), other wasps, primarily yellowjackets and bald-faced hornets (Vespidae), more frequently take prey of the sort that *Misumena* capture at the flowers. These wasps often visit flowers for nectar, where adult female *Misumena* (Morse 1981) prey upon them heavily at times. Not infrequently, however, yellowjackets and hornets capture insects at flowers as well, though much of their predation is carried out on caterpillars in the vegetation. We have recorded kills by several yellowjacket species (*Vespula* and *Dolichovespula* spp.), including the bald-faced hornet (*D. maculata*) on milkweed, where their prey consist primarily of large flies (sarcophagids, muscids, etc.).

Thus, although several spiders and predatory insects share prey species with *Misumena*, we have found little evidence that their predation lessens the foraging opportunities for *Misumena*. More likely, predation by certain of these species on *Misumena* could modify its behavior and thereby compromise its foraging repertoire, but we have no evidence of such an effect.

Interactions with Other Trophic Levels

As a small predator, *Misumena* preys on both herbivorous and nectivorous insects and other invertebrates, including small predators, given its ability to take prey as large or even larger than itself. However, it may be vulnerable to both the effects of lower trophic levels (bottom-up effects) and higher trophic levels (top-down effects). In addition to these direct effects, *Misumena* may be party to events that drastically modify its environment, yet over which it exerts little if any control. Some of these events may heavily impact *Misumena* populations at a local level. Later in the chapter I describe three instances in which herbivores have impacted plants that in turn would otherwise attract nectivores to their flowers. The drastic decline in nectivore numbers heavily impacts *Misumena*'s food supply.

Effect on Pollination Systems

Misumena's most obvious connections to pollination systems involve its direct role as a predator on pollinators and its possible indirect effect on plant seed production resulting from interactions with pollinators. These three trophic levels (predator, pollinator, plant) characterize *Misumena* and the essential aspects of its environment. If *Misumena* killed enough pollinators, or discouraged pollinators from visiting flowers by its presence, these effects could impinge on the long-term quality of the site for crab spiders. However, it remains far from clear whether these interactions routinely produce a measurable effect, or even whether a merely measurable effect will change the habitat enough to impinge on the quality of the site for the crab spiders.

Both of these issues are relatively complex because they involve flowers and their pollinators extending over the flowering season. In Chapter 3, I discussed the possible problem that absence of insect-attracting flower species for part of the growing season might prevent *Misumena* from establishing or maintaining a population. Further-more, heavy predation on the pollinators of such a plant could preclude the plant's continued presence. The disappearance of a single species of flower that bloomed when no others did could thus prevent the con-tinued presence of *Misumena*.

That being said, we have obtained little information to suggest that any such limitation occurs in our study areas, though most of our at-tention has focused on adult female *Misumena* hunting on milkweed umbels. Studies conducted at a population level (Morse 1986a) recorded no sign of the major flower visitors, bumble bees, avoiding spiders at the flowers. Furthermore, densities of spiders at these sites were so low (< 1 percent of flowering umbels occupied by *Misumena*) that bumble bees did not encounter spiders frequently. Neither did they affect the length of a bumble bee visit on an umbel. In fact, the two common bumble bee species, *Bombus terricola* and *B. vagans*, would, on average, only be attacked once every five to six days and be captured only after 70 to 110 days (Morse 1986a). Since worker bumble bees seldom forage on an inflorescence as much as five to six days, and never live as long as 70 to 110 days, the inconvenience of a spider for an individual bumble bee is small and the danger minimal. Thus, given the low success rates of capturing these extremely large prey, *Misumena* seem unlikely to present a threat that would select for these bees' avoidance behavior. In other words, precautionary behavior might entail a greater energetic cost on the bees (in terms of lost resources) than would ignoring the spiders. Since bumble bees are eusocial, losses of individuals to predators can be treated as losses in efficiency for the unit of selection, the hive of the queen mother.

Although these spider–bee interactions may take place infrequently from the viewpoint of the bee, individuals that do get attacked might respond to this exposure and avoid such sites in the future. In fact, Dukas and Morse (2003) found that some marked bumble bees did respond negatively to milkweed inflorescences occupied by *Misumena*. However, at the population level only the two smallest species, *Bombus ternarius* and honey bees exhibited a significant tendency to avoid occupied sites, either from earlier experience or from directly perceiving the spiders on the flowers. The larger *B. vagans*, one of the two species considered by Morse (1986a), did not avoid spider-occupied sites, and the even larger *B. terricola*, the other species considered by Morse (1986a), was not common enough to test at the sites used by Dukas and Morse (2003). Thus, the Dukas and Morse study, focusing on two smaller species, which the spiders capture far more readily than the larger bumble bees, demonstrated that highly vulnerable species are more likely to exhibit avoidance responses than are related larger species. The study thus does not contradict Morse (1986a), but suggests that highly vulnerable bees can develop the ability to avoid visiting sites occupied by dangerous predators. From a community perspective, however, an effect would be unlikely if the absence of previously attacked individuals merely resulted in other individuals subsequently moving into the foraging range previously occupied by the now vacating individuals.

Earlier experiments (Morse 1977a, 1977b) suggested that bumble bees do interact aggressively with each other. Combined with their tendency to develop trapline-like foraging routes (Williams and Thomson 1998), this trait suggests that other individuals might appropriate the routes formerly occupied by the now-departed individuals, thereby erasing a possible community-level effect on flower visitation by bumble bees. If the spider occupying this inflorescence were particularly active, it could convey this effect through a significant part of the bee populations. With obligately outcrossing flowers like the milk-

weed, enhanced turnover of pollinating bumble bees might, if anything, increase the fitness of the flowers. However, even the results obtained with the smaller bees may not characterize the world of flower
visitors at all times, for the densities of spiders in our experiments were
extremely high. Thus, some of these bees exhibited avoidance traits,
but it is unclear how often they avoided sites under other circumstances. It is also unclear whether the larger bumble bees ever exhibit
such avoidance strategies. Because *Misumena* do capture these large
species as well (Morse 1979), the large bumble bees undertake some
risk by visiting sites occupied by the spiders.

Misumena typically hunt concealed or partially concealed among the
flowers of milkweed umbels, but they are potentially much more visible when they hunt on such "open-faced" inflorescences as those of
ox-eye daisies (Morse 1999b) or black-eyed Susans, *Rudbeckia hirta*.
Unfortunately, we know much less about the spiders' relationships
with these flowers than we do for milkweed. Daisies are the preferred
hunting sites for penultimate females in our study areas, and adults also
use these flowers for hunting sites until the sites senesce. The later-
flowering black-eyed Susans are frequent hunting sites of adult females
where no milkweeds grow.

In some natural situations, adult female crab spiders significantly impact pollinators. Fautin (1946), Chew (1961), and Abraham (1983) all
reported high densities of adult *Misumenops* spp. on the spring flowers
of the herb and shrub strata of western North American deserts. Numbers were sometimes so high that the majority of the flowers of a com-
posite contained a spider. With densities of this magnitude, one would
expect flower visitors to avoid sites with predators, though these researchers apparently did not examine the possibility. Bristowe (1958)
reported that visitors avoided dandelion *Taraxacum officinale* flowers in
which he placed both yellow and contrasting objects as spider models.
He did not assess the effects on seed set, which would have been of
questionable significance, since this flower is largely apomictic. Subsequently, Suttle (2003) found that high densities of *Misumenops schlingeri*
caused pollinators to avoid ox-eye daisies *Chrysanthemum vulgare*
(=*leucanthemum*), upon which the spiders were highly conspicuous. He
experimentally detected a significant decrease in seed set associated
with the occupation of these flowers by the crab spiders.

Robertson and Maguire (2005) reported that peppergrass *Lepidium papilliferum* inflorescences with *Misumena* attracted significantly fewer visitors than those without the spiders, and that overall visitation rates increased when they removed the spiders. However, none of the results for individual species were significant, rendering the result equivocal. They did not examine seed set.

In a followup study to their 2003 paper, Dukas and Morse (2005) could detect no effect of bumble bees and honey bees on seed set in an experiment with common milkweed, using *Misumena* as the predator. They attributed this failure to the ambush hunting style of *Misumena* on milkweed, which made them relatively inconspicuous to the bees. In contrast, both Louda (1982) and Romero and Vasconcellos-Neto (2004) found that their spiders significantly enhanced seed set.

Heiling, Herberstein, and Chittka (2003) reported that honey bees were attracted to white Australian crab spiders *Thomisus spectabilis* on white daisies *Chrysanthemum frutescens*. They attributed this attraction to the bees responding to ultraviolet patterns of the spiders similar to those of their accustomed flowers. Neither the bees nor the flowers are native, however, which complicates the interpretation of this result. In a later experiment Heiling and Herberstein (2004) found that native bees avoided flowers occupied by *T. spectabilis*. This result suggests that response to the spider by prey species depends on a coevolutionary relationship between the players.

The early instars of *Misumena* often occupy flowers in far larger numbers than do adult females. Hence, it is of interest to see if they affect the success of their flowers, notably goldenrod. The major prey of these younger instars, particularly the second instars, consists of species, such as dance flies when available that probably play a relatively small role as pollinators of these flowers. However, early-instar spiders do capture modest numbers of somewhat larger hover flies, which are more likely pollinators of goldenrod. Nevertheless, large hairy visitors, primarily bumble bees (and honey bees in many areas), are probably far and away the most important pollinators of goldenrods (Heinrich 1979), and only adult female *Misumena* are large enough to capture them.

In other instances, however (see Chapter 3), *Misumena* probably play a passive role in the three-level relationship (flower, possible pollinator, predator). Until flower species colonize a site, *Misumena*'s pollinator prey will not visit that site, thereby providing no resources for *Misumena*, even if they could reach the area. The flowers on which they most often hunt differ widely in their dispersal capabilities. Since many prospective colonization sites are only transitory (patches in the forest, etc.), the relationship between the flower species, pollinators, and their environment may play a much greater role in plant fitness than the relationship between the spiders' presence and the success in seed production of these flowers.

As a slow-moving species often remaining for long periods at a site and possessing very few hairs that could trap pollen, *Misumena* are unlikely to perform any pollination services, in spite of the great amount of time spent on flowers. Since many of their principal hunting sites are self-incompatible flowers (milkweed, rose, goldenrod), the possibility of a pollinator role decreases yet further.

In sum, *Misumena* and other comparable small predators have at most a minor effect on both their prey populations and the flower species they occupy. Although some experiments have demonstrated a negative or even positive response of the flower visitors to the spiders, they involved modest changes obtained under high-density conditions. It is easier to envision them as a long-term directional force acting on plant visitors than as a strong ecological factor driving the current short-term characteristics of the community. Even their role as a selective force remains in question.

The Effect of Predators on Misumena

Misumena is subject to the attacks of several predators and parasitoids. The most predictable difference in its role would probably be a consequence of whether it had established itself in its hunting site or was searching for a hunting site, the time of likely greatest vulnerability.

As noted in Chapter 5, predators on adult *Misumena* are not common in the sites that we work and therefore seem unlikely to play an important role in these communities. However, one might expect them to play a larger role elsewhere, especially mud-dauber and spider wasps. The low density of *Misumena* observed in some more southerly areas where higher densities of these wasps are common (e.g., Rhode Island, Maryland) may result from the wasps' predation on the spiders.

Where known, these wasps prey heavily on late juvenile instars of *Misumenoides* and *Misumenops*, and one would expect a similar response to *Misumena*. Some mud-daubers fill their nests with relatively small spiders, and Muma and Jeffers (1945) and Dorris (1970) indicate that late juvenile crab spiders sometimes dominate the contents of these nests. One would therefore expect *Misumena*'s own putative effect on pollinators to diminish, with corresponding benefits to the flower populations (see Suttle 2003). This benefit should also enhance the success of the wasps, which frequently feed on the nectar of these flowers, the major energy source for their movements. Benefits as both a predator and nectar feeder should enhance the success of the wasps. However, it remains unclear whether any of these forces play a measurable role in the community.

Parasitoids

Parasitoids frequently attack *Misumena* clutches (Chapter 4) and at times are probably among the most important factors limiting their populations. In other situations, however, they exert a much more modest effect, and mortality in the early instars is then likely to be the major source of loss. Parasitoids, largely the ichneumonid egg-predator *Trychosis cyperia*, inflict a far greater impact on *Misumena* than do the predatory wasps in our study areas. With levels of parasitism of up to 75 percent, at least in field experiments, they have the ability to depress a target population severely. The frequency with which *Trychosis* actually can depress its prey species' populations is unclear. These ichneumonid spider-parasitoids tend to be specialists (Townes and Townes 1962; Shaw 1994). Our study areas provide few alternative species of crab spiders for them to use as alternative hosts, since they do not regularly parasitize the abundant *Xysticus* species, if at all. They may cycle with the numbers of *Misumena* available to them, although the results of our work are of far too short duration to make a strong case.

If real, a cyclical effect could have several ramifications at other trophic levels, if strong enough. It could result in enhanced numbers of unharassed pollinators two years following extremely heavy parasitism, with a correspondingly good seed set. However, it might put into movement a delayed cyclical fluctuation in numbers of pollinators, and then seed supplies. The effect on annual and perennial plant species should differ greatly.

A population with fluctuating numbers has a much smaller effective size than a steady-state population of the same size. Hence, the chances of an extinction should be considerably enhanced in a cycling population. This danger may be especially high for the parasitoid, which probably persists in low numbers to begin with, if totally or primarily dependent on *Misumena* for its livelihood. The local extinction of *Trychosis* would make the system revert to that of predator-pollinator-plant, which is likely a more stable system with more predictable patterns of abundance. Alternatively, the parasitoid might first exterminate the *Misumena* population, thereby enhancing the plant–pollinator relationships. Whether or not *Trychosis* eliminated *Misumena* would be greatly affected by whether or not *Trychosis* had an alternative host. If it did, it might survive at least until it had totally wiped out the *Misumena* population, as it could continue to reproduce even if the *Misumena* population could not sustain it. Following extinction of *Misumena*, the alternative host might not be able to sustain *Trychosis* on a permanent basis, in which instance it would likely go extinct after eliminating *Misumena*. Since many of these *Misumena* populations may simply be metapopulation-like groups, each extremely vulnerable to extinction, such events may account for why *Misumena* is absent from so many seemingly favorable sites.

We have not sampled *Trychosis* adequately to determine whether it, too, has a metapopulation-like distribution. We very seldom observe *Trychosis* in the field, even where it has a major impact on *Misumena* nests. Placing out sets of "bait" nests to sample its presence and potential impact is an extremely time- and resource-intensive activity. ("Bait" nests are nests made by gravid female *Misumena* placed in a cage in the field with vegetation suitable for making a nest. The nests are then exposed for parasitoid attack.) The literature is silent on the host relationships of *Trychosis cyperia*, consisting solely of its description (Townes and Townes 1962). Information on host relationships is not included in this literature, other than to note that it, in common with its closest relatives, probably parasitizes spiders. Henry Townes in fact confirmed to me that ours is the first host record for this species.

Trychosis is probably the only parasitoid in our study area that impacts *Misumena* numbers in a significant way. Our only other regularly encountered parasitoid, a scuttle fly *Megaselia* sp. (Phoridae), has never made up more than a small percentage ($\pm$ 10 percent) of the total parasitism on *Misumena*. In fact, we have yet to find it in our present primary study area, though we have sampled far less here than in our former primary site. Given its minute size, it is unlikely to be encountered during routine observations. As a group, scuttle flies, including members of the genus *Megaselia*, parasitize a prodigiously broad range of organisms (Disney 1994). Thus, *Megaselia* sp. probably is a generalist parasitoid, though we have no actual evidence for this supposition. If so, it is unlikely to be controlled by its interactions with *Trychosis*, which could be severe if it were so affected.

Thus, parasitoids of *Misumena* probably have a stronger community-wide effect, primarily via the action of the ichneumonid *Trychosis*, than do its predators, at least at the adult stage. The extent of these effects on other members of the community is not clear, however, and depends on such unknowns as the degree to which *Trychosis* relies on alternative hosts.

Interposition of Scavengers on Flower Communities

We have uncovered an interesting detrital pathway in which harvestmen *Phalangium opilio* feed upon the discarded carcasses of *Misumena*'s prey, largely bumble bees and moths (Morse 2001). No doubt the spiders leave measurable amounts of resources in the undamaged exoskeletons of their large prey (Chapter 5) that the harvestmen obtain in the process of breaking these largely hollow bodies apart. The harvestmen are unlikely to possess the symbionts that would permit them to break down the chitin in these structures.

Misumena drop their prey upon completing their processing of them. Often these prey litter the leaves below the spiders' hunting sites, especially on flower species with broad, horizontal leaves, such as common milkweed. These telltale signs often provide the best clues to the presence of an otherwise inconspicuous adult female *Misumena*. The carcasses often are blown off these leaves with high winds, but otherwise may accumulate for several days. However, we discovered that many of these carcasses disappeared from the leaves far more quickly than predicted by chance. If ants had recruited to the nectar supplies or aphid colonies on milkweed, they would sometimes remove these carcasses. However, the carcasses also disappeared in the absence of ant recruitment, often far more rapidly than accomplished by any of the ants that frequented these flowers.

Large numbers of harvestmen often occupy the milkweed plants at night, which would not be suspected by their general absence from the vegetation on clear, warm days, when they must seek cover to avoid excessive desiccation. The harvestmen feed on nectar at night (Morse 2001), but also consume any animal material they can process or capture, including the above-noted carcasses of large bees and moths discarded by the spiders. The harvestmen may even assume the role of facultative kleptoparasites, attempting to rob prey from the spiders while the spiders are still processing them, though we have not seen a successful "steal" to date (Morse 2001). We have no evidence that the spiders attack these harvestmen, which are likely to be extremely noxious (Cloudsley-Thompson 1968), though we have observed *X. emertoni* feeding upon them.

We obtained a measure of the harvestmen's ability to intercept this detrital pathway by lightly dusting insect carcasses (largely bumble bees) discarded by *Misumena* with powdered micronite dye. When we placed these dyed carcasses on the ground under the milkweeds, we found that they were heavily exploited on the first night, with some carried as high as 45 cm up into the plants and as far away as 95 cm from the cache site. Dyed insects were also removed from leaves of the plants, though not as rapidly as from the ground. By simultaneous pitfall trapping, we measured the frequency with which the mouthparts of the free-ranging harvestmen became colored with the dye. A sizable percentage of the harvestmen captured at this time had red-dyed mouthparts and surrounding areas. When we brought harvestmen into the laboratory and supplied them with dyed insects, they broke apart the carcasses, leaving remains identical to those found in the field.

Since the carcasses were frequently dismembered in the immediate vicinity of their posting, we often found these remains. Many of the sites frequented by hunting *Misumena* do not intercept the spent carcasses on leaves below the flowers as we have seen on the milkweed; instead they fall onto the ground layer, where the harvestmen also probably exploit them. In these places, however, other small detritivores or carnivores may more frequently feed on them before the harvestmen find them. Ants, some of which will defend these carcasses from the harvestmen, probably are the most frequent competitors on the ground. Omnivorous slugs *Arion subfuscus* also probably compete for the carcasses there.

Neither the harvestman nor the slug is a native member of the community. Although many of the flowers in these old-field communities are themselves not native, milkweed and most of the other major flowers of its community are natives, as is the majority of the insect visitors, with the exception of honey bees, which recruit to some of these sites. The effects of the harvestman and the slug on these communities are likely to be major, by virtue of their abundance. It is unlikely, however, that the spent-carcass detrital pathway described here has major effects on *Misumena* or the functioning of the community as a whole.

Ants, Nectar and Honeydew Feeding, and Spider Avoidance

Depending on the location of their nests, ants may recruit to flowers to gather nectar. There, they may have a strong effect on any *Misumena* hunting at these sites, for some (e.g., *Lasius neoniger*) will attack the spiders, first by biting their limbs. Although the spiders may attack a solitary ant when ants begin to recruit to a site, the spiders will typically abandon the sites as more ants arrive. Initially, they may avoid the ants by remaining suspended for considerable periods on lines below their previous foraging sites, where the ants cannot follow them. However, hanging from a line is hardly an effective foraging strategy, and if harassment by the ants continues, these spiders eventually leave. Because of the patchy distribution of the ant nests, only some patches of flowers, especially milkweeds, contain ants—and consequently few if any *Misumena*. As a result, ants may make sizable numbers of potential hunting areas off-limits for the spiders. Moreover, the ants will make the community far more patchy for the spiders, so that searching for satisfactory hunting sites becomes a yet more formidable task. The spiders may incur further costs from interacting with the ants, which may even entail some danger if the spiders are attacked en masse and unable to escape.

Other smaller, less aggressive ants *(Tapinoma sessile)* also recruit to flowers, especially milkweed, in large numbers. In the process, they remove considerable amounts of nectar and decrease the rate of insect visitation to the flowers and hunting opportunities for spiders occupying them. Since these spiders are generally food-limited, the resulting decrease in prey visitation may cause spiders to change sites, imposing a severe time expenditure. Maximizing the number of days hunting on the unusually rich milkweed source will largely determine the extent of the spiders' foraging success (Morse and Fritz 1982).

Ants also recruit to aphid (*Aphis asclepiadi* and *A. nerii*) colonies on the upper leaves of milkweed plants and tend the colony members located there. These sites frequently include leaves, often on nonflowering stems, that the spiders routinely appropriate as nest sites, and the spiders seldom use such sites if occupied by ants. Aphid colonies frequently form late in the summer after the spiders have already built their nests and laid their eggs. If the ants recruit to a colony near a spider nest, the spiders will frequently abandon their nests. Recruitment of spiderlings from such abandoned nests is extremely poor, because of subsequent ant predation on the eggs and spiderlings.

These interactions may play a major role in the spiders' appropriation of major resources. Given the central importance of milkweed to the spiders where they recruit to it, the consequence of being deprived of a premium site could decrease the size of their single egg mass. Furthermore, interference at the nest-guarding stage will almost certainly take place so late in the season that the displaced females will not have time to produce a replacement brood. Thus, usurpation of their sites by the ants will often condemn them to zero fitness.

The Effect of Spiderlings on Their Community

In common with most studies, we have thus far devoted inadequate attention to the early instars. Although adults may produce the greatest quantitative impacts on the community per individual, early instars are much more abundant and act at an earlier stage than the adults. Small impacts early in a sequence may skew results in a way that later takes on major significance. Here I comment on interactions likely to be of the greatest importance to their livelihood.

Spiderling *Misumena* are both numerous predators upon the smallest of herbivorous species and substantial resources for other, usually somewhat larger, predators. We have questioned whether they capture enough dance flies to impact those populations; the same may hold for the slightly older instars that feed heavily on small hover flies. As tiny, scantily haired insects, dance flies probably have minimal pollinating capabilities, in spite of their occasional great numbers. Because they spend long periods on single goldenrod clones, most of their putative pollinating doubtless consists of selfing. Since goldenrods are obligate outcrossers, heavy predation of the dance flies might enhance seed production of the goldenrods, if other pollinators are present. Given the greater movement rates of the small hover flies, especially between clones, any predation by the spiderlings on these flies might have the contrary effect of lowering outcrossing success. The actual levels of pollination services provided by these small flies are not known. However, if bumble bees or honey bees visit these flowers, they probably provide the bulk of pollination services.

The greatest community-level impact of these spiderlings as predators could be on the numbers of thrips (Thysanoptera) they capture. Thrips are small, often abundant in the flowers, and probably easy for the spiderlings to capture, making them important initial food items for the newly emerged tiny spiderlings, especially in the absence of dance flies. Thrips habitually inhabit flowers and feed heavily on them, largely by lacerating them with their mouthparts and sucking from the thus-damaged flowers. In the process they may severely impact flowering and fruiting success when present in high numbers (Dixon 1985). These interactions deserve more attention than they have received to date.

The spiderlings, in turn, provide a resource for other predators, which likely play an important role in diminishing the spiderling numbers. Jumping spiders are common small predators in the community that regularly capture *Misumena* spiderlings (Morse 2006), which sometimes may assume important proportions of their diet. Their prey otherwise consists heavily of small herbivores, especially those in their early instars. These herbivores may strongly impact plant species that are important flower sources for *Misumena* and other small predators. Thus, the jumping spiders may facilitate goldenrod productivity in this way. Aphids are likely to be among their more important prey. Accordingly, the role of jumping spiders relative to *Misumena* is an interesting one, which may depress spiderling numbers more than any other source. Countering that effect, jumping spiders may enhance the future success of surviving *Misumena* because of the jumping spiders' impact on herbivores of flowering species that provide hunting sites for the later *Misumena* instars.

The role that *Misumena* spiderlings play as cannibals should decrease the impact of any of the relationships discussed above. At the same time, one should recognize the possibility that this behavior, especially if carried on between members of the same brood, might act (however inefficiently) in a way analogous to trophic eggs. Nevertheless, since they often do not cannibalize each other when kept under crowded conditions, even if unrelated individuals are kept together, their relationships in the field warrant closer attention. If they exhibit any reluctance to prey upon each other, it would enhance their community-level impact. Barth (2002) noted that *Cuprennius salei* spiderlings did not become cannibalistic until they had exhausted their yolk sac at the end of nine days. However, *Misumena* spiderlings will capture *Drosophila* prey voraciously as soon as they have left their nests (Morse 2000b) while eschewing sibs and other conspecific spiderlings for three weeks or more when in close contact with them, including situations in which they will routinely capture *Drosophila*.

If early *Misumena* instars play an important role at the community level, it is of interest that these interactions take place only during a small part of the year, late summer/autumn. Therefore, they are seen on a relatively small proportion of the flowering plants likely to play an important role in establishing their lifetime fitness.

We thus know very little about the impacts of early *Misumena* instars on their community or of members of the community on them, a deficiency that holds for nearly all other species. Because *Misumena* experience strong phenological constraints, any significant impact of the young in community interactions will vary over time. Probably factors involving the young assume great importance for the maintenance of their populations, a time at which mortality appears to be greatest. It is unclear whether most losses during this period are nonrandom, such that they would differentially select for traits inherited from their parents (thereby maximizing their parents' fitnesses). It does appear unlikely that the spiderlings will measurably affect the dynamics of their major prey, which are likely to be present in large numbers at this time.

Consequences of Prey Foraging Strategies

Although I have concentrated thus far on the effects of *Misumena* on their prey (Chapter 5), the complete picture of the relationships between predator and prey is considerably more complex than the one given here. The actions of the prey probably influence this relationship as well, both by their responses to *Misumena* (Chapter 3) and their response to their own resources, which nevertheless may greatly affect *Misumena*'s success.

The bumble bees' delayed response to the bonanza of nectar produced by milkweed markedly decreases their potential intake of nectar (Morse and Fritz 1983), a resource that they subsequently exploit avidly. We have estimated that the bees may fail to recover as much as half of the total nectar production of milkweed clones we have studied (Morse 1982b). The bees' slow recruitment to milkweed almost certainly results in a significant loss of potential biomass to adult female *Misumena*. Not only do they lack convenient access to the bumble bees at this time, but they cannot begin to recruit to these prime hunting sites until the bees have already done so (Morse and Fritz 1982). Recruitment itself will delay their access to a concentrated population of prey by another two to three days. (Although the bees forage in the general vicinity and thus are potentially available to the spiders at that time, they are much more widely dispersed at flower sources visited much less frequently than milkweed.) Since the majority of spiders recruiting onto milkweed fail to reach their maximum possible mass in spite of unexcelled opportunities for capturing large prey there, this delay almost certainly results in smaller egg masses than if the bumble bees recruited to milkweed more rapidly (Morse and Fritz 1982). At the same time, the bees' slow recruitment should also affect the pollination success of the milkweeds themselves: maximizing the length of the pollination period maximizes the number of clones with which the obligately outcrossing milkweed can exchange pollinia, via bumble bees, by far the most important pollinators of milkweed (Morse and Fritz 1983). Seemingly, all participants would thus gain from earlier recruitment by the bees to milkweed. Pollination success should be enhanced, the bees should gain more resources than they lose from predation by the spiders, and the spiders will gain more resources as well.

The spiders' presence does not significantly affect the seed set of the flowers (Dukas and Morse 2005). This failure to recruit sooner, which has broad effects on the rest of the community, is thus somewhat of a paradox, though at this time pollen supply potentially limits colony growth of the bumble bees. Since all of its pollen is wrapped in pollinia, to which the pollinators do not have access, milkweed is not a one-stop shopping site (nectar and pollen), in contrast to flowers that make both resources available. This factor could account for why the bees continue to frequent yellow rattle *Rhinanthus crista-galli*, red clover *Trifolium pratense*, and cow vetch *Vicia cracca* after milkweed begins to bloom. Recruitment to milkweed is rapid once it begins, suggesting that the bees can shift resources as soon as it becomes economical to do so.

Impacts of Some Herbivores and Their Effects on Misumena

With the exception of my comments on bumble bee recruitment to nectar supplies (see preceding section), I have not focused on community-level effects that impact on *Misumena* populations, but in which it is not a major player. However, seemingly unrelated major changes in a community may affect most or all of its species pool indirectly, including *Misumena*. I will discuss three such events that have occurred recently in our primary study area and that indirectly but strongly affect the spiders. For the most part this work is qualitative, for it did not make up part of our ongoing research. However, the events were striking and bound to impinge on the well-being of *Misumena*. In each instance they greatly decreased the availability of satisfactory foraging sites for the spiders' prey, and consequently the ability of the spiders to access them. Over a slightly longer term, these changes seem likely to affect the spiders' prey populations, the flower-visiting insects, leading to a change in their own population size (and viability where populations are small).

Weevils and Milkweeds. In the late spring of 2002, we first noticed the wholesale death of young milkweed stems in our major study clone, which decreased the number of flowering stems from several hundred in 2001 to 15 that year, 3 in 2003, and only scattered vegetative growth in 2004 and 2005. The only remaining flowering stems in the immediate vicinity were isolated outliers, likely members of the same clone, but well separated from its center. Bumble bees never recruited in significant numbers to these scattered remaining stems, with a maximum of three individuals counted over the area at any time, as opposed to dozens previously. No *Misumena* recruited to the few flowering stems in 2002 or 2003.

Close inspection revealed that the stems of these young plants had been drilled with a number of holes, each marked with white spots denoting the exudation of latex from the attack sites, action of the weevil

Rhyssomatus lineaticollis (Curculionidae) (Fordyce and Malcolm 2000). These drill-sites sever the latex ducts of the leaves above them, causing them to wilt and eventually die if not eaten first by slugs *Arion subfuscus*, which also attacked young milkweed undamaged by the weevils. The weevils lay their eggs in these holes. Although the weevil is a native species, we have not previously noted this highly conspicuous damage in 30 years of intensive field work. Much of this work has focused on milkweed, either in smaller clones widely separated from this clone in the 3.5 ha study field or elsewhere. Initially, these attacks were confined to the largest milkweed stand at one end of the study area, but in subsequent years these effects have spread to smaller clones in other parts of the study area.

This behavior bears many similarities to behavior exhibited by several milkweed specialists that cut the veins of milkweed leaves and then feed on the parts distal to the cut. Monarch butterfly larvae *Danaus plexippus* (Dussourd and Eisner 1987), milkweed bugs *Lygaeus kalmii* and *Oncopeltus fasciatus*, and milkweed beetles *Tetraopes tetraophthalmus* all cut veins in this way. They seldom kill a plant in the process, however, in contrast to the weevils, which kill most young shoots that they attack. As a result, milkweed clones are quickly decimated, leading to the disappearance of their regular associates, including bumble bees and *Misumena*.

Goldenrod Beetles and Goldenrod. Goldenrod beetles *Trirhabda* spp. are common herbivores on goldenrods (e.g., Hufbauer and Root 2002), where they sometimes seriously inhibit the growth of these plants. In the study area, their numbers appear to fluctuate widely. In 2002 their larvae produced highly noticeable herbivory, such that the leaves of many *Solidago canadensis* clones took on a noticeably "shredded" appearance by the time that the beetle larvae metamorphosed. Flowering was suppressed in the clones most heavily grazed. In 2003 the impact of the beetles became so severe that several clones were totally denuded of their leaves. Five percent of the clones did not survive the following winter, and even more produced only a few small, nonflowering stems in 2004. Noticeable herbivory took place in 2004, though no clones lost all of their foliage. Negligible herbivory occurred in 2005, but most surviving clones still bore few if any flowers, and other clones had died. Although it is too early to determine the long-term effect of this massive attack, goldenrod defoliators may in the long term sup-

press these plants to generate striking differences in old-field communities (Carson and Root 2000).

The loss of flowering goldenrods cut back considerably the resources available to bumble bees and in turn the hunting sites and prey
for adult spiders that had not yet laid a clutch. Perhaps most importantly, it severely curtailed the number of recruitment and feeding sites
available to the spiderlings. The loss should thus have strong effects on
the quality of this area as a site that can sustain a spider population.
With a much more limited number of local hunting sites, spiderlings of
the earlier-laying spiders would presumably balloon from the area with
increased frequency, decreasing the next local generation. The later
adults would experience an increasing probability of not capturing
enough prey to lay a clutch of eggs, or if they did succeed, the clutch
would be considerably smaller, and probably later, than otherwise expected.

On a longer-term basis, the shortage of feeding sites for bumble bees
contributed to decreases in their populations, already at risk because of
the demise of the milkweeds. By 2004, one of the most common
species of bumble bee, *Bombus ternarius*, may have gone locally extinct
in the study area, as we observed only a single worker, which could
have been a vagrant, during a summer of intensive work. This species,
which depends heavily on goldenrod (Morse 1977a; Heinrich 1979),
was the most common bumble bee in 2002 and earlier years and an important prey item of adult female *Misumena*. Similarly, numbers of *B.
terricola*, a common species in earlier years (Morse 1986a), declined so
severely that only occasional workers were observed. It, too, may not
have nested in the study area.

Combined with these attacks to the foliage and developing inflorescences of *S. canadensis*, we noted an unprecedented high frequency of
spindle galls on the stems of these plants, which far exceeded any numbers we have previously seen, here or elsewhere. In some of the totally
devastated clones, as many as 60 percent of the large stems had been
galled (mean = 40.0% $\pm$ 4.74 SE, N = 10 clones). The effects of these
galls, caused by the gelichiid moth *Gnorimoschema gallaoesolidaginis*, in
themselves may be strong and may largely or completely curtail flowering of the stem parasitized (Abrahamson and Weis 1997). Whether
the moths had prospered on these goldenrods as a consequence of their
altered state is unknown, but they further depressed the resources for

Misumena's important bumble bee prey, as well as the dipteran prey of the spiderlings.

Harris' Checkerspot Butterflies and Asters. Flat-topped white asters *Aster umbellatus* provide alternative hunting sites for recently emerged spiderlings. This fact takes on particular importance for individuals whose parents laid in areas with many asters and few goldenrods, although they are suboptimal hunting sites (Morse 2005). These plants grow in relatively damp meadows and forest edges and thus exhibit a rather patchy distribution. The Harris' checkerspot butterfly *Chlosyne harrisii*, a species of special concern, is an obligate herbivore on this aster (Robakiewicz and Robbins 2001). Notwithstanding its vulnerable status, this species may locally become an important defoliator of the aster, in some instances totally consuming the leaves and buds of these plants. Although some of these stems recover to produce adventitious buds and leaves, they do not produce flowers that year. As a consequence during the summers of 2000 and 2001, almost no asters flowered in the study area, making one-fifth of the entire study area an unsatisfactory site for the second-instar spiderlings. Infestation declined in 2002 and was even lighter in 2003. Numbers of flying checkerspots in 2003 still stood at less than one-tenth that of the 2001 density, consistent with long-term observations of this species in the study area and in other comparable nearby areas. Although these secondary sites of *Misumena* probably did not support enough spiders to affect population size seriously in the study, if the checkerspot outbreak had coincided with that of the milkweed and goldenrod attacks, the effect on *Misumena* might have been seriously exacerbated.

Synopsis—Interactions within and among Trophic Levels

The perceived importance of other community members to a focal individual or population has shifted markedly over recent decades. Indirect effects are likely to play a critically important role in determining success, and no single factor seems likely to act in the absence of others. As suggested, the above-noted, plant-herbivore interactions may have a major effect on *Misumena*, even though *Misumena* does not appear to play a leading or direct role in any of these interactions.

The strength of between-trophic level interactions thus considerably exceeds that of within-trophic level interactions in *Misumena*, at least during the adult stage, and probably the late instars as well. We have found little evidence of interspecific competition between *Misumena* and other species, in keeping with many other spider species, including closely related species traditionally treated as competitors (Wise 1993). Although one might argue that the scarcity of *Misumenoides formosipes* in our northern study areas results from past interspecific competition, the difference in their geographical ranges suggests that sensitivity to climatic factors plays a much more important role in determining their present ranges than does competition (see Schmalhofer 1999). The small predatory species frequenting the same sites as *Misumena* are scarce relative to *Misumena* under most circumstances. Even if present in far greater numbers, all species combined would occupy only a small minority of the satisfactory hunting sites. Only very seldom do adult female *Misumena* come in contact on a random basis (Morse and Fritz 1982); modifying that calculation to include other sit-and-wait predators on the flowers will not change this frequency of flower occupation significantly.

There is strong evidence, however, for food limitation in spider populations, including *Misumena*. This condition likely represents the rule rather than the exception (Wise 1993). In *Misumena* the paradox of abundant resources and food limitation results from the difficulties of finding superior hunting sites, which may not be rare but often are difficult to find. This factor may control the numbers of individuals, as well as their size and subsequent fecundity. Although cannibalism, which might be considered an aspect of interference competition, is at times intense among early instars of wolf spiders (Wagner and Wise 1996; Morse 1997), we have no convincing information to suggest that it plays an important role in *Misumena*'s population size. Even though early-instar *Misumena* sometimes initially reach high densities immediately after leaving their natal nests, they quickly space themselves more widely. This action in itself minimizes the possibility of cannibalism.

In contrast to the apparent minor significance of putative competitors, some of *Misumena*'s interactions with members of other trophic levels play an important role in its lifestyle. Bumble bees are the most important herbivore/nectivores directly impacting *Misumena*'s livelihood in our study areas, though honey bees may assume this position where present. Adult female *Misumena* depend on captures of extremely large prey to produce a clutch of eggs, and large bees, perhaps supplemented by noctuid moths, provide the only regularly available resources that allow that opportunity.

The ichneumonid parasitoid *Trychosis cyperia* appears to be the most important predator/parasitoid impacting *Misumena*, through its attacks on *Misumena*'s egg masses. As an apparent specialist on *Misumena* in this community, *Trychosis*'s actions on *Misumena* should be simple and direct and should not draw *Misumena*'s numbers down as low as if it had alternate hosts. This apparent one-on-one relationship may account for the possible signs of cycling in their numbers. Nevertheless, given the modest likely sizes of both the spider and the wasp populations, they should be vulnerable to local extinctions.

Any decrease of *Misumena* numbers might enhance the insects' impact as pollinators, although the absence of predators could have the opposite effect if pollinators then remain longer on flowers or clones, increasing inbreeding. However, a test on milkweed did not lead to enhanced seed production (Dukas and Morse 2005). Overall, however, any such effects of *Misumena* on its prey appear to be minor and of questionable impact. The spiders may also change the behavior of their prey species, a pattern sometimes exhibited by bumble bees in this system (Morse 1982a; Dukas and Morse 2003).

Mud-dauber and spider wasps are potentially dominating factors acting on *Misumena* populations. We attribute the near absence of crab-spider-hunting members of these groups in our study areas to other, independent factors (likely climate). The absence of mud-dauber and spider wasps may allow *Misumena* to hunt at sites that would otherwise expose it to the wasps, thereby enhancing its hunting ability in our study areas.

Thus far we have noted little evidence of mutualistic relationships involving *Misumena*, including species that depressed the success of other species that competed with, preyed on, or parasitized *Misumena*. Whether such indirect effects exert a measurable impact on the community, as a whole, remains open to question. However, since competition does not appear to play an important role in its life, we might not expect major indirect effects of this sort to develop about *Misumena*. Indirect effects involving predators and parasitoids might appear to be the most promising ground, but since predators on the adult spiders appear scarce in our study area, impacts involving parasitoids appear the most likely candidates. Here we draw a near blank because we know so little about *Misumena*'s principal parasitoid, *Trychosis*. Parasitoids are likely to be attacked by hyperparasitoids, most likely chalcidoid wasps. Although *Trychosis* may be too uncommon to support specialist hyperparasitoids, many hyperparasitoids are generalists (Godfray 1994; Sullivan and Völkl 1998). Thus, *Trychosis* should be vulnerable, even if uncommon. To date, however, we have yet to find a hyperparasitoid that attacks *Trychosis*.

The standard predators of *Misumena*-like spiders do, however, appear to be wasps—spider wasps and, especially mud-daubers. They seem unlikely to affect our populations of *Misumena* significantly, and, by extension, they do not present a significant resource for other organisms, especially parasitoids that might secondarily enhance *Misumena*'s success by suppressing the wasps. One might inquire whether our spider populations are restrained by such parasitoids at this moment, but we have no evidence for such an effect. At the same time, the parasitoids could have an important positive indirect effect on *Misumena*'s populations where the impact of mud-daubers is high, but we have not studied such systems. In sum, we have little basis for attributing a significant role to mutualism in *Misumena*'s lifestyle.

Like the adults, early instars also probably have little impact at a community level. We have no evidence to suggest that they affect their small dipteran prey to the degree that it modifies their probably modest impact as pollinators or nectar thieves. However, young *Misumena* may provide resources for larger predators, such as older jumping spiders. We thus ask whether species such as *Misumena* regularly make a "difference" in community functioning, or can a certain number insinuate themselves into a community without severely altering its balance? And if so, under what conditions, and how many can accomplish this feat?

On *Misumena*'s Place in the Community

The picture at present thus suggests that *Misumena* seldom if ever exert a strong effect on other members of their community, though potentially interacting with a host of other factors, which involve organisms both at higher and lower trophic levels. Not infrequently, a single species (keystone species) may have a dominating effect on other members of a community that imposes a structure on that community, as in the starfish *Pisaster ochraceus* in some intertidal communities of western North America (Paine 1966) and the sea otter *Enhydra lutris* in near-shore marine communities of western North America (Estes and Palmisano 1974). *Misumena* does not appear to possess such attributes, probably in common with most of the members of the community in which it lives. Nevertheless, detailed knowledge of such species is critical, if often ignored, providing needed insight into how communities are assembled, as well as what traits generate keystone effects.

𝒳 11

Xysticus emertoni,
a Cohabiting Crab Spider

SELDOM IS *MISUMENA* THE SOLE crab spider inhabiting the areas it frequents. In addition to the occasional *Misumenoides* and *Misumenops* encountered at our study sites, members of the genus *Xysticus* are ubiquitous in the old fields and roadsides frequented by *Misumena*. Of these *Xysticus* species, only *X. emertoni* (Figure 11.1) regularly occupies flowers frequented by *Misumena*. I will thus concentrate my comments on *X. emertoni*, since only it has the opportunity to interact regularly with *Misumena*, even though some of the other *Xysticus* species may considerably exceed *X. emertoni* in abundance. However, they are virtually exclusive inhabitants of the litter and seldom visit flowers. In the largest of the *Xysticus* species in our study area, *X. emertoni*, females roughly match *Misumena* in size, while most of the other *Xysticus* species more closely match penultimate female *Misumena* in size. One species, *X. triguttatus*, markedly smaller than the rest, even more closely resembles an antepenultimate female *Misumena* in size than a penultimate one. We do not understand why so many species of *Xysticus* coexist in this community. The mean size of both adult males and females differs significantly between some species, but since they almost completely overlap each other if the earlier instars are considered, the problem is a complex one. If competition is generally absent among these species, as Wise (1993) has argued for other spider assemblages of this sort, the various species may not exclude each other. The limiting factor remains unclear, however.

Forces acting on the early instars could prove to be an important key. Early instars of *X. emertoni* that place their nests in the herbaceous canopy recruit downward on lines into the litter upon leaving their

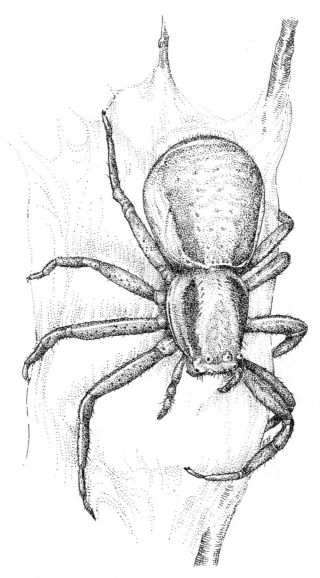

Figure 11.1. Adult female *Xysticus emertoni.* Illustration by Elizabeth Farnsworth.

natal nests, in contrast to the upward-moving behavior of *Misumena* spiderlings (Morse 1992b). Living primarily in the litter layer exposes *X. emertoni* to a totally different suite of coexisting species than those associated with *Misumena* in its abode well up in the vegetation.

Hunting Behavior

Given their similarities in size, one might anticipate extensive interaction between *Misumena* and *X. emertoni*. Yet, we have seen but few interactions, probably as a result of the relatively low densities of both species occupying flowers in the vegetation in relation to the number of hunting sites available on the flowers. In addition, adult female *X. emertoni*, in contrast to *Misumena*, spend but a small part of their time hunting on flowers, which further diminishes the likelihood of interactions between them. Not only do they occupy flowers less frequently than *Misumena*, but *X. emertoni* occupy them for shorter periods of time—less than two days per flower, with more than half remaining less than one day (Morse 1983). Furthermore, we almost never reencounter any *X. emertoni* that we mark after observing them on flowers (Morse 1983). Either these individuals are part of a very large population whose members only infrequently occupy flowers, or they are extremely vagile and do not remain in the vicinity. Catches from pitfall traps (D. H. Morse, unpublished data) suggest that populations of *X. emertoni* are far larger than anticipated simply from their appearance on flowers.

Xysticus emertoni usually confines its activity to the underside of the umbels on milkweed, the flower upon which we have most frequently seen it foraging. This pattern might account for the difference in its prey capture rates from those of *Misumena*, which spends much of its time on or in the top of the umbels. However, *X. emertoni* attacks many of the same prey as *Misumena*; thus, location is unlikely to be the primary basis for their difference in success. Although the similarity of the two species might suggest that they compete, the two species occupy only a small percentage of the high-quality hunting sites in a clone and capture only a small percentage of the constantly renewing insect supply at these sites (estimated at under 1 percent: Morse 1986a).

Xysticus emertoni may only occupy flowers infrequently because they do not capture prey as frequently as *Misumena* do, especially the largest prey available (bumble bees) that form the important staple in *Misumena*'s diet. Moths and intermittently available honey bees make up the most important prey of *X. emertoni* when on these flowers. *Misumena* that do not ingest biomass faster than *X. emertoni* usually leave their hunting sites shortly (Morse 1983); thus, these *X. emertoni* most likely leave because the sites are unprofitable for them. In that *X. emertoni*'s raptorial limbs match those of *Misumena* for size, their failure to capture bumble bees with any regularity is somewhat surprising. If they even captured bumble bees at a fraction of *Misumena*'s rate, these sites should be profitable ones for *X. emertoni*. In contrast, *X. emertoni* take a variety of prey that we have never seen *Misumena* feed upon, including monarch butterfly caterpillars *Danaus plexippus* and ladybird beetles (Coccinellidae). These species are potentially distasteful or toxic to them (Figure 11.2: Morse 1983).

Only on one occasion have we seen *X. emertoni* frequently occupying flowers. During a single spring they hunted on common buttercups *Ranunculus acris* visited in unprecedented numbers by a medium-sized, brown muscoid fly. We often found 12 or more adult female *X. emer-*

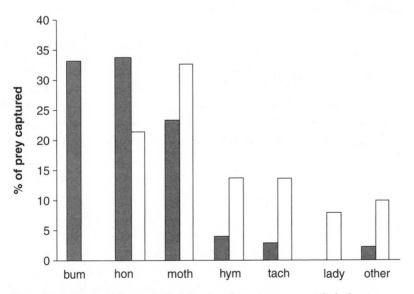

Figure 11.2. Prey of *Misumena* (dark bars) and *Xysticus emertoni* (light bars). Bum = bumble bee, hon = honey bee, moth = noctuid and geometrid moths, hym = other wasps or bees, tach = tachinid fly, lady = ladybird beetle, other = other. Data from Morse (1983).

toni within about a 50 m² patch of buttercups, which, however, had several thousand flowers in bloom at the time. The spiders enjoyed good success hunting these flies, with most individuals capturing one or more of them over the period of a day. In no previous or subsequent year have we observed more than an occasional *X. emertoni* using these sites, though we have never observed an abundance of these flies approaching the magnitude seen during that year. We did not observe *Misumena* on these buttercups during this time. Given the number of flower sites attracting flies, however, it is questionable whether the two species would have regularly engaged each other at even this site.

Most likely *X. emertoni*, and probably the other *Xysticus* species as well, capture most of their food in the litter layer. We have no information on this part of their life cycle, nor do we know that they adopt a sit-and-wait hunting strategy there as well. Although in common with other crab spiders their two pairs of forelimbs are markedly larger than their hind limbs, *X. emertoni*'s hind legs are considerably longer and more robust than those of *Misumena*. Furthermore, their abdomen does not enlarge to the extent seen in some gravid *Misumena*, which correlates with a lower reproductive effort. Both of these factors suggest that its cursorial capabilities exceed those of adult female *Misumena*.

Other Co-occurring Species

Living primarily in the litter layer, *X. emertoni* is exposed to a group of species rather dissimilar to those encountered by *Misumena* in the herbaceous canopy. As a result, they also probably capture a rather different group of prey species from those obtained when hunting in the canopy.

Other Crab Spiders

The most similar species encountered by *X. emertoni* will be the other species of *Xysticus*, although none of these congeners yet found in the study areas approaches the size of adult female *X. emertoni*. We have already found eight species in our old-field study area, mostly by pitfall trapping. A speciose genus in general (Dondale and Redner 1978), many *Xysticus* coexist in other areas as well (Nyffeler and Benz 1979),

but to the best of our knowledge their relationships have not been rig-
orously analyzed. We work at a heterogeneous site, with wet and dry
areas, as well as patches of various plant species, but do not yet know
whether any of the *Xysticus* species are confined to small parts of the
area. However, we sometimes capture males of up to four of these
species at one time in a single pitfall trap baited with *d*-2-octenal,
which apparently effectively mimics pheromones emitted by female
Xysticus (see Aldrich and Barros 1995). The other *Xysticus* species generally resemble *X. emertoni* and range in size from *X. triguttatus*, whose
gravid females may not much exceed 20 mg, to *X. punctatus*, whose
adult mass generally falls between one-half and two-thirds that of *X.
emertoni*.

At present we know little about the relationship between *X. emertoni*
and its congeners, or the relationships among the other congeners.
The gradations in size between certain subsets of these *Xysticus* species
somewhat resemble those of some congeneric birds (Diamond 1978)
and mammals (McNab 1971). However, we have no evidence for character displacement. Differences in adult size of these species are
breached by juveniles (which often make up a high proportion of a
population). The result is a complete gradation in size that varies both
between and within species over much of the season. Not only that, but
some species even closely resemble each other in adult size.

The strong association of our *Xysticus* species with the litter layer has
been striking. As opposed to the females, which we periodically or
rarely find in the higher vegetation, males virtually never occupy
higher vegetation. Thus, we never encounter them during casual observation. However, we can readily capture them in pitfall traps, especially those baited with *d*-2-octenal. Reports elsewhere suggest that
(other) *Xysticus* species frequently occupy the herbaceous canopy, including flowers, as well as the litter, and that they capture a wide variety of prey species in all of these areas (Palmgren 1972; Nyffeler and
Benz 1979; Nyffeler and Breene 1990). However, these papers do not
contain detailed time budget information, making it difficult to compare them with our results from *X. emertoni* and, by inference, the
other *Xysticus* species.

The relationships of all of these species during their earliest instars
bear attention, since they reach densities that greatly exceed those of
the older instars. Yet, their total biomass probably never exceeds that of

the smaller number of large conspecifics surviving until later in ontogeny. Thus, their demands may not exceed those of the later stages,
although they would depend on different kinds of prey, which might
engender conditions totally unlike those experienced by adults of their
species. They would also likely encounter each other more frequently,
solely as a consequence of their own numbers, which might form a pre-
text to cannibalism or predation. Density-dependent predation (cannibalism) is sometimes high in other juvenile spiders (Wagner and Wise
1996; Morse 1997).

Other Spiders

In addition to the closely related crab spiders, *X. emertoni* shares this
habitat with a large number of wandering spiders. The most common
of these wandering spiders are wolf spiders, especially the small, long-
legged wolf spiders of the genus *Pardosa*. *Pardosa* are abundant, also
consisting of several species at a site. They provide another potentially
interesting question of coexistence and of a size completely overlap-
ping some *Xysticus* species, including the earlier instars of *X. emertoni*.
Pardosa are highly vagile and thus probably capable of capturing many
prey that would be difficult for similar-sized *Xysticus* to capture. However, several workers (Ford 1978; Nakamura 1982) have characterized
wolf spiders as sit-and-wait predators rather than cursorial predators as
traditionally envisioned (Comstock 1940; Bristowe 1958), in which
case their foraging habits and prey capture may closely resemble those
of the co-occurring *Xysticus*. Nevertheless, one species of *Pardosa* that I
have studied intensively in a different habitat, *P. lapidicina*, varied its
hunting style with the contexts experienced. Styles ranged all the way
from a sit-and-wait predator to a far more cursorial style (Morse 1997).
It is currently unclear how the *Pardosa* species coexisting with the *Xysticus* assemblage hunt, but the pattern used could drive the amount of
interaction between them.

Probably the most important interaction between the two groups
will be to act as predators or prey of the other, largely a matter of the
smaller or younger being exploited by larger members. Overall, the direction of these relationships could be a consequence of the hunting
style adopted by wolf spiders. Nyffeler and Benz (1979) indicated that
Xysticus species in field ecosystems prey heavily on *Pardosa*, providing

about 10 percent of their resources when the mown vegetation forces *Xysticus* to the ground level where they are in regular contact with these wolf spiders.

Other larger wolf spiders (*Hogna, Schizocosa* spp., etc.) routinely inhabit the litter community as well. As adults, they likely interact with late instars of the *Xysticus* group. These large wolf spiders also prey heavily on the smaller *Pardosa* species (Persons et al. 2001) and presumably prey on earlier *Xysticus* instars as well. A wide range of other spiders occupies the litter layer, with sac spiders (Clubionidae) probably the most frequent of species likely to overlap with *Xysticus* species.

Harvestmen (Opiliones) form a group that is concentrated in the litter layer but that ventures into the canopy as well. Although primarily scavengers (Bishop 1949), they will voraciously attack any potential prey they can handle, albeit they have but modest attack capabilities, consisting of a pair of chelicerae with no poison and no silk to wrap prey. Nevertheless, they probably prey heavily on early *Xysticus* instars (as well as *Misumena* spiderlings in the canopy, which they frequent at night and at other times when the humidity is high). As roaming predators they cover considerable ground, and they often attain high densities (Morse 2001). The overwhelmingly numerically dominant harvestman in our study areas is *Phalangium opilio*, an introduced palaearctic species.

Beyond these spiders and harvestmen, a wide range of small invertebrates live in the litter layer, most of which do not venture into the herbaceous canopy. Small centipedes, often common in these habitats, are important components of this litter-level community. However, I do not know enough of their foraging activities to comment upon their interactions with *X. emertoni*. However, as mobile predators (Cloudsley-Thompson 1968), they would probably prey on small *Xysticus*, as would larger *Xysticus* upon them.

Of the other species occupying these areas, Collembola may assume the greatest significance, at least to the earlier instars of *X. emertoni*, which probably depend heavily on them as basic food items. Probably they interact little with millipedes because of the extremely toxic defensive compounds of the millipedes (although *X. emertoni* does take a number of seemingly toxic species when it hunts in the canopy (Morse 1983). We know little about other inhabitants of these sites, as they relate to *X. emertoni*.

Xysticus emertoni and its congeners probably are more vulnerable to both meadow voles *Microtus pennsylvanicus* and garter snakes *Thamnophis sirtalis* than are *Misumena*. Although we hypothesized that during an outbreak of meadow voles the voles had taken several gravid female *Misumena* as their lines had sagged to the understory of the plants they were traversing (Morse 1985b), exposure of the *Xysticus* species would probably greatly exceed that of *Misumena*. Leopard frogs *Rana pipiens*, gray treefrogs *Hyla versicolor*, and spring peepers *H. crucifer* are other likely occasional predators on these spiders in the litter layer.

In short, most *X. emertoni* live in an environment markedly different from *Misumena*, and its co-occupants of the litter layer overlap but little with those of *Misumena*. In addition to the different kinds of animals encountered, it differs in terms of the substantial number of closely related species with which it shares its habitat. This is in marked contrast to *Misumena*, which has its lifestyle largely to itself in our study areas, sharing it only intermittently with *X. emertoni* and a few true bugs.

Predation

We have observed only one direct case of predation of *Misumena* and *X. emertoni* upon each other. This act involved an adult female *Misumena* guarding its egg sac when an adult female *X. emertoni* happened upon it while moving through the crown of a milkweed plant. The highly emaciated *Misumena* held its ground, but the roaming *X. emertoni* quite systematically approached, attacked, and killed the guarding female and fed upon it at the spot where it killed it. This attack was probably a highly unusual event. The *Misumena* was in a postpartum, starved condition and unlikely to be a formidable assailant or to leave its egg sac; under those conditions, the *X. emertoni* could easily overpower the *Misumena*. Furthermore, this *Misumena* had placed its nest in the midst of milkweed flowers, which may have attracted the *X. emertoni* to the site in the first place. *Misumena* seldom place their nests in flowering milkweeds, and when they do, these flowers have usually senesced. We have not conducted encounter experiments between the two species, and thus we do not know whether *X. emertoni* normally dominate *Misumena* in interactions between evenly matched individuals (same size and energetic condition). *Xysticus emertoni*'s robust carapace and legs suggest that it would prevail over *Misumena*. However, *Misumena*'s far higher success at capturing large, powerful prey items, especially bumblebees, would predict the opposite outcome.

This *X. emertoni* may have attempted to claim the nest of the guarding mother *Misumena*, even though it killed and ate the guarding *Misumena* and did not remain to claim the nest. In work with *Misumena* nest guarding, we have observed five instances of *X. emertoni* guarding *Misumena* nests, and two of them had earlier killed and were feeding on the guarding *Misumena* when discovered. This behavior may occur regularly, since we have demonstrated experimentally that female *Misumena* that have already laid will readily guard nests of other species that build similar nests (*X. emertoni, Pelegrina insignis, Enoplognatha ovata*) (Morse 1989). We have also observed a large female jumping spider (*Phidippus clarus*) feeding on a (previously) guarding *Misumena*.

Reproductive Behavior

Although *X. emertoni* (and other *Xysticus* as well) apparently make major use of pheromones in mating behavior, the basis for this insight is currently indirect. Cotton swabs containing a few drops of *d*-2-octenal have proved highly effective in attracting large numbers of male *Xysticus* of several species, including *X. emertoni*, to pitfall traps, and capture of several species at one site suggests that they heavily or completely overlap spatially. These swabs will attract male *Xysticus* over the period of a few days without recharging. Whether female pheromones are similarly long lasting is unclear, though I would not expect them to have that much lasting power if they function primarily in signaling their presence to males. We have not conducted tests to determine how far these males will recruit to these sites, and we do not know the effectiveness of *d*-2-octenal in relation to the presumed pheromones released by a penultimate female molting into the adult stage. Aldrich and Barros (1985) initially demonstrated that *d*-2-octenal attracted large numbers of *Xysticus*, especially *X. ferox*. Efforts to attract male *Misumena* with this chemical have repeatedly proved unsuccessful, and numbers of other spiders (primarily wolf spiders) captured did not exceed those captured in unbaited controls.

We have not studied the mating behavior of *X. emertoni*, though such an effort should be rewarding. Male *X. emertoni* are relatively much larger in relation to the females than are males of *Misumena*, though still significantly smaller than the females, so the females should be highly dangerous potential mates. One of the arguments advanced for the dwarfism of some male spiders is that they will not be a worthwhile prey item for females (Elgar 1992). Nonetheless, comparisons of the two genera, as well as that of male *X. emertoni* of varying sizes, would be worthwhile efforts.

Males of at least some *Xysticus* species exhibit an unusual courtship behavior, in which they "tie down" the female with threads before mating (Bristowe 1958). This network of threads seems unlikely to restrain the females. Thus, the act is probably largely symbolic and may be a modification of the rapid movements performed by some male *Misumena* on the dorsum of the females' abdomen after mounting.

Nest Sites

As noted earlier, *X. emertoni* build nests in the herbaceous canopy that resemble those of *Misumena*, except that they are not covered with as much stiff silk as *Misumena*'s and thus feel far less firm when palpated. Neither do they attain the consequent shiny nature of *Misumena* nests when exposed to sunlight. Since they guard from the inside of their nest, rather than the outside as *Misumena* do, they presumably cannot add this silk. We do not know whether they all place nests in the canopy, but nests of other *Xysticus* species are habitually placed in the litter in such locations as the underside of a loose stone. We did not expect to find nest sites of *X. emertoni* in the canopy, since they spend the majority of their time in the litter layer, although we might sooner expect them in the canopy than for any of their congeners in the study area. This placement suggests that nests may be safer at this elevation than at lower levels, though we have not compared levels of parasitism of *Xysticus* nests from the canopy and the litter.

We have found a majority of *X. emertoni*'s nests on milkweed in positions similar to those of *Misumena*, possibly because we have concentrated on *Misumena*. We have found other nests on aster *Aster* spp., chokecherry *Prunus virginiana*, and raspberry *Rubus* spp. All of the nests except one on flat-topped white aster *A. umbellatus* resembled the standard configuration of *Misumena* nests on milkweed, except for their failure to cover the exterior with large amounts of silk. The one strikingly different nest occupied a narrow leaf that was folded two times, thereby providing a third side from plant material. We have never seen a similar nest constructed by *Misumena*.

Nest Guarding

Nest-guarding *X. emertoni* are spared from predation at their nests by *Misumena* because *X. emertoni*, in common with other *Xysticus* species, takes up abode within rather than outside the nest sac. We do not know which one of these strategies provides superior protection for the egg mass. *Misumena* females are more exposed to possible danger; at the same time, they may occupy a better position to respond to some contingencies. We know through experimental manipulations that unguarded *Misumena* nests suffer a higher predation rate than do guarded ones (Morse 1988c). However, we cannot run a similar removal experiment with *X. emertoni* because it would require opening the nest, and it would be difficult to reconstruct ("resew") the nest in a way that simulated the original. We do know that *X. emertoni* nests are vulnerable to parasitic wasps, but we do not know whether they differ from *Misumena* as a result of being unable to confront the parasitoids directly. Different species of wasps parasitize *X. emertoni* and *Misumena* nests.

The *Xysticus* strategy of occupying the nest chamber while guarding resembles that of both jumping spiders nesting commonly in the study area and the candy-stripe spider. Parasitoids of these species also differ from those that attack *Misumena*.

Misumena may provide better protection for their broods than do *X. emertoni*, though *X. emertoni* may not have to contend with a parasitoid comparable to *Trychosis*. A key issue here would be whether *X. emertoni* (and its congeners, which guard their nests in a similar way) typically produce more than one clutch. If so, they should place a higher priority on their own survival and a relatively lower one on the survival of any single clutch. Since we have no evidence that *X. emertoni* adults overwinter, the defense of first broods and any hypothetical second broods might differ markedly.

Alternatively, *X. emertoni* may be more vulnerable to desiccation than *Misumena*. If so, remaining on the outside of the nest is perhaps not an option available to *X. emertoni*. (When rearing broods in the laboratory, we have formed the impression that *Xysticus* females suffer more severe rates of desiccation than do guarding *Misumena*.) On the other hand, remaining outside the nest is the exception, and thus perhaps *Misumena*'s guarding behavior represents a derived condition.

Dispersal

We have exposed newly emerged second instars of *X. emertoni* to dispersal experiments (Figure 11.3: Morse 1992b) similar to those run on *Misumena* (Morse 1993b). Their behavior differs strikingly from that of *Misumena* spiderlings in that they exhibit much less of a tendency to remain on prime hunting sites (goldenrod flowers) than do *Misumena* spiderlings, although remaining longer on goldenrod inflorescences, potential hunting sites, than on milkweed leaves. When they do leave, they very seldom proceed by ballooning, only infrequently even using horizontal lines to other vegetation. Most of these horizontal lines are probably intentionally vertical lines blown by the wind. Instead, these spiders show a strong tendency to drop on lines or walk down the stems into the undercover, a behavior consistent with that of the adults. However, these individuals are ones whose parents placed and guarded their egg sacs in the canopy of the herbaceous cover. As suggested earlier, the elevated nest site may provide them with protection from parasitoids and predators frequenting the litter area. Early instars do, however, occasionally hunt in the flower canopy of the goldenrods, and when there they manage to capture small dance flies and other minute Diptera. However, in contrast to their general abundance in the habitat, they never constitute more than a minute fraction of the very young thomisids foraging in the flowers of the canopy (2 out of over 600 in one census: Morse 1992b).

The striking difference in dispersal behavior of *X. emertoni* and *Misumena* suggests that both the patchiness of their environments and their population structures differ markedly from each other. In support of this argument, Greenstone (personal communication in Morse 1992b) reported that *Xysticus* were extremely rare in aerial captures he carried out in Missouri, and that most of the abundant crab spiderling ballooners belonged to the *Misumena*-like genera *Misumenops* and *Misumenoides*. One would predict that local populations of *X. emertoni* (and

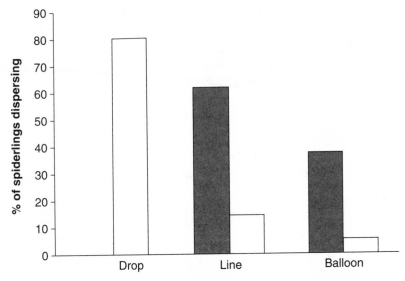

Figure 11.3. Dispersal of *Misumena* (dark bars) and *Xysticus emertoni* (light bars) spiderlings. Drop = on line or down stem into litter, Line = proceed horizontally on line, Balloon = become airborne on line. Data from Morse (1992b).

those of other *Xysticus* species as well) were more strongly differentiated genetically than those of *Misumena* (Morse 1992b), and perhaps those of many other species of spiders prone to disperse by ballooning. In order to pursue this question further, however, we would need to establish whether the *Xysticus* species have habitat requirements similar enough to those of *Misumena* that other factors, such as the distances between favorable sites for populations to establish and maintain themselves, are similar to each other. Given the apparent high density of the *Xysticus* populations, as well as the tendency of the young not to disperse widely, we would expect favorable sites to be occupied more consistently by the *Xysticus* species than by *Misumena*, and for their metapopulation structures to differ accordingly.

Conclusions

Though a member of the same family as *Misumena*, *Xysticus emertoni* has a lifestyle that does not overlap heavily with that of *Misumena*. Only sporadically does *X. emertoni* come in contact with it as an adult in the canopy on flowers, as well as placing its nests in sites similar to those of *Misumena*. *Xysticus emertoni* is primarily a member of the litter community, and several of its differences from *Misumena* are related to its life in the litter. These traits include robust hind legs that probably make it considerably more agile than adult female *Misumena;* a lower reproductive effort than *Misumena;* apparent use of pheromones in sexual interactions; and a strong tendency as newly emerged second instars to descend to the litter layer on lines rather than to move upward like *Misumena* and to leave that site on lines or by ballooning. *Xysticus emertoni* males are much larger than male *Misumena,* the likely consequence of passing through one more instar than male *Misumena.* In contrast to *Misumena,* *X. emertoni* coexists in the litter layer with several congeneric species of unknown ecological relationships.

12

Conclusions and Future Directions

$\mathcal{I}$ HAVE USED THE CRAB spider *Misumena* as a model species to evaluate the role that differences in behavior play in affecting lifetime fitness, and I have focused on how these variables may affect both males and females. These efforts have concentrated on matters related to foraging and predation, life history variables, and sexual selection. I have also placed considerable emphasis on constraints that prevent a substantial minority of individuals from approximating the optimal strategy for the situation experienced. Perhaps more importantly, I have investigated why many members of a population fall far short of the most successful individuals. In the process I have attempted to present an integrated picture of a species' life cycle and to compare it with other species that are either related or share similar lifestyles. To put these variables into broader perspective, I have also attempted to place them into the community in which the spiders live.

Probably the greatest insight I have gained from this work with *Misumena* has been to achieve a detailed impression of the importance of certain key acts in an animal's lifetime. In most other studies efforts to evaluate comparable phenomena have rested on "dodgy" assumptions based on highly indirect information. Although a long and arduous task, working out the importance of the many intermediate steps between acts of patch choice and the fitness consequences resulting from these acts has allowed us to measure the "value" of decisions with a degree of accuracy denied to most others. I credit this accomplishment to the choice of an organism uniquely suited for many of the tasks I set out to accomplish. The contributions of my many associates over the years have been responsible for getting me to this point. Apart from this central unifying theme, the work itself has provided the opportunity to explore several important subthemes. Some have been indelibly part of the narrow focus of addressing lifetime fitness, and others have offered highly worthwhile (to me, at least) diversions from the main task at hand. I would be in error to suggest that I perceived my total agenda upon beginning this line of research, for I was initially driven by certain desirable shorter-term goals that progressively revealed themselves. Only slowly did I begin to realize that I could achieve broader goals with this system.

Misumena as a Model Species

Misumena is an ideal subject through which we can explore questions of the economics of resource exploitation, life history variation, sexual dimorphism, sexual selection, and, more generally, lifetime fitness. In particular, its semelparous condition and prey-capturing strategy provide an outstanding combination of features for such studies. Females can be readily studied and manipulated under naturalistic field situations and easily adapted to simple laboratory conditions as needed. Furthermore, any manipulations that we must run in the laboratory can be closely approximated in the field. In this way we can control for most of the conditions in question and provide a ready check on the possible artificiality of any laboratory work. Several additional features of *Misumena*'s life history make it possible to obtain realistic lifetime fitness estimates that would be virtually impossible to obtain from most species. In addition to its semelparity, it sustains a remarkably consistent reproductive effort; as an adult it captures a few large prey, so that we can readily evaluate the basis for its reproductive output; and it has a relatively simple behavioral repertoire. Thus, we can cancel several potentially critical life history variables and reduce others to simple differences.

Although the work initially focused on adult females (especially on patch choice of an animal under strong selection to maximize food intake), success in obtaining progressively more direct estimates of lifetime fitness meant that the value of obtaining comparable information on the males increased greatly, although the males are considerably more difficult to study than the females. The highly dimorphic nature of the two sexes immediately raises questions about what factors have prompted this great disparity of size, from a presumably much more modest dimorphism.

The issue of sexual dimorphism takes on importance at several levels: (1) How does dimorphism takes place through changes in the number of instars (by one sex or both)? (2) What factors resulted in selection for these changes? (3) How does a system in which size change is dictated (at least in major part) by instar number differ from one in which any such change is continuous, as in vertebrates? (4) Does its potential for change differ as a result of this condition, and if so, in what way?

Misumena is one of the most highly dimorphic terrestrial species in which both sexes exist as free-living organisms. It has a female-biased sex ratio, a feature of potential significance as it relates to lifetime fitness and of inherent interest in its own right. Complementing this great disparity of size, considerable variation exists in the size of both males and females, part but not all of which is clearly environmental.

These differences imply a definite "tension" between the sexes and encourage evaluation of how the relations between the two sexes are resolved, especially to compare circumstances in which their interests are consistent with circumstances in which their interests are likely to be in conflict. Such factors could fall under the control of direct selection or merely be the consequences of selection on other traits, as seen in several characteristics of *Misumena*.

Fitness Considerations

A major advantage of working with *Misumena* is the ability to obtain accurate estimates of lifetime fitness, as seen in measuring the patch choice of adult females. Arguably this ability is the single most important factor dictating female success, inasmuch as it is the key factor affecting fecundity selection in these individuals. The number of virgin females found is a factor of comparable importance to the males, though it is not as tidy a surrogate for lifetime fitness as adult patch choice turns out to be for the females. Although one might argue that this difficulty of dealing with the males disqualifies *Misumena* as an out-standing subject for studying lifetime fitness, the very fact that fitness can be readily estimated in one of the sexes argues strongly for using it as a model species. In few species can one obtain such precise information for even one of the sexes, and the ability to accomplish this task with one sex justifies extra effort on the other sex.

Work with female *Misumena* has permitted more independent direct measures and experimental tests of lifetime fitness consequences of foraging traits under field conditions than have been otherwise obtained to date. They provide an important testing system for key aspects of foraging theory, which is based on the assumption that organisms optimize resource use. In practice, however, they can only maximize their efficiency of foraging subject to such possible constraints as predator avoidance and other aspects of risk avoidance. Our detailed experimental analysis of fitness variables over several key points in *Misumena*'s life history contrasts with the criterion frequently used to measure fitness, energy gain per unit time. This criterion is a logical, but extremely indirect, choice, often separated from a direct measure by nearly an entire life cycle. Although estimates of partial fitness are frequently used, particularly as they relate to reproduction or some other discrete part of the life cycle, these factors, too, are of questionable adequacy, especially in the absence of more comprehensive studies such as this one.

Using the fitness criterion of offspring contribution to the next breeding generation clarifies the point that energy gain/time is greatly separated from a direct estimate of lifetime fitness. Our work with *Misumena* allows us to obtain a realistic estimate of this factor. Lifetime fitness as I have defined it must in turn be merged with inclusive fitness and success of subsequent generations in order to obtain a comprehensive sense of an individual's long-term contribution to a population. However, it is questionable whether these variables could be exhaustively studied in any free-ranging population.

Optimization theory is concerned with obtaining effectively phenotypic measures of verification. However, although it ultimately depends on the criterion of lifetime fitness, one may be able to perform useful work without making this exhaustive verification. Since the formal introduction of optimality theory into ecology (and evolutionary biology) by MacArthur and Pianka (1966) and Emlen (1966), though, the provisional nature of such indirect information has often been ignored and not been compared against more direct estimates. Some investigators have thus read considerably more into their conclusions than they could adequately justify. The results from our work with *Misumena* permit us to compare and evaluate relatively direct and extremely indirect fitness estimates. In this regard it is encouraging to find that our indirect estimates of patch choice provided a very accurate estimate of lifetime fitness. We followed the initial success of testing for patch choice with a string of positively correlated results over subsequent life-cycle stages, some probably caused by this initial result and others mere consequences of it. We did not expect enhancement of the original patch-choice advantage at the subsequent life-cycle stage, for one might have instead predicted tradeoffs depreciating any initial gain emanating from patch-choice decisions.

Using a single species cannot in itself provide an adequate evaluation of general theory. As a small sit-and-wait predator, *Misumena* is unlikely to mimic perfectly the analogous behavior of widely divergent lifestyles, such as grazing herbivores or parasites. In many ways, however, a sit-and-wait predator provides a relatively simple, unspecialized entity that may more closely approach the ideal model than most species, in addition to its feasibility for study. Testing of theory related to fitness issues must start somewhere, and this model species may provide suggestions about how next to proceed in this quest.

Information on key factors from related species may help to obtain both a broader appreciation of *Misumena*'s life history and its use as a model species for testing theory. Problems exist on two levels here. First, comparable information on other crab spiders is for the most part scarce, though there has recently been a heartening increase in research on this group, especially on North American *Misumenoides* and *Misumenops*. A second problem, especially if we wish to apply meaningful modern comparative analysis (Harvey and Pagel 1991), is that the taxonomy of the crab spiders requires revision. This deficiency somewhat compromises the usefulness of available information from these other crab spiders.

The few comparable efforts to test the lifetime fitness consequences of foraging traits in field and laboratory-based systems provide further insight, namely, the studies of Ritchie (1990, 1991) and Lemon (1991) and Lemon and Barth 1992). Ritchie also obtained a good fit between the foraging and reproductive success of ground squirrels *Spermophilus columbianus* and found that it, too, correlated with growth rate, survivorship of young, and litter size. Lemon obtained a similar relationship between foraging patch choice and reproductive success in zebra finches *Taeniopygia guttata* in the laboratory. Carefully chosen and manipulated laboratory systems may contribute to analysis of this problem, though they cannot match the reality of working in a field-based system that uses extremely naturalistic manipulations. However, the close relationship between fitness and foraging in the three studies is heartening.

Working on an entire life cycle enhances the opportunity to compare multiple traits that are potentially relevant to fitness, a goal frequently stressed (Prout 1971; Ehrman and Parsons 1981) but seldom accomplished. Our crab spider system presents a virtually unique experimental insight into the importance of several life history traits to fitness in multiple populations over multiple years. Additional studies on *Misumena* will further enrich the understanding resulting from this study, but work completed thus far enhances our ability to predict outcomes arising under other circumstances. Key life history variables of other *Misumena* populations doubtless differ from those of our populations and should provide further insight into fitness variables in the wild.

The Use of Cues over Ontogeny, between Sexes and among Taxa

The method by which *Misumena* choose patches changes between the newly emerged young and experienced adult. The shift from innate patterns to responding directly to visiting prey follows a logical development that takes advantage of the progressively greater experience that an individual attains over its lifespan. However, the various difficulties suffered by both the newly venturing young (e.g., not incorporating some types of experience) and the adults (e.g., not exploiting innate traits used early in life) raise interesting questions about what constraints prevent these spiders from adopting broader "game plans" that combine advantages of the innate and the learned. It is easier to understand why the spiderlings would follow a conservative strategy, since initially they have had no experience, than why the adults largely or completely eschew innate abilities. "Simple" animals with "limited brain power" have traditionally been thought to be limited to either-or constraints, but the importance of learning is now appreciated. Nevertheless, the numerical dominance of specialist species in many groups of animals suggests the role that cognitive constraints may play in such circumstances (Egan and Funk 2006). Determining the form of this behavioral change (whether gradual or stepped), and the relative importance of innate and learned traits over this transition—a problem that we hope to resolve shortly—will greatly enhance our insight into this issue.

The use of cues varies among groups and in many instances includes modes that humans cannot monitor accurately without the use of specialized equipment (Huber 2005). However, differences within groups may at times receive less attention, notwithstanding that the relative importance of various cues may differ even between closely related forms. General examples include the likely importance of visual courtship cues between isolates in depauperate habitats such as islands and in mainland communities containing additional closely related species. Several ducks of the genus *Anas*, mallards *A. platyrhynchos* and their close relatives (Lack 1974), present a familiar example: those on isolated islands often are monomorphic and lack the striking plumage associated with drakes of mainland situations. Under these circumstances, visual factors likely assume less importance than in more diverse communities, perhaps indirectly facilitating selection for other traits in the process. Such factors as these might contribute to some of the more extreme radiations seen in isolation, even though they are generally perceived to have been direct responses to the absence of potential competitors for food and comparable resources (e.g., extinct giant flightless birds of New Zealand or Hawaii).

Xysticus emertoni and congeners may use chemical cues in species recognition as a way that avoids potential wasteful hybridization. The presence of several congeners, all of which appear to use female pheromones of a similar basic nature (judging from their common response to *d*-2-octenal) should select for special prowess in chemical sensation, as opposed to *Misumena*, which does not use pheromones in these circumstances and does not normally have closely related species in its environment. The failure of male *Misumena* to respond to putative female pheromones came as a surprise, and its basis prompts speculation. We have suggested that habitat conditions are key to this result, since *Misumena* is a habitual denizen of the herbaceous canopy at all stages of its life cycle. This factor alone would provide an adaptive basis for differing from their predominately litter-layer *Xysticus* relatives, but it does not accommodate for why aerial orb-weavers of both the genera *Argiope* and *Nephila* use pheromones so extensively in a reproductive context (see Elgar 1998). Assuming that the pheromonal trait is primitive, *Misumena*'s condition should be a relatively recently derived one, and include factors in addition to the canopy locations it frequents. We do not know whether *Misumenoides* or *Misumenops* species use pheromones in this context—an environmental explanation would predict that they did not use them in this way, but knowledge of their role would greatly enhance our understanding of the basis for *Misumena*'s apparent failure to use them.

It would be of interest to know whether the absence of common close relatives has played a role leading to *Misumena*'s apparent failure to use pheromones in mate search. It would be of further interest to investigate this characteristic in other parts of its range where it may experience substantial numbers of closely related species. Alternatively, the normal environment of *Misumena* in the canopy of herbaceous vegetation, where chemical cues might not play as great a role, may drive its responses. However, *Misumena*'s concentration on the canopy could result from other species (e.g., *Xysticus* spp.) occupying the litter and lower vegetation, where the majority of crab spiders live (Gertsch 1939). Any of these factors may assume considerable importance in determining the success of these species, as well as their evolutionary trajectories, which may vary in different environmental situations as well as geographically. However, they will probably be difficult to explore in the absence of an up-to-date phylogeny of the family.

Implications of Size for *Misumena* as a Predator and as Prey

In contrast to the focus on adult behavior of many types, juvenile behavior has received relatively little attention. I emphasized this dearth of attention in a similar concluding chapter over 25 years ago (Morse 1980), but it remains true today as well. In some groups, which reach full size before beginning to forage (some holometabolous insects, birds, etc.), differences may relate largely to experience, allowing separation of experience and size. However, in many other extremely important groups, of which the spiders we have featured serve as a typical example, the size of the forager may vary a hundredfold or more over its lifetime. A female may progress from 0.6 mg to over 400 mg, pushing the envelope to nearly three orders of magnitude. Over this *Misumena* period, some potential prey may range from hard-to-capture, relatively gigantic items to ones so relatively small that the spiders lose mass while feeding on them. The 4-mg hover fly *Toxomerus marginatus* is such an item—an item that may be six times as large as a newly emerged spiderling and only one hundredth as large as some extremely gravid adult females. Many other prey are only accessible to part of the size range of female *Misumena*, but the same variables will hold. To what extent will experiences early in life affect their later foraging? In contrast, males will change only tenfold at most, and as adults use small hover flies as a prime food item. Are these male–female differences of any further (indirect) significance?

Certain other species prey on small *Misumena* but in turn become prey of late-instar and adult females. Some of the jumping spiders that share the vegetation with *Misumena* provide among the clearest examples of this sort. They may even capture spiderlings emerging from their natal nests (Morse 1992a). We have further evidence of the early impact of jumping spiders on *Misumena* populations from their common initial hunting areas on goldenrod. Presumably danger from jumping spiders or other small predators accounts for *Misumena* spiderlings' behavior, and their initial prey capture patterns on their first foraging substrates, sparse and dense goldenrod inflorescences (Morse 2006). The spiderlings' behavior on different substrates strongly suggests that predator avoidance initially affects their responses, but that foraging concerns soon modify their behavior. We do not know whether these events affect their behavioral repertoire later in life.

Dwarf Males and Sexual Dimorphism

As one of the most strikingly dimorphic species among free-living terrestrial animals, *Misumena* invites comparison with other strongly dimorphic species, as well as a consideration of Ghiselin's (1974) predictions about the conditions under which extreme dimorphism should occur and where it should be found. I have associated *Misumena*'s extreme dimorphism in part with its major resource requirements. Comparison with the intensively studied orb-weaver spiders, which exhibit a very different lifestyle, and the deep-sea organisms discussed by Ghiselin, present a profitable opportunity. Ghiselin drew spiders and deep-sea organisms together with the analogy of orb-weavers as plankton feeders, in common with some extremely sexually dimorphic deep-sea forms, noting that these insect "plankton" flew in a (dilute) viscous medium, namely, air. He further drew attention to the fact that female crab spiders concentrated at sites with extremely large prey, allying them to deep-sea ceratioid angler fishes, which also take prey whose size may even exceed theirs. Furthermore, since most female *Misumena* do not obtain enough food to lay maximally large clutches (Morse and Stephens 1996), effectively they live in a food-scarce environment. This is notwithstanding the illusion of abundant prey about them, which are difficult to catch, probably in part because of the spiders' cognitive limitations (Chapter 6). This condition may drive *Misumena* toward extreme dimorphism in the same way as that suggested by Ghiselin for the deep-sea forms he discussed, namely, that of relatively large sedentary females accumulating resources while minimizing energetic costs and males minimizing expenditures associated with size to the extent possible while retaining the ability to seek out females.

Most of the evidence suggests that the greatest size changes occur in male *Misumena*, inasmuch as males are the active sex. Taking males to be the flexible sex makes sense, since whether tiny or huge (crab spiders or seals), they make no subsequent contribution to the welfare of their offspring. However, Coddington, Hormiga, and Scharff's (1997) argument that the size differences of male and female spiders often result from female giantism rather than male dwarfism complicates any simple general explanation. Higgins (2000) has also presented independent biological arguments in support of female giantism as the driving force in the tetragnathan genus *Nephila*, the group focused on by Coddington, Hormiga, and Scharff. Adult female *Misumena* may also exhibit a derived condition (relatively large and rather immobile for part of their adult stage), according to Ghiselin's argument. Relative to other crab spiders, such as *Xysticus emertoni*, they do in fact appear to have become less mobile, though as adults much of that difference results from their extremely gravid condition. The frequent inability of adult female *Misumena* to reach maximum size due to food limitation probably minimizes selective pressure to become yet larger. Although I do not believe that Coddington, Hormiga, and Scharff's explanation holds for *Misumena* (they allowed for exceptions), the likelihood that it holds for many grossly dimorphic spiders (female giantism, not male dwarfism) argues against a scheme of one explanation fits all.

It is profitable to compare *Misumena*'s dimorphic attributes with the more frequently studied ones, such as those of mammals. Here, in the most extreme cases, the male/female size ratio may reach 8:1 (some seals), which, however, does not approach that of *Misumena* and is male rather than female-biased. The large seals present some of the most striking examples of sexual dimorphism driven by intrasexual competition (fighting over females). One is tempted to ask what factors have prevented size ratios among these forms from achieving the degree of difference seen in *Misumena* and the orb-weavers. Nonsexual factors may place the brakes on such growth, though the difference could reflect phylogenetic constraints as well. Although thermal considerations may place limits on the minimum size of aquatic mammals (Calder 1984), females of the most highly dimorphic seals do not approach the minimal sizes of this group.

Other aquatic mammals, whales, have also become vastly larger than the largest seals. However, whales have become completely aquatic, so that they no longer have any requirements for functioning on land. Ghiselin noted that the most highly dimorphic forms were typically aquatic, but this conclusion could have been affected by his sampling regime. If terrestrial constraints do limit the extent of dimorphism, it makes the spiders' extreme dimorphism, especially the crab spiders' "non-planktivorous" dimorphism, all the more remarkable.

In this regard, it may be profitable to compare systems in which males are the fighting sex or the larger sex with those in which females are the fighting sex or the larger sex. For an example concerning males, the traits associated with finding females versus fighting over females are fundamentally different: scramble competition versus interference competition and require a totally different array of traits, which may be completely noncomplementary and drive the morphological and behavioral characteristics of these species in different directions. Male spiders probably cannot be both large and fast. Although that constraint sounds reasonable (probably a consequence of our familiarity with mammalian systems), if they maintain large size only at a great energetic cost, and if finding females yields more quality matings than fighting over them, one would expect any extra bulk to disappear.

Misumena females could be hard to find because of their numbers, their behavior, their size, or the habitat they occupy. If any of these traits serve to make them inconspicuous, then selection occurs for effectively searching males—and dimorphism is born or enhanced, in a way distinct from male contest competition and resulting in forms fundamentally different from the large fighting specialists associated with contests. And, Ghiselin argues, if those conditions include extreme difficulty for females accumulating enough resources to reproduce, the extent of dimorphism may be extreme. Curiously, these small, mobile males bear some resemblance to the sneaker or satellite males often associated with various communal breeding systems of fish dominated by large males (Dominey 1980; Gross 1982). Both of these auxiliary males and the scramble competitor spiders are small and highly mobile.

Ghiselin's examples show that in situations in which one sex, the male, becomes literally a parasite of the other one, as in some of the deep-sea anglerfishes, sexual dimorphism has proceeded far beyond the boundaries seen in free-living forms. Moreover, this boundary seems to have been crossed time and again, as seen by the regular presence of this condition among several lines of barnacles (Darwin 1854). Developing this secondarily parasitic relationship does not appear to have presented unusual difficulties for them. For instance, even closely related barnacle species may differ strikingly in this regard, ranging from males that are complete parasites on the females to situations in which the two sexes are similar in size, totally separate, and free-living. No comparable terrestrial animals have crossed that boundary and moved into the area of intrasexual parasitism, and one is tempted to ask why this is so.

What special conditions have permitted some, but not other, deep-sea organisms to cross this line to parasitism? Comparison with the terrestrial dimorphic species may provide useful cues. Is the difference a phylogenetic one, in which few terrestrial forms, spiders being the prime example, have approached a condition in which extreme dimorphism may occur, but, because of peculiar phylogenetic constraints, they have been unable to cross this boundary? Or do the problems lie more basically in surmounting environmental problems? Since the most extreme terrestrial examples of free-living dimorphism occur among the spiders, they should provide excellent insights into the bases for variation and extremes in dimorphism.

Based on 16 species of orb-weaving spiders, Ramos et al. (2005) have proposed that the allometric growth patterns of male and female genitalia could limit the extent of sexual size dimorphism in that the allometry of male and female genitalia progresses at a rate that exceeds that of overall body size. Thus, the forces acting on female fecundity selection and male mate-finding would, at the same time that they drove dimorphism to increase, generate an even stronger dimorphism between male and female genitalia. Mismatches would then occur, thus initiating stabilizing selection that would preclude such mismatches. Such a limit might also be relevant to *Misumena*. The relatively similar extremes in dimorphism of crab spiders and orb-weavers are consistent with such an interpretation. It should be immediately apparent, however, that this factor would only apply to forms with internal fertilization. Ramos et al. further note that it might play a particularly strong role in taxa with sclerotized genitalia.

Ghiselin's (1974) speculations about extreme sexual dimorphism and dwarf males argue for the importance of environmental factors in sexual dimorphism and the extent to which they will initiate and drive this phenomenon, orient it, and secondarily, dictate the breeding systems associated with it as well. Other aspects of sexual dimorphism may also be associated with depauperate communities, though not necessarily ones with extremely low resource availability (e.g., islands). In the case of some island bird species (e.g., huia, Hispaniolan woodpecker: Selander 1966), specialization to particular types of resources—a way by which they expand their resource base in the absence of likely competitors—appears to drive this change. The presumable result is superior access to food for their young. The birds' beaks have diverged so greatly that the sexes can access food unavailable to the other sex, although differences in body size are not great. Sexual dimorphism probably developed as a direct consequence of resource conditions in the bird species, while in those involving the dwarf male syndrome, dimorphism arose as an indirect consequence of the resource condition.

It would be worthwhile to examine whether dimorphism in spider species showed analogous shifts in isolated or otherwise depauperate circumstances. Densities of some island populations, including spiders

(Schoener and Toft 1983), may greatly exceed those to be found in comparable mainland environments. It is also of interest that island-type environments have led toward secondary monomorphism of certain nontrophic, sexually dimorphic characters, such as the plumage colors of ducks noted earlier. The latter traits are likely acted on by both sexual and natural selection.

Male *Misumena* achieve their tiny adult size by reaching maturity in fewer instars than the females, seemingly a very convenient way of reaching this condition. However, which sex has actually changed its numbers of instars from a presumably less dimorphic condition to achieve this striking dimorphism requires more attention, since rates of molting may differ between the sexes, as seen in *Misumena* (D. H. Morse, unpublished data) and other species. A sensitive analysis of *Misumena* itself will require an adequate phylogenetic analysis of the crab spiders. However, the number of male instars in *Misumena* seems low for a spider of this size, and especially for one that produces females of the observed size. Furthermore, the apparent variation in number of instars required for the males to reach maturity (see Gabritschevsky 1927) suggests that vagility still exists at the low end of the spectrum, which supports the argument of dwarf males. Where known, triggering of adulthood is governed hormonally in arthropods by the concentration of juvenile hormone produced at time of molt. Although juvenile hormone has not to date been found in spiders (Trabalon, Pourié, and Hartmann 1998; Pourié and Trabalon 2003), it nevertheless remains likely that an analogous hormonal switch controls the onset of maturity in *Misumena* and other species. Changing the timing of that trigger might result in a difference comparable to that seen in male *Misumena*.

The sensitivity of maturation to hormonal control in arthropods has been known since the pioneering work of Kopeć (1922) and Wigglesworth (1934). Experimental analysis of extremely large males, perhaps derived by hormonal manipulation, might go a long way in terms of evaluating the adaptive significance of size in the males. Particularly likely sources for investigation include the performance of large males in locomotory rate, ability to find females (presumably through increased locomotory rate, but further "developed" males might have sensory capabilities that the earlier instars [including present-day adults] lack), and vulnerability to cannibalism by females during mating attempts. One predicts that these individuals will not perform as well as the present-day adults. Since we have found two "giant" males over the period of these studies, we know that they can survive in the wild. However, one would expect their frequency to be far higher if current-day conditions favored them.

Different selective regimes may have led to dwarf males and giant females in crab spiders and orb-weavers: the driving force in male dwarfism of finding scarce females and the primary force acting on female giantism to produce the most offspring possible in a single clutch. These differences may in turn lead to a rich range of responses if the fitness considerations of the two sexes do not perfectly complement each other. The selective regimes driving these two parts to extreme dimorphism may have differed fundamentally, even though the two groups exhibit strikingly comparable levels of difference. This conclusion raises questions about the origins and consequences of these lifestyles, which exhibit marked similarities (sit-and-wait predators) as well as differences (use of snare and silk vs. stealth and potent venom). It is of interest, however, that both of these striking lines evolved within the same major taxonomic group (entelegyne spiders).

Comparison of some of these traits with ambushing bugs (Nabidae, Phymatidae, Reduviidae, Pentatomidae), as well as mantids, should expand our understanding of this problem, given the similarity of their hunting patterns. None of these insects exhibits the extreme size dimorphism of *Misumena* and certain other species. One major difference is that since insects can fly, though not agile afoot, they may disperse to new hunting sites more readily than spiders. Thus, it would be of interest to determine whether food routinely limits these insects as well.

Sex Ratios, Population Density, and Dimorphism

We do not know the basis for the significantly, though not dramatically, female-biased sex ratios found in newly emerged young *Misumena*. Gunnarsson and Andersson (1992) found that a solitary sheet-web weaver *(Pityohyphantes phrygianus)* had a primary sex ratio of about one-third males, which apparently the females can control through a sperm-sorting technique. Through this technique, they acquire sperm while in different mating positions, thereby probably ensuring that the sperm are directed into different spermathecae (Gunnarsson, Uhl, and Wallin 2004). Sex ratios of social spiders, the only other spiders currently known to have a strongly female-biased primary sex ratio, have been related to inbreeding concerns (Buskirk 1981; Vollrath 1986).

More recently, Avilés et al. (2000) have provided evidence suggesting that at least some of the social spiders use a sperm-sorting mechanism as well. These biases supplement the usual high mortality of adult male spiders, thus skewing the secondary sex ratio even more severely. However, since male *Misumena* do not suffer severe mortality as adults, this factor (adult male mortality) should not intervene in determining the secondary ratio. Male *Misumena* do suffer a disproportionately high overwintering mortality, another factor that may substitute for adult male mortality. Gunnarsson (1987, 1988) also reported this pattern in his sheet-web weavers. It appears that younger spiders also suffer disproportionately high overwintering losses, which we have attributed to their small size, as also suggested by Gunnarsson. Males might similarly be at risk overwintering as a consequence of their small size, though we have no independent evidence in support of this point.

A female-biased sex ratio might be predicted if sib mating takes place regularly. Given the apparently small sizes of many populations, some of which are almost certainly far smaller than the adult populations of roughly 100 adults found in the study areas (Holdsworth and Morse 2000), sib mating might often occur. On the other hand, the dispersal abilities of the young should minimize sib mating in large populations. However, in low-density populations whose females are all nevertheless mated, females investing heavily in daughters should be rewarded in the same way as those from sib mating. Although we do not know whether *Misumena* can manipulate its sex ratio, the wide variance in individual sex ratios (LeGrand and Morse 2000) warrants attention. Wide variations about the mean are by no means unknown (Frank 1990), and selection for sex ratio should take place at the level of the local population unit.

Low population densities should favor small males, or at least minimize the premium of fighting for females. The same selective factor should act to drive sex ratios and dimorphism, presumably to the point that substantial numbers of females go unmated. In this regard, large females (though not as extreme as *Misumena*), are generally the rule among arachnids (Foelix 1996a) and animals other than mammals and birds (Ghiselin 1974).

Sexual Selection

Misumena provides excellent opportunities to work on factors directly related to sexual selection, especially through scramble competition, which may well have contributed to its highly dimorphic condition. In *Misumena*, scramble competition plays an important role in sexual selection (LeGrand and Morse 2000). Scramble competition is an understudied aspect of sexual selection, even though a fundamental part of (male) success may lie in finding the opposite sex. In sexual selection, emphasis has usually focused on male combat, with selection for large, aggressive males (the mammal model) or male display, with highly conspicuous traits (the bird model). However, large or showy males constitute the minority among animals; contest competition probably plays a much less important role in the majority of species, and scramble competition a much more important role, than the spectacular examples of contest competition that occupy most textbook space.

Misumena's highly dimorphic condition provides an excellent opportunity to investigate the role of female aggression in facilitating extreme dimorphism. Predatory females can alternatively treat a male as a potential mate or prey item. The range of within-sex size in *Misumena* provides an opportunity to test the contribution of dimorphism to sexual selection, and, conversely, to test the possibility that sexual selection has contributed to the high level of dimorphism observed.

The females might favor relatively large males as prey by virtue of their greater food value, but these males might also be the highest quality mates. Males may vary in the ease with which the females may capture them; the larger individuals might provide easier targets by virtue of their size alone, unless they are more agile. However, small males are also extremely agile, and we have not found a size-related differential vulnerability to cannibalism among the males. All of these males are extremely small, and none of them may provide a worthwhile meal for the large females. In fact, we know that adult females cannot maintain their weight on a diet of hover flies of similar size to the smaller males, and the reward of the largest may not be much better.

All males may be so small that they do not provide worthwhile food resources. This observation raises the question of whether minimizing
cannibalism has played a significant role in generating the extreme dimorphism that we observe in *Misumena*.

Elgar and Nash (1988) have suggested that females may obtain important nutrients by feeding on males. However, this explanation does
not hold for the *Misumena* we have studied, for we have demonstrated
experimentally that females that did not feed on males had clutches as
large and as successful as those that fed on one or more males. With
few exceptions, investigations of the role of males in female nutrition
have found that males do not provide a significant source of resources.
One recent study that did report such a gain (Johnson 2005) involved
fishing spiders *Dolomedes triton*, whose males are relatively (to their females) far larger than are male *Misumena* to their females. However, female *Misumena* probably do not make a decision to attack based on
food quality. We have found that adult female *Misumena* do not discriminate in the prey they capture and that they readily attack unprofitable prey.

Conflict and Cooperation between the Sexes

Traditionally, it was assumed that interactions between the two sexes
evolved under selection to become as compatible as possible (e.g., Allee
1951), thereby maximizing the production of offspring to the benefit of
both parents. However, this line of thinking is covertly group-selective
(Williams 1966; Ghiselin 1969). Natural selection dictates that members of a pair should interact in a way that maximizes each one's own
individual fitness. (Individuals that pair for life and produce multiple
broods provide a possible exception to this rule.)

Highly dimorphic systems such as *Misumena*'s raise immediate questions about relative advantage to the two sexes. Often the resulting interactions between pair members will prove far less advantageous to
one or even both members of the pair than the formerly predicted mutualistic interaction. At the same time, both individuals do have
common goals that overlap to a certain degree: to recruit as many offspring as possible into the next breeding generation. This factor alone
will ensure a minimal level of cooperation, especially if demands of the
young require that both members of a pair contribute to rearing the

young. It may also define the limits of selfishness that selection will allow. Nevertheless, the most effective way of maximizing one's recruitment of young into the next breeding generation may differ considerably between males and females.

High reproductive effort requires much energy and many resources, which in themselves may result in stressful conditions. Even if both parents care for their young, if one member minimizes its contribution to the upkeep of the dependent young to the extent consistent with their survival, this action may force an increased effort by the exploited parent to compensate for the cheater's sloth. As a result, the survivorship of the exploited parent is compromised, and that of the cheater is enhanced. If the cheater lives one year longer as a result, during which it can parent another clutch and so measurably enhance its fitness, then its tactics are rewarded. If its present mate lives a year less, or exists in substandard condition as a result of the experience, and suffers a poor breeding year, its loss will be of little consequence to its former mate.

Unsuccessful or relatively unproductive matings were formerly assumed to result from either genetic or behavioral incompatibility. (Behavioral incompatibility could in itself result from genetic deficiencies.) Such discussions seldom addressed the contribution of competition for advantage between the sexes, yet a substantial part of reproductive failure might be a consequence of competition between the sexes for their own personal lifetime advantages.

Arnqvist and Rowe (2005) argue that the sexually selected traits of males and females facilitate their ability to compete against each other. Examples of pre-mating cannibalism by female *Misumena* illustrate this point well. Our detailed work on both sexes of this species (LeGrand and Morse 2000; Morse 2004; Morse and Hu 2004), which we have pursued largely to address fitness and dimorphism issues, provide useful insights for addressing male–female cooperation and exploitation.

To what degree do sexual competition and sexual selection play off against natural selection? This is an important question to be addressed in understanding the complex relationships between the two sexes. According to theory, factors operating through natural selection need not inevitably be in conflict. In fact, selection should favor this type of cooperation under certain conditions, though the degree to which selection can act is unclear, and it remains unclear whether one could separate potentially synergistic and antagonistic effects from natural selection per se. Situations in which a sexually selected trait in one sex may be balanced by a naturally selected trait in the other sex remain an important problem to be resolved. *Misumena* provides a potentially useful system in which to test some of these ideas and in the process may help to point the direction toward a more general understanding of competition between the sexes.

Plasticity of Breeding Systems

Although the low density of our *Misumena* populations appears to assign little selective advantage to male contest competition, the "willingness" of males to fight when placed in appropriate settings suggests that *Misumena* have a plastic breeding system. This condition should work well in systems regularly subject to striking fluctuations in numbers, especially unpredictable ones. Our *Misumena* populations actually maintain a relatively constant population size. However, small populations in successional sites probably result from regular colonization events and declines toward extinction. *Misumena*'s plastic traits seem ideal for forms with high probabilities of experiencing conditions different from those of their parents. Retaining the ability to respond aggressively to other males may favor the members of a population whose neighbors have traditionally experienced a wide variety of conditions, owing in part to the successional nature of many of their sites. However, the retention of extremely small males suggests that male fighting does not strongly drive selection for body size, as opposed to species such as the sheet-web weaver *Linyphia litigiosa*, in which male contests are the rule and males are larger than their females (Watson 1990). The high levels of aggression in dense populations of the closely related *M. formosipes* (Dodson and Beck 1993) demonstrate the fighting abilities retained by extremely small males.

Cause and Effect

Although direct causal factors may play a dominant role in sculpting an individual's life history traits, their consequences may play an equally important role. Results from our work with *Misumena* suggest that these correlated factors may dominate important aspects of this species' life cycle. Patch-choice decisions of foraging adult females play a key role in their subsequent reproductive success and the early success of their offspring, the latter through maternal effects resulting from maternal size. Large size is in turn a consequence of the superior success of individuals that experienced great foraging success resulting from good patch-choice decisions earlier in the life cycle.

The fitness of female *Misumena* is thus the consequence of a long string of events (Chapter 4) that follow from their hunting patch-choice decisions. Selection apparently acts directly on some of these events, but other results follow as a consequence of the earlier events. These results, which derive from selection on other traits, have doubtless gone underreported, underappreciated, and often under the heading of direct effects in other systems. This issue prompted Gould and Lewontin (1979) to decry what they referred to as the "adaptationist programme" and forced the field to take stock of important effects that do not occur directly as a result of an adaptation, but may merely be the byproduct of some other factor that is being independently selected.

This mechanism may not make the result any less important, but it may have clear implications for the possibilities of subsequent change. Thus, current scholars have been forced to treat the subject of adaptation more precisely than before (Rose and Lauder 1996b). Gould and Lewontin expressed particular concern about situations in which workers assumed the presence of an adaptation, most often of a morphological nature, though the critique has much broader applicability. Because adaptation was the matter at issue, their concerns also apply to other "decisions" of animals that may not attract the attention that morphological traits do inasmuch as they are perceived to be "ecological choices" rather than adaptations. Yet, they involve decisions likely to be reined in by stabilizing selection, and thus they have a fundamentally similar evolutionary nature.

Misumena provide an excellent opportunity to evaluate the impact of these indirect effects in a natural system, since they play such an important role in its life cycle. They can be clearly seen from the hunting-site choices of newly molted adult females to the dispersal of the newly emerged offspring. Success of the mature female's foraging efforts will accurately determine her prepartum mass, the size of her clutch, and her postpartum mass. The latter two factors play an important role in determining the eventual lifetime fitness of the mother (and, indirectly, the father as well).

The size of the egg mass follows from the size of the mother and forms the primary basis of her lifetime fitness: she produces only a single clutch, thereby greatly simplifying this variable. In this instance, a closely correlated factor, postpartum mass, plays a key role. Simply as a result of her greater size after having laid her clutch, a female can better guard it at a time of high losses. Furthermore, since the prepartum mass and egg mass are nearly linearly related over most of the range of female mass at egg-laying, the spider does not compromise her initial advantage of large size. By virtue of her larger size alone, a successful postpartum female can more effectively guard her eggs up to emergence of the young. She is also more likely to be alive when the young emerge from their nests, at which point she may provide further significant protection from small predators that would otherwise capture the young as they leave their nests. In this system, not only does no tradeoff occur, but large broods enjoy enhanced success over this period. This is the result of the lingering advantages of foraging success, here dictated by differences in the mother's size. As a result of a few fortunate events, one or two of which go "right," the rest of the spider's life falls nicely into place, even though several of the subsequent variables do not appear to involve direct selective events— hence the accuracy of Grafen's (1988) reference to such results as the "silver spoon effect."

After the young disperse, differences in their success should vary randomly in relation to their mother's patch-choice decisions. Since the mother's fitness-related benefits result from the quantity of offspring she has reared to this point, rather than differences in their quality, we should have identified all of the stages in the life cycle that can yield differences in parental fitness resulting from their patch-choice decisions. Now the young are on their own! Nevertheless, we continue to explore the juvenile stages to test this tentative conclusion.

Sorting out the respective roles of cause, correlation, and effect is an important task for the study of life histories. Our work suggests the importance of looking for critical points in a life cycle that will strongly affect how subsequent parts of the life cycle will play out. Perturbations of this stage may produce particularly strong effects on the individual involved, which will ripple through its life cycle relatively impermeable to subsequent variables. In addition to the effort needed to assess the respective contributions of direct and indirect factors to animal life histories, evaluation of how these factors may vary in importance at different stages, among different lifestyles, and among different taxonomic groups warrants attention. Identification of critical stages, manipulation of the relevant variables to test predictions, and phylogenetic analyses of well-known groups would all be most profitable goals. They should provide insight into the ubiquity and importance of such life-dominating stages among animals; at what parts of the life cycle they take place; if more often at some than at others, what consequences such positioning has on their impact; and the relative importance of innate and environmental factors in determining their recurring presence, strength, and temporal position in the life cycle.

This brief analysis of cause and consequence in *Misumena* provokes some general questions. Are some taxonomic groups more likely to exhibit certain traits, including the ones noted above, than others? Are these traits more likely to occur in some situations than in others, and if so, why? What are the traits' underlying bases? Do they represent an example of pleiotropic genes inadvertently at work, or are the acting genes randomly placed relative to each other? Presumably, selection acts on the entire package of traits of concern here. That is, the size of the female as well as the size of the egg mass, and only certain combinations of such traits would manage to survive. If the genetic basis for such traits is complex, indirect effects might be common. Although our studies with *Misumena* cannot answer the latter questions, they do provide a test case that other studies may subsequently use as a basis for comparison. Thus, the work I have reviewed here suggests several interesting venues for future work.

References

Abraham, B. J. 1983. Spatial and temporal patterns in a sagebrush steppe spider community (Arachnida, Araneae). *Journal of Arachnology* 11:31–50.

Abrahamson, W. G., and A. E. Weis. 1997. Evolutionary ecology across three trophic levels. *Monographs in Population Biology* 29:1–456.

Abramson, C. L. 1994. *A primer of invertebrate learning.* Washington, DC: American Psychological Association.

Aguinaldo, A. M. A., J. M. Turbeville, L. S. Linford, M. C. Rivera, J. M. Garey, R. A. Raff, and J. A. Lake. 1997. Evidence for a clade of nematodes, arthropods and other moulting animals. *Nature* 387:489–493.

Aldrich, J. R., and T. M. Barros. 1995. Chemical attraction of male crab spiders (Araneae, Thomisidae) and kleptoparasitic flies (Diptera, Milichiidae and Chloropidae). *Journal of Arachnology* 23:212–214.

Allee, W. C. 1951. *Cooperation among animals.* New York: Henry Schuman.

Anderson, J. F. 1974. Responses to starvation in the spiders *Lycosa lenta* Hentz and *Filistata hibernalis* (Hentz). *Ecology* 55:576–585.

Anderson, J. T., and D. H. Morse. 2001. Pick-up lines: Cues used by male crab spiders to find reproductive females. *Behavioral Ecology* 12:360–366.

Andersson, M. 1994. *Sexual selection.* Princeton, NJ: Princeton University Press.

Andrade, M. C. B. 1996. Sexual selection for male sacrifice in the Australian redback spider. *Science* 271:70–72.

———. 1998. Female hunger can explain variation in cannibalistic behavior despite male sacrifice in redback spiders. *Behavioral Ecology* 9:35–42.

Arnold, S. J. 1983. Morphology, performance and fitness. *American Zoologist* 23:347–361.

———. 1992. Constraints on phenotypic evolution. *American Naturalist* (Supplement) 140:S85–S107.

Arnqvist, G., and S. Henriksson. 1997. Sexual cannibalism in the fishing spider and a model for the evolution of sexual cannibalism based on genetic constraints. *Evolutionary Ecology* 11:255–273.

Arnqvist, G., and L. Rowe. 2005. *Sexual conflict.* Princeton, NJ: Princeton University Press.

Austad, S. N. 1982. First male sperm priority in the bowl and doily spider, *Frontinella pyramitela* (Walckenaer). *Evolution* 36:777–785.

———. 1984. First male sperm priority patterns in spiders. In *Sperm competition and the evolution of animal mating systems,* ed. R. L. Smith, 223–249. New York: Academic Press.

Avilés, L. 1986. Sex ratio bias and possible group selection in the social spider *Anelosimus eximus. American Naturalist* 128:1–12.

Avilés, L., J. McCormack, A. Cutter, and T. Bukowski. 2000. Precise, highly female-biased sex ratios in a social spider. *Proceedings of the Royal Society of London B* 267:1445–1449.

Balduf, W. V. 1939. Food habits of *Phymata pennsylvanica americana* Melin (Hemip.). *Canadian Entomologist* 71:66–74.

———. 1941. Life history of *Phymata pennsylvanica americana* Melin. *Annals of the Entomological Society of America* 34:204–214.

Balfour, R. A., C. M. Buddle, A. L. Rypstra, S. E. Walker, and S. D. Marshall. 2003. Ontogenetic shifts in competitive interactions and intra-guild predation between two wolf spider species. *Ecological Entomology* 28:25–30.

Bartels, M. 1930. Uber den Fressmechanismus und chemischen Sinn einiger Netzspinnen. *Revue Suisse Zoologie* 37:1–42.

Barth, F. G. 1985. *Insects and flowers.* Princeton, NJ: Princeton University Press.

———. 2002. *A spider's world.* Berlin: Springer-Verlag.

Bartos, M. 2004. The prey of *Yllenus arenarius* (Araneae, Salticidae). *Bulletin of the British Arachnological Society* 13:83–85.

Beck, M. W., and E. F. Connor. 1992. Factors affecting the reproductive success of the crab spider *Misumenoides formosipes:* The covariance between juvenile and adult traits. *Oecologia* 92:287–295.

Bell, G. 1984. Measuring the cost of reproduction. I. The correlation structure of the life table of a plankton rotifer. *Evolution* 38:300–313.

Ben-Shahar, Y., H.-T. Leung, W. L. Pak, M. B. Sokolowski, and G. E. Robinson. 2003. cGMP-dependent changes in phototaxis: A possible role for the foraging gene in honey bee division of labor. *Journal of Experimental Biology* 206:2507–2515.

Birkhead, T. R., and A. P. Møller, eds. 1998. *Sperm competition and sexual selection.* London: Academic Press.

Blakely, N. 1981. Life history significance of size-triggered metamorphosis in milkweed bugs *(Oncopeltus). Ecology* 62:57–64.

Blanckenhorn, W. U. 2000. The evolution of body size: What keeps organisms small? *Quarterly Review of Biology* 75:385–407.

———. 2005. Behavioral causes and consequences of sexual size dimorphism. *Ethology* 111:977–1016.

Bonner, J. T. 1988. *The evolution of complexity.* Princeton, NJ: Princeton University Press.

Bonte, D., N. Vandenbroecke, L. Lens, and J.-P. Maelfait. 2003. Low propensity for aerial dispersal in specialist spiders from fragmented landscapes. *Proceedings of the Royal Society of London B* 270:1601–1607.

Briscoe, A. D., and L. Chittka. 2001. The evolution of color vision in insects. *Annual Review of Entomology* 46:471–510.

Bristowe, W. S. 1958. *The world of spiders.* London: Collins.

Brueseke, M. A., A. L. Rypstra, S. E. Walker, and M. H. Persons. 2001. Leg autotomy in the wolf spider *Pardosa milvina:* A common phenomenon with few apparent costs. *American Midland Naturalist* 146:153–160.

Burger, M., W. Nentwig, and C. Kropf. 2003. Complex genital structures indicate cryptic female choice in a haplogyne spider (Arachnida, Araneae, Oonopidae, Gamasomorphinae). *Journal of Morphology* 255:80–93.

Buskirk, R. E. 1981. Sociality in the Arachnida. In *Social insects,* vol. 2, ed. H. R. Hermann, 281–367. New York: Academic Press.

Buskirk, R. E., C. Frohlich, and K. G. Ross. 1984. The natural selection of sexual cannibalism. *American Naturalist* 123:612–625.

Calder, W. A., III. 1984. *Size, function, and life history.* Cambridge, MA: Harvard University Press.

Carrel, J. E., H. K. Burgess, and D. M. Shoemaker. 2000. A test of pollen feeding by a linyphiid spider. *Journal of Arachnology* 28:243–244.

Carson, W. P., and R. B. Root. 2000. Herbivory and plant species coexistence: community regulation by an outbreaking phytophagous insect. *Ecological Monographs* 70:73–99.

Chapman, T., G. Arnqvist, J. Bangham, and L. Rowe. 2003. Sexual conflict. *Trends in Ecology and Evolution* 18:41–47.

Chapman, T., L. F. Little, J. M. Kalb, M. F. Wolfner, and L. Partridge. 1995. Cost of mating in *Drosophila melanogaster* females is mediated by accessory-gland products. *Nature* 373:241–244.

Charnov, E. L. 1976. Optimal foraging: The marginal value theorem. *Theoretical Population Biology* 9:129–136.

Chew, R. M. 1961. Ecology of the spiders of a desert community. *Journal of the New York Entomological Society* 69:5–41.

Chien, S. A., and D. H. Morse. 1998. Foraging patterns of male crab spiders *Misumena vatia* (Araneae: Thomisidae). *Journal of Arachnology* 26:238–243.

Chittka, L. 2001. Camouflage of predatory crab spiders on flowers and the colour perception of bees (Aranida: Thomisidae/Hymenoptera: Apidae). *Entomologia Generalis* 25:181–187.

Clark, C. W., and R. Dukas. 1994. Balancing foraging and anti-predator demands: An advantage of sociality. *American Naturalist* 144:542–548.

Clayton, N. C., and D. W. Lee. 1998. Memory and the hippocampus in food-storing birds. In *Animal cognition in nature,* ed. R. P. Balda, I. M. Pepperberg, and A. C. Kamil, 99–118. San Diego, CA: Academic Press.

Cloudsley-Thompson, J. L. 1968. *Spiders, scorpions, centipedes and mites.* Oxford: Pergamon Press.

Clutton-Brock, T. H., ed. 1988. *Reproductive success.* Chicago: University of Chicago Press.

Clutton-Brock, T. H., F. E. Guinness, and S. D. Albon. 1982. *Red deer: Behavior and ecology of two sexes.* Chicago: University of Chicago Press.

Coddington, J. A., G. Hormiga, and N. Scharff. 1997. Giant female or dwarf male spiders. *Nature* 385:687–688.

Coddington, J. A., and H. W. Levi. 1991. Systematics and evolution of spiders (Araneae). *Annual Review of Ecology and Systematics* 22:565–592.

Cohen, A. C. 1995. Extra-oral digestion in predaceous terrestrial Arthropoda. *Annual Review of Entomology* 40:85–103.

Collatz, K.-G. 1987. Structure and function of the digestive tract. In *Ecophysiology of spiders*, ed. W. Nentwig, 229–238. Berlin: Springer-Verlag.

Comstock, J. H. 1940. *The spider book.* Revised and edited by W. J. Gertsch. Ithaca, NY: Comstock Publishing Associates.

Cordaux, R., A. Michel-Salzat, and D. Bouchon. 2001. *Wolbachia* infection in crustaceans: Novel hosts and potential routes for horizontal transmission. *Journal of Evolutionary Biology* 14:237–243.

Cuthill, I. C., H. C. Partridge, A. T. D. Bennett, S. C. Church, N. S. Hart, and N. Hunt. 2000. Ultraviolet vision in birds. *Advances in the Study of Behavior* 29:159–215.

Darwin, C. 1854. *A monograph on the sub-class Cirripedia, with figures of all the species. The Balanidae (or sessile cirripedes); the Verrucidae, etc., etc.* London: Ray Society.

————. 1871. *The descent of man, and selection in relation to sex.* London: Murray.

Diamond, J. M. 1975. Assembly of species communities. In *Ecology and evolution of communities*, ed. M. L. Cody and J. M. Diamond, 343–444. Cambridge, MA: Harvard University Press.

————. 1978. Niche shifts and the rediscovery of interspecific competition. *American Scientist* 66:322–331.

Dickinson, J. K. 1992. Egg cannibalism by larvae and adults of the milkweed leaf beetle (*Labidomera clivicollis*, Coleoptera: Chrysomelidae). *Ecological Entomology* 17:209–218.

Disney, R. H. L. 1994. *Scuttle flies: The Phoridae.* London: Chapman and Hall.

Dixon, A. F. G. 1985. *Aphid ecology.* Glasgow: Blackie.

Dodson, G. N., and N. W. Beck. 1993. Pre-copulatory guarding of penultimate females by male crab spiders, *Misumenoides formosipes. Animal Behaviour* 46:951–959.

Dodson, G. N., and A. T. Schwaab. 2001. Body size, leg autotomy, and prior experience as factors in the fighting success of male crab spiders, *Misumenoides formosipes. Journal of Insect Behavior* 14:841–855.

Dominey, W. J. 1980. Female mimicry in male bluegill sunfish—A genetic polymorphism? *Nature* 284:546–548.

Dondale, C. D., and J. H. Redner. 1978. *The insects and arachnids of Canada, Part 5. The crab spiders of Canada and Alaska.* Hull, Quebec: Canada Department of Agriculture.

Dong, Q., and G. A. Polis. 1992. The dynamics of cannibalistic populations: A foraging perspective. In *Cannibalism: Ecology and evolution among diverse taxa*, ed. M. A. Elgar and B. J. Crespi, 13–37. New York: Oxford University Press.

Dorris, P. R. 1970. Spiders collected from mud-dauber nests in Mississippi. *Journal of the Kansas Entomological Society* 43:10–11.

Double, M. C., and A. Cockburn. 2003. Subordinate superb fairy-wrens (*Malurus cyaneus*) parasitize the reproductive success of attractive dominant males. *Proceedings of the Royal Society of London B* 270:379–384.

Dudley, R. 2000. *The biomechanics of insect flight.* Princeton, NJ: Princeton University Press.

Dugatkin, L. A., and H. K. Reeve, eds. 1998. *Game theory and animal behavior.* New York: Oxford University Press.

Dukas, R., ed. 1998a. *Cognitive ecology.* Chicago: University of Chicago Press.

———. 1998b. Introduction. In *Cognitive ecology,* ed. R. Dukas, 1–19. Chicago: University of Chicago Press.

———. 1998c. Evolutionary ecology of learning. In *Cognitive ecology,* ed. R. Dukas, 129–174. Chicago: University of Chicago Press.

———. 1999. Ecological relevance of associative learning in fruit fly larvae. *Behavioral Ecology and Sociobiology* 45:195–200.

———. 2001. Effects of perceived danger on flower choice by bees. *Ecology Letters* 4:327–333.

———. 2002. Behavioural and ecological consequences of limited attention. *Philosophical Transactions of the Royal Society of London B* 357:1539–1547.

———. 2005. Bumblebee predators reduce bumblebee density and plant fitness. *Ecology* 86:1401–1406.

Dukas, R., and A. O. Mooers. 2003. Environmental enrichment improves mating success in fruit flies. *Animal Behaviour* 66:741–749.

Dukas, R., and D. H. Morse. 2003. Crab spiders affect flower visitation by bees. *Oikos* 101:157–163.

———. 2005. Crab spiders show mixed effects on flower visitation by bees and no effect on plant fitness. *Ecoscience* 12:244–247.

Dukas, R., D. H. Morse, and S. Myles. 2005. Experience levels of individuals in natural bee populations and their ecological implications. *Canadian Journal of Zoology* 83:492–497.

Dussourd, D. E., and R. E. Denno. 1994. Host-range of generalist caterpillars—trenching permits feeding on plants with secretory canals. *Ecology* 75:69–78.

Dussourd, D. E., and T. Eisner. 1987. Vein-cutting behavior: Insect counterplay to the latex defense of plants. *Science* 237:898–901.

Dyer, F. C. 1998. Spatial cognition: Lesson from central-place foraging insects. In *Animal cognition in nature,* ed. R. P. Balda, I. M. Pepperberg, and A. C. Kamil, 119–154. San Diego, CA: Academic Press.

Eaton, M. D., and S. M. Lanyon. 2003. The ubiquity of avian ultraviolet plumage reflectance. *Proceedings of the Royal Society of London B* 270:1721–1726.

Eberhard, W. G. 1996. *Female control: Sexual selection by cryptic female choice.* Princeton, NJ: Princeton University Press.

Edgar, W. D. 1971. Seasonal weight changes, age structure, natality and mortality in the wolf spider *Pardosa lugubris* Walck. in central Scotland. *Oikos* 22:84–92.

Egan, S. P., and D. H. Funk. 2006. Individual advantages to ecological specialization: Insights on cognitive constraints from three conspecific taxa. *Proceedings of the Royal Society of London B* 273:843–848.

Ehrman, L., and P. A. Parsons. 1981. *Behavior genetics and evolution.* New York: McGraw-Hill.

Elgar, M. A. 1992. Sexual cannibalism in spiders and other invertebrates. In *Cannibalism: Ecology and evolution among diverse taxa,* ed. M. A. Elgar and B. J. Crespi, 128–155. Oxford: Oxford University Press.

———. 1995. Duration of copulation in spiders: Comparative patterns. *Records of the Western Australia Museum* (Supplement) 52:1–11.

———. 1998. Sperm competition and sexual selection in spiders and other arachnids. In *Sperm competition and sexual selection*, ed. T. R. Birkhead and A. P. Møller, 307–339. San Diego, CA: Academic Press.

Elgar, M. A., and B. J. Crespi, eds. 1992. *Cannibalism: Ecology and evolution among diverse taxa*. Oxford: Oxford University Press.

Elgar, M. A., and D. R. Nash. 1988. Sexual cannibalism in the garden spider *Araneus diadematus*. *Animal Behaviour* 36:1511–1517.

Elgar, M. A., and J. M. Schneider. 2004. Evolutionary significance of sexual cannibalism. *Advances in the Study of Behavior* 34:135–163.

Emlen, J. M. 1966. The role of time and energy in food preference. *American Naturalist* 100:611–617.

Emlen, S. T., and L. W. Oring. 1977. Ecology, sexual selection, and the evolution of mating systems. *Science* 197:215–223.

Enders, F. 1974. Vertical stratification in orb-web spiders (Araneidae, Araneae) and a consideration of other methods of coexistence. *Ecology* 55:317–328.

Endler, J. A. 1986. Natural selection in the wild. *Monographs in Population Biology* 21:1–336.

Erickson, K. S., and D. H. Morse. 1997. Predator size and the suitability of a common prey. *Oecologia* 109:608–614.

Estes, J. A., and J. F. Palmisano. 1974. Sea otters: Their role in structuring nearshore communities. *Science* 185:1058–1060.

Evans, T. A. 1995. Two new species of social crab spiders of the genus *Diaea* from eastern Australia, their natural history and distribution. *Records of the Western Australian Museum* (Supplement) 52:151–158.

———. 1998. Factors influencing the evolution of social behaviour in Australian crab spiders (Araneae: Thomisidae). *Biological Journal of the Linnean Society* 63:205–219.

Evans, T. A., and M. A. D. Goodisman. 2002. Nestmate relatedness and population genetic structure of the Australian social crab spider *Diaea ergandros* (Araneae: Thomisidae). *Molecular Ecology* 11:2307–2316.

Falconer, D. S. 1989. *Introduction to quantitative genetics*. 3rd ed. London: Longman.

Fautin, R. W. 1946. Biotic communities of the northern desert shrub biome in western Utah. *Ecological Monographs* 16:251–310.

Fink, L. S. 1986. Costs and benefits of maternal behaviour in the green lynx spider (Oxyopidae, *Peucetia viridans*). *Animal Behaviour* 34:1051–1060.

Fisher, R. A. 1930. *The genetical theory of natural selection*. Oxford: Oxford University Press.

Fitzpatrick, M. J., Y. Ben-Sharar, H. M. Smid, L. E. M. Vet, G. E. Robinson, and M. B. Sokolowski. 2005. Candidate genes for behavioural ecology. *Trends in Ecology and Evolution* 20:96–104.

Foelix, R. F. 1985. Mechano- and chemoreceptive sensilla. In *Neurobiology of arachnids*, ed. F. G. Barth, 118–137. Berlin: Springer-Verlag.

———. 1996a. *Biology of spiders*. 2nd ed. New York: Oxford University Press.

———. 1996b. How do crab spiders (Thomisidae) bite their prey? *Revue Suisse de Zoologie, hors serie:* 203–210.

Foellmer, M. W., and D. J. Fairbairn. 2003. Spontaneous male death during copulation in an orb-weaving spider. *Proceedings of the Royal Society of London B* (Supplement) 270:S183–S185.

————. 2005a. Competing dwarf males: Sexual selection in an orb-weaving spider. *Journal of Evolutionary Biology* 18:629–641.

————. 2005b. Selection on male size, leg length and condition during mate search in a sexually highly dimorphic orb-weaving spider. *Oecologia* 142:653–662.

Ford, M. J. 1978. Locomotory activity and the predation strategy of the wolf-spider *Pardosa amentata* (Clerck) (Lycosidae). *Animal Behaviour* 26:31–35.

Fordyce, J. A., and S. B. Malcolm. 2000. Specialist weevil, *Rhyssomatus lineaticollis*, does not spatially avoid cardenolide defenses of common milkweed by ovipositing into pith tissue. *Journal of Chemical Ecology* 26:2857–2874.

Formanowicz, D. R. 1990. Antipredator efficacy of spider leg autotomy. *Animal Behaviour* 40:400–401.

Forrest, T. G. 1987. Insect size tactics and development strategies. *Oecologia* 73:178–184.

Forster, R. R., and L. M. Forster. 1973. *New Zealand spiders: An introduction*. Auckland: Collins.

Framenau, V. W., and C. J. Vink. 2002. Review of the wolf spider genus *Venatrix* Roewer (Araneae: Lycosidae). *Invertebrate Taxonomy* 15:927–970.

Frank, S. A. 1990. Sex allocation theory for birds and mammals. *Annual Review of Ecology and Systematics* 21:13–55.

Freeland, W. J. 1976. Pathogens and the evolution of primate sociality. *Biotropica* 8:12–24.

Fritz, R. S., and D. H. Morse. 1981. Nectar parasitism of *Asclepias syriaca* by ants: Effect on nectar levels, pollinia insertion, pollinaria removal and pod production. *Oecologia* 50:316–319.

————. 1985. Reproductive success, growth rate and foraging decisions of the crab spider *Misumena vatia*. *Oecologia* 65:194–200.

Fujiwara, M., P. Sengupta, and S. L. McIntire. 2002. Regulation of body size and behavioral state of *C. elegans* by sensory perception and the EGL-4cGMP-dependent protein kinase. *Neuron* 36:1091–1102.

Gabritschevsky, E. 1927. Experiments on color change and regeneration in the crab-spider, *Misumena vatia*. *Journal of Experimental Zoology* 47:251–267.

Gaskett, A. C., M. E. Herberstein, B. J. Downes, and M. A. Elgar. 2004. Changes in male mate choice in a sexually cannibalistic orb-web spider (Araneae: Araneidae). *Behaviour* 141:1197–1210.

Gauld, I., and B. Bolton, eds. 1988. *The Hymenoptera*. Oxford: Oxford University Press.

Gavrilets, S., G. Arnqvist, and U. Friberg. 2001. The evolution of female mate choice by sexual conflict. *Proceedings of the Royal Society of London Series B—Biological Sciences* 268:531–539.

Gertsch, W. J. 1939. A revision of the typical crab spiders (Misumeninae) of America north of Mexico. *Bulletin of the American Museum of Natural History* 76:277–442.

————. 1979. *American spiders*. 2nd ed. New York: Van Nostrand Reinhold.

Ghiselin, M. T. 1969. *The triumph of the Darwinian method*. Berkeley: University of California Press.

————. 1974. *The economy of nature and the evolution of sex*. Berkeley: University of California Press.

Gigerenzer, G., P. M. Todd, and the ABC Research Group. 1999. *Simple heuristics that make us smart.* New York: Oxford University Press.

Gillespie, R. G., and T. Caraco. 1987. Risk-sensitive foraging strategies of two spider populations. *Ecology* 68:887–899.

Giurfa, J., and M. Lehrer. 2001. Honeybee vision and floral displays: From detection to close-up recognition. In *Cognitive ecology of pollination*, ed. L. Chittka and J. D Thomson, 61–82. Cambridge: Cambridge University Press.

Godfray, W. C. F. 1994. *Parasitoids.* Princeton, NJ: Princeton University Press.

Gomulkiewicz, R. 1998. Game theory, optimization and quantitative genetics. In *Game theory and animal behavior*, ed. L. A. Dugatkin and H. K. Reeve, 283–303. New York: Oxford University Press.

Gould, S. J., and R. C. Lewontin. 1979. The spandrels of San Marco and the Panglossian paradigm: A critique of the adaptationist programme. *Proceedings of the Royal Society of London B* 205:581–598.

Grafen, A. 1988. On the uses of data on lifetime reproductive success. In *Reproductive success*, ed. T. H. Clutton-Brock, 454–471. Chicago: University of Chicago Press.

Gray, R. D. 1987. Faith and foraging: A critique of the "paradigm argument from design." In *Foraging behavior*, ed. A. C. Kamil, J. R. Krebs, and H. R. Pulliam, 69–142. New York: Plenum.

Greco, C. F., and P. G. Kevan. 1994. Contrasting patch choosing by anthophilous ambush predators: Vegetation and floral cues for decisions by a crab spider *(Misumena vatia)* and males and females of an ambush bug *(Phymata americana)*. *Canadian Journal of Zoology* 72:1583–1588.

Greco, C. F., P. Weeks, and P. G. Kevan. 1995. Patch choice in ambush predators: Plant height selection by *Misumena vatia* (Araneae, Thomisidae) and *Phymata americana* (Heteroptera, Phymatidae). *Ecoscience* 2:203–205.

Greenstone, M. H. 1990. Meteorological determinants of spider, ballooning: The roles of thermals vs. the vertical windspeed gradient in becoming airborne. *Oecologia* 84:164–168.

Gross, M. R. 1982. Sneakers, satellites and parentals: Polymorphic mating strategies in North American sunfishes. *Zeitschrift für Tierpsychologie* 60:1–26.

Grubb, T. C., Jr., and L. Greenwald. 1982. Sparrows and a brushpile: Foraging responses to different combinations of predation risk and energy cost. *Animal Behaviour* 30:637–640.

Gunnarsson, B. 1987. Sex ratio in the spider *Pityohyphantes phrygianus* affected by winter severity. *Journal of Zoology* 213:609–619.

———. 1988. Body size and survival: Implications for an overwintering spider. *Oikos* 52:274–282.

Gunnarsson, B., and A. Andersson. 1992. Skewed primary sex ratios in the solitary spider *Pityohyphantes phyrgianus. Evolution* 46:841–845.

Gunnarsson, B., G. Uhl, and K. Wallin. 2004. Variable female mating positions and offspring sex ratio in the spider *Pityohyphantes phrygianus* (Araneae: Linyphiidae). *Journal of Insect Behavior* 17:129–144.

Hagar, R., and R. A. Johnstone. 2003. The genetic basis of family conflict resolution in mice. *Nature* 421:533–535.

Haig, D. 2000. The kinship theory of genomic imprinting. *Annual Review of Ecology and Systematics* 31:9–32.

———. 2004. Genomic imprinting and kinship: How good is the evidence? *Annual Review of Genetics* 38:553–585.

Hamilton, W. D., and M. Zuk. 1982. Heritable true fitness and bright birds: A role for parasites. *Science* 218:384–387.

Hanlon, R. T., and J. B. Messenger. 1996. *Cephalopod behaviour*. Cambridge: Cambridge University Press.

Hanski, I., and D. Simberloff. 1997. The metapopulation approach, its history, conceptual domain, and application to conservation. In *Metapopulation biology*, ed. I. A. Hanski and M. E. Gilpin, 5–26. San Diego, CA: Academic Press.

Harland, D. P., and R. R. Jackson. 2004. *Portia* perceptions: The *Umwelt* of an araneophagic jumping spider. In *Complex worlds from simpler nervous systems*, ed. F. R. Prete, 5–40. Cambridge, MA: MIT Press.

Harvey, P. H., and M. D. Pagel. 1991. *The comparative method in evolutionary biology*. Oxford: Oxford University Press.

Haynes, D. L., and P. Sisojevic. 1966. Predatory behavior of *Philodromus rufus* Walckenaer (Araneae: Thomisidae). *Canadian Entomologist* 98:113–133.

Head, G. 1995. Selection on fecundity and variation in the degree of sexual size dimorphism among spider species (Class Araneae). *Evolution* 49:776–781.

Hedrick, A. V., and S. E. Riechert. 1989. Genetically-based variation between two spider populations in foraging behavior. *Oecologia* 80:533–539.

Heiling, A. M., K. Cheng, and M. E. Herberstein. 2004. Exploitation of floral signals by crab spiders (*Thomisus spectabilis*, Thomisidae). *Behavioral Ecology* 15:321–326.

Heiling, A. M., and M. E. Herberstein. 2004. Predator-prey coevolution: Australian native bees avoid their spider predators. *Proceedings of the Royal Society of London B* (Supplement) 271:S196–S198.

Heiling, A. M., M. E. Herberstein, and L. Chittka. 2003. Crab spiders manipulate flower signals. *Nature* 421:334.

Heinrich, B. 1979. *Bumblebee economics*. Cambridge, MA: Harvard University Press.

Herberstein, M. E., J. M. Schneider, and M. A. Elgar. 2002. Costs of courtship and mating in a sexually cannibalistic orb-web spider: Female mating strategies and their consequences for males. *Behavioral Ecology and Sociobiology* 51:440–446.

Hieber, C. S. 1992. Spider cocoons and their suspension systems as barriers to generalist and specialist predators. *Oecologia* 91:530–535.

Higgins, L. E. 1992. Developmental plasticity and fecundity in the orb-weaving spider *Nephila clavipes*. *Journal of Arachnology* 20:94–106.

———. 2000. The interaction of season length and development time alters size at maturity. *Oecologia* 122:51–59.

Higgins, L. E., and M. A. Rankin. 1996. Different pathways in arthropod postembryonic development. *Evolution* 50:573–582.

———. 2001. Mortality risk of rapid growth in the spider *Nephila clavipes*. *Functional Ecology* 15:24–28.

Hinton, H. E. 1976. Possible significance of the red patches of the female crabspider, *Misumena vatia*. *Journal of Zoology (London)* 180:35–39.

Hippa, H., and I. Oksala. 1979. Colour polymorphism of *Enoplognatha ovata* (Clerck) (Araneae, Theridiidae) in western Europe. *Hereditas* 90:203–212.

————. 1981. Polymorphism and reproductive strategies of *Enoplognatha ovata* (Clerck) (Araaneae, Theridiidae) in northern Europe. *Annales Zoologici Fennici* 18:179–190.

Hocking, B. 1968. Insect-flower associations in the high Arctic with special reference to nectar. *Oikos* 19:359–388.

Hoefler, C. D. 2002. Is contest experience a trump card? The interaction of residency status, experience, and body size on fighting success in *Misumenoides formosipes* (Araneae: Thomisidae). *Journal of Insect Behavior* 15:779–790.

Holdsworth, A. R., and D. H. Morse. 2000. Frequencies of male guarding and female aggression in the crab spider *Misumena vatia*. *American Midland Naturalist* 143:201–211.

Holl, A. 1987. Coloration and chromes. In *Ecophysiology of spiders*, ed. W. Nentwig, 16–25. Berlin: Springer-Verlag.

Holland, B., and W. R. Rice. 1998. Perspective: Chase-away sexual selection: Antagonistic seduction versus resistance. *Evolution* 52:1–7.

Homann, H. 1934. Beiträge zur Physiologie der Spinnenaugen. IV. Das Sehvermögen der Thomisiden. *Zeitschrift für Vergleichende Physiologie* 20:420–492.

————. 1949. Uber das Wachstum und die mechanischen Vorgänge bei der Häutung von *Tegenaria agrestis* (Araneae). *Zeitschrift für Vergleichende Physiologie* 31:413–440.

Hormiga, G., N. Scharff, and J. A. Coddington. 2000. The phylogenetic basis of sexual size dimorphism in orb-weaving spiders (Araneae, Orbiculariae). *Systematic Biology* 49:435–462.

Houle, D. 1991. Genetic covariance of fitness correlates: What genetic correlations are made of and why it matters. *Evolution* 45:630–648.

Houston, A. I., and J. M. McNamara. 1999. *Models of adaptive behaviour.* Cambridge: Cambridge University Press.

Hu, H. H., and D. H. Morse. 2004. The effect of age on encounters between male crab spiders. *Behavioral Ecology* 15:883–888.

Huber, B. A. 2005. Sexual selection research on spiders: Progress and biases. *Biological Reviews* 80:363–385.

Hufbauer, R. A., and R. B. Root. 2002. Interactive effects of different types of herbivore damage: *Trirhabda* beetle larvae and *Philaenus* spittlebugs on goldenrod *(Solidago altissima)*. *American Midland Naturalist* 147:204–213.

Hukusima, S., and M. Miyafuji. 1970. Life histories and habits of *Misumenops tricuspidatus* Fabricius (Araneae: Thomisidae). *Annual Report of the Society for Plant Protection of Northern Japan* 21:5–12. In Japanese, English summary.

Ingram, K. K., P. Oefner, and D. M. Gordon. 2005. Task-specific expression of the *foraging* gene in harvester ants. *Molecular Ecology* 14:813–818.

Jackson, R. R., and S. D. Pollard. 1996. Predatory behavior of jumping spiders. *Annual Review of Entomology* 41:287–308.

Jackson, R. R., P. W. Taylor, A. S. McGill, and S. D. Pollard. 1995. The web and prey-capture behaviour of *Diaea* sp., a crab spider (Thomisidae) from New Zealand. *Records of the Western Australian Museum* (Supplement) 52:33–37.

Johnson, J. C. 2005. Cohabitation of juvenile females with mature males promotes sexual cannibalism in fishing spiders. *Behavioral Ecology* 16:269–273.

Johnson, M. L., and S. Johnson. 1982. Voles. In *Wild mammals of North America,* ed. J. A. Chapman and G. A. Feldhamer, 326–354. Baltimore, MD: Johns Hopkins University Press.

Johnson, S. A., and E. M. Jakob. 1999. Leg autotomy in a spider has minimal costs in competitive ability and development. *Animal Behaviour* 57:957–965.

Kahneman, D. 2003. A perspective on judgment and choice. *American Psychologist* 58:697–720.

Kahneman, D., and A. Tversky. 1979. Prospect theory: An analysis of decisions under risk. *Econometrica* 47:263–291.

Kamil, A. C., J. R. Krebs, and H. R. Pulliam. 1987. The reproductive consequences of foraging. In *Foraging behavior*, ed. A. C. Kamil, J. R. Krebs, and H. R. Pulliam, 415. New York: Plenum.

Kareiva, P., D. H. Morse, and J. Eccleston. 1989. Stochastic prey arrivals and crab spider giving-up times: Simulations of spider performance using two simple "rules of thumb." *Oecologia* 78:542–549.

Kaston. B. J. 1948. Spiders of Connecticut. *Connecticut State Geological and Natural History Survey Bulletin* 70:1–874.

Keasar, T., U. Motro, Y. Shir, and A. Shmida. 1996. Overnight memory retention of foraging skills by bumblebees is imperfect. *Animal Behaviour* 52:95–104.

Kerr, D., and M. W. Feldman. 2003. Carving the cognitive niche: Optimal learning strategies in homogeneous and heterogeneous environments. *Journal of Theoretical Biology* 220:169–188.

Kevan, P. G. 1972. Heliotropism in some Arctic flowers. *Canadian Field-Naturalist* 86:41–44.

Kevan, P. G., and C. F. Greco. 2001. Contrasting patch choice behaviour by immature ambush predators, a spider *(Misumena vatia)* and an insect *(Phymata americana)*. *Ecological Entomology* 26:148–153.

Kingsolver, J. G., and D. W. Schemske. 1991. Path analyses of selection. *Trends in Ecology and Evolution* 6:276–280.

Kopeć, S. 1922. Studies on the necessity of the brain for the inception of insect metamorphosis. *Biological Bulletin* 42:323–342.

Krebs, J. R., J. C. Ryan, and E. L. Charnov. 1974. Hunting by expectation or optimal foraging? A study of patch choice by chickadees. *Animal Behaviour* 22:953–964.

Krebs, J. R., D. W. Stephens, and W. J. Sutherland. 1983. Perspectives in optimal foraging. In *Perspectives in ornithology*, ed. A. H. Brush and G. A. Clark, Jr., 165–216. New York: Cambridge University Press.

Krell, F.-T., and F. Krämer. 1998. Chemical attraction of crab spiders (Araneae, Thomisidae) to a flower fragrance component. *Journal of Arachnology* 26:117–119.

Kurczewski, F. E. 2003. Comparative nesting behavior of *Crabro monticola* (Hymenoptera: Sphecidae). *Northeastern Naturalist* 10:425–450.

Lack, D. 1968. *Ecological adaptations for breeding in birds*. London: Methuen.

———. 1974. *Evolution illustrated by waterfowl*. Oxford: Blackwell.

Land, M. F. 1985. The morphology and optics of spider eyes. In *Neurobiology of arachnids*, ed. F. G. Barth, 53–78. Berlin: Springer-Verlag.

Leech, R. E. 1966. The spiders (Araneida) of Hazen Camp 81°49′N, 71°W. *Quaestiones Entomologicae* 2:153–212.

LeGrand, R. S., and D. H. Morse. 2000. Factors driving extreme sexual size dimorphism of a sit-and-wait predator under low density. *Biological Journal of the Linnean Society* 71:643–664.

Lemon, W. C. 1991. Fitness consequences of foraging behaviour in the zebra finch. *Nature* 352:153–155.

———. 1993. Heritability of selectively advantageous foraging behavior in a small passerine. *Evolutionary Ecology* 7:421–428.

Lemon, W. C., and R. H. Barth, Jr. 1992. The effects of feeding rate on reproductive success in the zebra finch, *Taeniopygia guttata. Animal Behaviour* 44:851–857.

Leonard, A. S., and D. H. Morse. 2006. Line-following preferences of male crab spiders *Misumena vatia. Animal Behaviour* 71:717–724.

Lewontin, R. C. 1979. Fitness, survival, and optimality. In *Analysis of ecological systems*, ed. D. J. Horn, G. R. Stairs, and R. D. Mitchell, 3–21. Columbus: Ohio State University Press.

Li, C.-C. 1975. *Path analysis, a primer.* Pacific Grove, CA: Boxwood Press.

Li, D., and J. Zhao. 1991. Life history and biology of the spider, *Misumenops tricuspidatus*, and outlines for the classification of the life history of spiders. *Journal of Hubei University (Natural Science)* 13:170–174. In Chinese, English summary.

Lima, S. L. 1986. Predation risk and unpredictable feeding conditions: determinants of body mass in birds. *Ecology* 67:377–385.

———. 2002. Putting predators back into behavioral predator-prey interactions. *Trends in Ecology and Evolution* 17:70–75.

Lima, S. L., and L. M. Dill. 1990. Behavioral decisions made under the risk of predation: A review and prospectus. *Canadian Journal of Zoology* 68:619–640.

Lockley, T. C., O. P. Young, and J. L. Hayes. 1989. Nocturnal predation by *Misumena vatia* (Araneae, Thomisidae). *Journal of Arachnology* 17:249–251.

Louda, S. M. 1982. Inflorescence spiders: A cost/benefit analysis for the host plant, *Haplopappus venetus* (Asteraceae). *Oecologia* 55:185–191.

Lutzy, R. M., and D. H. Morse. MS. Effects of leg loss on male *Misumena vatia.*

MacArthur, R. H. 1955. Fluctuations of animal populations and a measure of community stability. *Ecology* 36:533–536.

MacArthur, R. H., and E. C. Pianka. 1966. On optimal use of a patchy environment. *American Naturalist* 100:603–609.

Magnhagen, C. 1991. Predation risk as a cost of reproduction. *Trends in Ecology and Evolution* 6:183–186.

Main, B. Y. 1976. *Spiders.* Sydney: Collins.

———. 1988. The biology of a social thomisid spider. *Miscellaneous Publication, Australian Entomological Society* 5:55–74.

Masumoto, T. 1993. The effect of the copulatory plug in the funnel-web spider *Agelena limbata* (Araneae, Agelenidae). *Journal of Arachnology* 21:55–59.

Maynard Smith, J. 1978. Optimization theory in evolution. *Annual Review of Ecology and Systematics* 9:31–56.

———. 1982. *Evolution and the theory of games.* Cambridge: Cambridge University Press.

Maynard Smith, J., and G. R. Price. 1973. The logic of animal conflict. *Nature* 246:15–18.

Mayntz, D., and S. Toft. 2006. Nutritional value of cannibalism and the role of starvation and nutrient imbalance for cannibalistic tendencies in a generalist predator. *Journal of Animal Ecology* 75:288–297.

McNab, B. K. 1971. On the ecological significance of Bergmann's rule. *Ecology* 52:845–854.

Mesnick, S. L., and B. J. LeBoeuf. 1991. Sexual behavior of male northern elephant seals. II. Female response to potentially injurious encounters. *Behaviour* 117:262–280.

Miller, R. S. 1967. Pattern and process in competition. *Advances in Ecological Research* 4:1–74.

Miyashita, K. 1969. Seasonal changes of population density and some characteristics of overwintering nymphs of *Lycosa T-insignita* BOES. et STR. (Araneae: Lycosidae). *Applied Entomology and Zoology* 4:1–8.

Mommsen, T. P. 1978. Digestive enzymes of a spider (*Tegenaria atrica* Koch). I. General remarks, digestion of proteins. *Comparative Biochemistry and Physiology* 60A:365–370.

Morse, D. H. 1968. A quantitative study of foraging of male and female spruce-woods warblers. *Ecology* 49:779–784.

———. 1970. Ecological aspects of some mixed-species foraging flocks of birds. *Ecological Monographs* 40:119–168.

———. 1973. Interactions between tit flocks and sparrowhawks *Accipiter nisus*. *Ibis* 115:591–593.

———. 1977a. Resource partitioning in bumble bees: The role of behavioral factors. *Science* 197:678–680.

———. 1977b. Foraging of bumble bees: The effect of other individuals. *Journal of the New York Entomological Society* 85:240–248.

———. 1979. Prey capture by the crab spider *Misumena calycina* (Araneae, Thomisidae). *Oecologia* 39:309–319.

———. 1980. *Behavioral mechanisms in ecology.* Cambridge, MA: Harvard University Press.

———. 1981. Prey capture by the crab spider *Misumena vatia* (L.) (Thomisidae) on three common native flowers. *American Midland Naturalist* 105:358–367.

———. 1982a. The turnover of milkweed pollinia on bumble bees, and implications for outcrossing. *Oecologia* 53:187–196.

———. 1982b. Behavior and ecology of bumble bees. In *Social insects*, vol. 3, ed. H. R. Hermann, 245–322. New York: Academic Press.

———. 1983. Foraging patterns and time budgets of the crab spiders *Xysticus emertoni* Keyserling and *Misumena vatia* (Clerck) (Araneae: Thomisidae) on flowers. *Journal of Arachnology* 11:87–94.

———. 1984. How crab spiders (Araneae, Thomisidae) hunt at flowers. *Journal of Arachnology* 12:307–316.

———. 1985a. Milkweeds and their visitors. *Scientific American* 253(1):112–119.

———. 1985b. Nests and nest-site selection of the crab spider *Misumena vatia* (Araneae, Thomisidae) on milkweed. *Journal of Arachnology* 13:383–390.

———. 1986a. Predatory risk to insects foraging at flowers. *Oikos* 46:223–228.

———. 1986b. Foraging behavior of crab spiders (*Misumena vatia*) hunting on inflorescences of different quality. *American Midland Naturalist* 116:341–347.

————. 1986c. Inflorescence choice and time allocation by insects foraging on milkweed. *Oikos* 46:229–236.

————. 1987. Attendance patterns, prey capture, changes in mass, and survival of crab spiders *Misumena vatia* (Araneae, Thomisidae) guarding their nests. *Journal of Arachnology* 15:193–204.

————. 1988a. Relationship between crab spider *Misumena vatia* nesting success and earlier patch-choice decisions. *Ecology* 69:1970–1973.

————. 1988b. Cues associated with patch-choice decisions by foraging crab spiders *Misumena vatia*. *Behaviour* 107:297–313.

————. 1988c. Interactions between the crab spider *Misumena vatia* (Clerck) (Araneae) and its ichneumonid egg predator *Trychosis cyperia* Townes (Hymenoptera). *Journal of Arachnology* 16:132–135.

————. 1989. Nest acceptance by the crab spider *Misumena vatia* (Araneae, Thomisidae). *Journal of Arachnology* 17:49–57.

————. 1990. Leaf choices of nest-building crab spiders *(Misumena vatia)*. *Behavioral Ecology and Sociobiology* 27:265–267.

————. 1991. Homing by crab spiders *Misumena vatia* (Araneae, Thomisidae) separated from their nests. *Journal of Arachnology* 19:111–114.

————. 1992a. Predation on dispersing spiderlings *Misumena vatia* and its relationship to their mothers' earlier patch-choice decisions. *Ecology* 73:1814–1819.

————. 1992b. Dispersal of the spiderlings of *Xysticus emertoni* (Araneae, Thomisidae), a litter-dwelling crab spider. *Journal of Arachnology* 20:217–221.

————. 1993a. Placement of crab spider *(Misumena vatia)* nests in relation to their spiderlings' hunting sites. *American Midland Naturalist* 129:241–247.

————. 1993b. Some determinants of dispersal by crab spiderlings. *Ecology* 74:427–432.

————. 1993c. Choosing hunting sites with little information: Patch-choice responses of crab spiders to distant cues. *Behavioral Ecology* 4:61–65.

————. 1994. Numbers of broods produced by the crab spider *Misumena vatia* (Araneae, Thomisidae). *Journal of Arachnology* 22:195–199.

————. 1995. Changes in biomass of penultimate-instar crab spiders *Misumena vatia* (Araneae, Thomisidae) hunting on flowers late in the summer. *Journal of Arachnology* 23:85–90.

————. 1997. Distribution, movement, and activity patterns of an intertidal wolf spider *Pardosa lapidicina* population (Araneae, Lycosidae). *Journal of Arachnology* 25:1–10.

————. 1998. The effect of wounds on desiccation of prey: Implications for a predator with extra-oral digestion. *Oecologia* 115:184–187.

————. 1999a. Location of successful strikes on prey by juvenile crab spiders *Misumena vatia* (Araneae, Thomisidae). *Journal of Arachnology* 27:171–175.

————. 1999b. Choice of hunting site as a consequence of experience in late-instar crab spiders. *Oecologia* 120:252–257.

————. 2000a. Flower choice by naive young crab spiders and the effect of subsequent experience. *Animal Behaviour* 59:943–951.

————. 2000b. The effect of experience on the hunting success of newly emerged spiderlings. *Animal Behaviour* 60:827–835.

————. 2000c. The role of experience in determining patch-use by adult crab spiders. *Behaviour* 137:265–278.

———. 2001. Harvestmen as commensals of crab spiders. *Journal of Arachnology* 29:273–275.

———. 2004. A test of sexual cannibalism models, using a sit-and-wait predator. *Biological Journal of the Linnean Society* 81:427–437.

———. 2005. Initial responses to substrates by naive spiderlings: Single and simultaneous choices. *Animal Behaviour* 70:319–328.

———. 2006. Fine-scale substrate use by a small sit-and-wait predator. *Behavioral Ecology* 17:405–409.

Morse, D. H., and R. S. Fritz. 1982. Experimental and observational studies of patch-choice at different scales by the crab spider *Misumena vatia*. *Ecology* 63:172–182.

———. 1983. Contributions of diurnal and nocturnal insects to the pollination of common milkweed (*Asclepias syriaca* L.) in a pollen-limited system. *Oecologia* 60:190–197.

———. 1987. The consequences of foraging for reproductive success. In *Foraging behavior*, ed. A. C. Kamil, J. R. Krebs, and H. R. Pulliam, 443–455. New York: Plenum.

———. 1989. Milkweed pollinia and predation risk to flower-visiting insects by the crab spider *Misumena vatia*. *American Midland Naturalist* 121:188–193.

Morse, D. H., and H. H. Hu. 2004. Age-dependent cannibalism of male crab spiders *Misumena vatia*. *American Midland Naturalist* 151:318–325.

Morse, D. H., and E. G. Stephens. 1996. The consequences of adult foraging success on the components of lifetime fitness in a semelparous, sit and wait predator. *Evolutionary Ecology* 10:361–373.

Morton, E. S., and M. D. Shalter. 1977. Vocal response to predators in pair-bonded Carolina wrens. *Condor* 79:222–227.

Moya-Laraño, J., J. Halaj, and D. H. Wise. 2002. Climbing to reach females: Romeo should be small. *Evolution* 56:420–425.

Muma, M. H., and W. F. Jeffers. 1945. Studies of the spider prey of several mud-dauber wasps. *Annals of the Entomological Society of America* 38:245–255.

Muniappan, R., and H. L. Chada. 1970. Biology of the crab spider, *Misumenops celer*. *Annals of the Entomological Society of America* 63:1718–1722.

Nakamura, K. 1982. Prey capture tactics of spiders: An analysis based on a simulation for spider's growth. *Researches in Population Ecology* 24:302–317.

Nentwig, W. 1986. Non-webbuilding spiders: Prey specialists or generalists? *Oecologia* 69:571–576.

Newman, J. A., and M. A. Elgar. 1991. Sexual cannibalism in orb-weaving spiders: An economic model. *American Naturalist* 138:1372–1395.

Nijhout, H. F. 1994. *Insect hormones*. Princeton, NJ: Princeton University Press.

Nishimura, K., and Y. Isoda. 2004. Evolution of cannibalism: Referring to costs of cannibalism. *Journal of Theoretical Biology* 226:291–300.

Noordwijk, A. J. van, and G. De Jong. 1986. Acquisition and allocation of resources, their influence on variation in life history tactics. *American Naturalist* 128:137–142.

Nowak, M. A., and K. Sigmund. 2004. Evolutionary dynamics of biological games. *Science* 303:793–799.

Nyffeler, M., and G. Benz. 1979. Nischenüberlappung bezüglich der Raum-und Nahrungskomponenten bei Krabbenspinnen (Araneae: Thomisidae) und

Wolfspinnen (Araneae: Lycosidae) in Mähwiesen. *Revue Suisse Zoologie* 86:855–865.

Nyffeler, M., and R. G. Breene. 1990. Spiders associated with selected European hay meadows, and the effects of habitat disturbance, with the predation ecology of the crab spiders, *Xysticus* spp. (Araneae, Thomisidae). *Journal of Applied Entomology* 110:149–159.

Oaten, A. 1977. Optimal foraging in patches: A case for stochasticity. *Theoretical Population Biology* 12:263–285.

O'Brien, W. N., N. A. Slade, and G. L. Vinyard. 1976. Apparent size as the determinant of prey selection by bluegill sunfish *(Lepomis macrochirus)*. *Ecology* 57:1304–1310.

Oh, H. W., M. G. Kim, S. W. Shin, K. S. Bae, Y. S. Ahn, and H. Y. Park. 2000. Ultrastructural and molecular identification of a *Wolbachia* endosymbiont in a spider, *Nephila clavata*. *Insect Molecular Biology* 9:539–543.

Ollason, J. G. 1980. Learning to forage—optimally? *Theoretical Population Biology* 18:44–56.

Osaki, S. 1996. Spider silk as mechanical lifeline. *Nature* 384:419.

———. 1999. Is the mechanical strength of spider's drag-lines reasonable as lifeline? *International Journal of Biological Macromolecules* 24:283–287.

Osborne, K. A., A. Robichon, E. Burgess, S. Butland, R. A. Shaw, A. Coulthard, H. S. Pereira, R. J. Greenspan, and M. B. Sokolowski. 1997. Natural behavior polymorphism due to a cGMP-dependent protein kinase of *Drosophila*. *Science* 277:834–836.

Oxford, G. S. 1976. The colour polymorphism in *Enoplognatha ovata* (Clerck) (Araneae: Theridiidae)—temporal stability and spatial variability. *Heredity* 36:369–381.

———. 1983. Genetics of colour and its regulation during development in the spider *Enoplognatha ovata* (Clerck) (Araneae: Theridiidae). *Heredity* 51:621–634.

———. 1985. Geographical distribution of phenotypes regulating pigmentation in the spider *Enoplognatha ovata* (Clerck) (Araneae: Theridiidae). *Heredity* 55:37–45.

———. 2005. Genetic drift within a protected polymorphism: Enigmatic variation in color-morph frequencies in the candy-stripe spider, *Enoplognatha ovata*. *Evolution* 59:2170–2184.

Oxford, G. S., and P. R. Reillo. 1994. The world distribution of species within the *Enoplognatha ovata* group (Araneae: Theridiidae): Implications for their evolution and for previous research. *Bulletin of the British Arachnological Society* 9:226–232.

Oxford, G. S., and M. W. Shaw. 1986. Long-term variation in color-morph frequencies in the spider *Enoplognatha ovata* (Clerck) (Araneae: Theridiidae): Natural selection, migration and intermittent drift. *Biological Journal of the Linnean Society* 27:225–249.

Packard, A. 1905. Change of color and protective coloration in *Misumena vatia*. *Journal of the New York Entomological Society* 13:85–96.

Paine, R. T. 1966. Food web complexity and species diversity. *American Naturalist* 100:65–75.

———. 1980. Food webs: Linkage, interaction strength, and community infrastructure. *Journal of Animal Ecology* 49:667–685.

Palmgren, P. 1972. Studies on the spider populations of the surroundings of the Tvärminne Zoological Station, Finland. *Commentationes Biologicae* 52:1–133.

Papaj, D. R., and A. C. Lewis, eds. 1993. *Insect learning.* New York: Chapman and Hall.

Parker, G. A. 1970. Sperm competition and its evolutionary consequences in the insects. *Biological Reviews* 45:525–567.

———. 1974. Courtship persistence and female guarding as male time investment strategies. *Behaviour* 48:157–184.

———. 1979. Sexual selection and sexual conflict. In *Sexual selection and reproductive competition in insects,* ed. M. S. Blum and N. A. Blum, 123–166. New York: Academic Press.

Partridge, L., and P. H. Harvey. 1988. The ecological context of life history evolution. *Science* 241:1449–1455.

Partridge, L., and L. D. Hurst. 1998. Sex and conflict. *Science* 281:2003–2008.

Persons, M. H., S. E. Walker, A. L. Rypstra, and S. D. Marshall. 2001. Wolf spider predator avoidance tactics and survival in the presence of diet-associated predator cues (Araneae: Lycosidae). *Animal Behaviour* 61:43–51.

Petrie, M. 1994. Improved growth and survival of offspring of peacocks with more elaborate trains. *Nature* 371:598–599.

Pfennig, D. W. 1997. Kinship and cannibalism: Understanding why animals avoid preying on relatives offers insights into the evolution of social behavior. *Bioscience* 47:667–675.

Platnick, N. I. 2004. *The world spider catalog, version 5.0.* http://research.amnh.org/entomology. New York: American Museum of Natural History.

Pollard, S. D. 1989. Constraints affecting partial prey consumption by a crab spider, *Diaea* sp. indet. (Araneae: Thomisidae). *Oecologia* 81:392–396.

———. 1990. The feeding strategy of a crab spider, *Diaea* sp. indet. (Araneae: Thomisidae): Post-capture decision rules. *Journal of Zoology* 222:601–615.

Pollard, S. D., M. W. Beck, and G. N. Dodson.1995. Why do male crab spiders drink nectar? *Animal Behaviour* 49:1443–1448.

Pollard, S. D., A. M. MacNab, and R. R. Jackson. 1987. Communication with chemicals: Pheromones and spiders. In *Ecophysiology of spiders,* ed. W. Nentwig, 133–141. Berlin: Springer-Verlag.

Poole, R. W., and B. J. Rathcke. 1979. Regularity, randomness, and aggregation in flowering phenologies. *Science* 203:470–471.

Pound, J. M., and J. H. Oliver, Jr. 1979. Juvenile hormone: Evidence of its role in the reproduction of ticks. *Science* 206:355–357.

Pourié, G., and M. Trabalon. 2003. The role of 20-hydroxyecdysone on the control of spider vitellogenesis. *General and Comparative Endocrinology* 131:250–257.

Preisser, E. L., D. I. Bolnick, and M. F. Benard. 2005. Scared to death? The effects of intimidation and consumption in predator-prey interactions. *Ecology* 86:501–509.

Prenter, J., R. W. Elwood, and W. I. Montgomery. 1999. Sexual size dimorphism and reproductive investment by female spiders: A comparative analysis. *Evolution* 53:1987–1994.

Prenter, J., C. MacNeil, and R. W. Elwood. 2006. Sexual cannibalism and mate choice. *Animal Behaviour* 71:481–490.

Proctor, M., and P. Yeo. 1973. *The pollination of flowers*. London: Collins.

Prout, T. 1971. The relation between fitness components and population prediction in *Drosophila*. I: The estimation of fitness components. *Genetics* 68:127–149.

Pulliam, H. R., and G. S. Mills. 1977. Use of space by wintering sparrows. *Ecology* 58:1393–1399.

Punzo, F. 1997. Leg autotomy and avoidance behavior in response to a predator in the wolf spider, *Schizocosa avida* (Araneae, Lycosidae). *Journal of Arachnology* 25:202–205.

Pyke, G. H. 1984. Optimal foraging theory: A critical review. *Annual Review of Ecology and Systematics* 15:523–575.

Pyke, G. H., H. R. Pulliam, and E. L. Charnov. 1977. Optimal foraging: A selective review of theory and tests. *Quarterly Review of Biology* 52:137–154.

Ramos, M., J. A. Coddington, T. E. Christenson, and D. J. Irschick. 2005. Have male and female genitalia coevolved? A phylogenetic analysis of genitalic morphology and sexual size dimorphism in web-building spiders (Araneae: Araneoidea). *Evolution* 59:1989–1999.

Ratner, R. K., B. E. Kahn, and D. Kahneman. 1999. Choosing less-preferred experiences for the sake of variety. *Journal of Consumer Research* 26:1–15.

Read, D., G. Antomides, L. van den Ouden, and H. Trienekens. 2001. Which is better: Simultaneous or sequential choice? *Organizational Behavior and Human Decision Processes* 84:54–70.

Rehnberg, B. G. 1987. Selection of spider prey by *Trypoxylon politum* (Say) (Hymenoptera: Sphecidae). *Canadian Entomologist* 119:189–194.

Reillo, P. R., and D. H. Wise. 1988a. Genetics of color expression in the spider *Enoplognatha ovata* (Araneae, Theridiidae) from coastal Maine. *American Midland Naturalist* 119:318–326.

———. 1988b. An experimental evaluation of selection on color morphs of the polymorphic spider *Enoplognatha ovata* (Araneae: Theridiidae). *Evolution* 42:1172–1189.

Reissland, A., and P. Görner. 1985. Trichobothria. In *Neurobiology of arachnids*, ed. F. G. Barth, 138–161. Berlin: Springer-Verlag.

Reznick, D. 1985. Costs of reproduction—an evaluation of the empirical evidence. *Oikos* 44:257–267.

Rice, W. R. 1996. Sexually antagonistic male adaptation triggered by experimental arrest of female evolution. *Nature* 381:232–234.

———. 2000. Dangerous liaisons. *Proceedings of the National Academy of Sciences USA* 97:12953–12955.

Richter, M. R., and K. D. Waddington. 1993. Past foraging experience influences honey bee dance behaviour. *Animal Behaviour* 46:123–128.

Riechert, S. E. 1982. Spider interaction strategies: Communication versus coercion. In *Spider communication: Mechanisms and ecological significance*, ed. P. N. Witt and J. Rovner, 281–313. Princeton, NJ: Princeton University Press.

———. 1991. Prey abundance vs. diet breadth in a spider test system. *Evolutionary Ecology* 5:327–338.

———. 1993. Investigation of potential gene flow limitation of behavioral adaptation in an aridlands spider. *Behavioral Ecology and Sociobiology* 32:355–363.

Riechert, S. E., and R. F. Hall. 2000. Local population success in heterogeneous habitats: Reciprocal transplant experiments completed on a desert spider. *Journal of Evolutionary Biology* 13:541–550.

Riechert, S. E., and A. V. Hedrick. 1990. Levels of predation and genetically based anti-predator behaviour in the spider, *Agelenopsis aperta*. *Animal Behaviour* 40:679–687.

Ritchie, M. E. 1988. Individual variation in the ability of Columbian ground-squirrels to select an optimal diet. *Evolutionary Ecology* 2:232–252.

———. 1990. Optimal foraging and fitness in Columbian ground squirrels. *Oecologia* 82:56–67.

———. 1991. Inheritance of optimal foraging behaviour in Columbian ground squirrels. *Evolutionary Ecology* 5:146–159.

Robakiewicz, P., and J. E. Robbins. 2001. Oviposition site choice in Harris' checkerspot, *Charidryas harrisii* (Nymphalidae). *Northeastern Naturalist* 8:293–300.

Roberts, J. A., P. W. Taylor, and G. W. Uetz. 2003. Kinship and food availability influence cannibalism tendency in early-instar wolf spiders (Araneae: Lycosidae). *Behavioral Ecology and Sociobiology* 54:416–422.

Robertson, I. C., and D. K. Maguire. 2005. Crab spiders deter insect visitations to slickspot peppergrass flowers. *Oikos* 109:577–582.

Roff, D. A. 1992. *The evolution of life histories*. New York: Chapman and Hall.

Romero, G. Q., and J. Vasconcellos-Neto. 2004. Beneficial effects of flower-dwelling predators on their host plant. *Ecology* 85:446–457.

Rose, M. R., and G. V. Lauder. 1996. Post-spandrel adaptationism. In *Adaptation*, ed. M. R. Rose and G. V. Lauder, 1–8. San Diego, CA: Academic Press.

Rowley, S. M., R. J. Raven, and E. A. McGraw. 2004. *Wolbachia pipientis* in Australian spiders. *Current Microbiology* 49:208–214.

Salmon, J. T., and N. V. Horner. 1977. Aerial dispersion of spiders in north central Texas. *Journal of Arachnology* 5:153–157.

Sasaki, T., and O. Iwahashi. 1995. Sexual cannibalism in an orb-weaving spider *Argiope aemula*. *Animal Behaviour* 49:1119–1121.

Schaefer, M. 1977. Winter ecology of spiders (Araneida). *Zeitschrift für Angewandte Entomologie* 83:113–134.

Schausberger, P., and B. A. Croft. 2001. Kin recognition and larval cannibalism by adult females in specialist predaceous mites. *Animal Behavior* 61:459–464.

Scheiner, S. M., R. L. Caplan, and R. F. Lyman. 1989. A search for trade-offs among life history traits in *Drosophila melanogaster*. *Evolutionary Ecology* 3:51–63.

Schmalhofer, V. R. 1999. Thermal tolerances and preferences of the crab spiders *Misumenops asperatus* and *Misumenoides formosipes* (Araneae, Thomisidae). *Journal of Arachnology* 27:470–480.

———. 2000. Diet-induced and morphological color changes in juvenile crab spiders (Araneae, Thomisidae). *Journal of Arachnology* 28:56–60.

Schmalhofer, V. R., and T. M. Casey. 1999. Crab spider hunting performance is temperature insensitive. *Ecological Entomology* 24:345–353.

Schmitz, O. J. 1997. Commemorating 30 years of optimal foraging theory. *Evolutionary Ecology* 11:631–632.

Schmitz, O. J., V. Krivan, and O. Ovadia. 2004. Trophic cascades: The primacy of trait-mediated indirect interactions. *Ecology Letters* 7:153–163.

Schneider, J. M., and Y. Lubin. 1997. Does high adult mortality explain semelparity in the spider *Stegodyphus lineatus* (Eresidae)? *Oikos* 79:92–100.

Schneider, J. M., M. Salomon, and Y. Lubin. 2003. Limited adaptive life-history plasticity in a semelparous spider, *Stegodypus lineatus* (Eresidae). *Evolutionary Ecology Research* 5:731–738.

Schoener, T. W. 1987. A brief history of optimal foraging ecology. In *Foraging behavior*, ed. A. C. Kamil, J. R. Krebs, and H. R. Pulliam, 5–67. New York: Plenum.

Schoener, T. W., and C. A. Toft. 1983. Spider populations—extraordinarily high densities on islands without top predators. *Science* 219:1353–1355.

Seger, J., and J. W. Stubblefield. 1996. Optimization and adaptation. In *Adaptation*, ed. M. R. Rose and G. V. Lauder, 93–123. San Diego, CA: Academic Press.

Selander, R. K. 1966. Sexual dimorphism and differential niche utilization in birds. *Condor* 68:113–151.

Serventy, D. L. 1971. Biology of desert birds. In *Avian biology*, vol. 1, ed. D. S. Farner and J. R. King, 287–339. New York: Academic Press.

Shaw, M. R. 1994. Parasitoid host ranges. In *Parasitoid community ecology*, ed. B. A. Hawkins and W. Sheehan, 111–144. Oxford: Oxford University Press.

Shear, W. A. 1986. The evolution of web-building behavior in spiders: A third generation of hypotheses. In *Spiders: Webs, behavior, and evolution*, ed. W. A. Shear, 364–400. Stanford, CA: Stanford University Press.

Shettleworth, S. J. 1998. *Cognition, evolution, and behavior*. New York: Oxford University Press.

Shuster, S. M., and M. J. Wade. 2003. *Mating systems and strategies*. Princeton, NJ: Princeton University Press.

Simmons, L. W. 2001. *Sperm competition and its evolutionary consequences in the insects*. Princeton, NJ: Princeton University Press.

Simmons, L. W., T. Llorens, M. Schinzig, D. Hosken, and M. Craig. 1994. Sperm competition selects for male mate choice and protandry in the bushcricket, *Requena verticalis* (Orthoptera, Tettigoniidae). *Animal Behaviour* 47:117–122.

Simonson, I. 1990. The effect of purchase quantity and timing on variety seeking behavior. *Journal of Marketing Research* 32:150–162.

Simpson, G. G. 1944. *Tempo and mode in evolution*. New York: Columbia University Press.

Smith, R. B., and T. P. Mommsen. 1984. Pollen feeding in an orb-weaving spider. *Science* 226:1330–1332.

Smithe, F. B. 1975. *Naturalist's color guide*. New York: American Museum of Natural History.

Sokolowski, M. B. 1980. Foraging strategies of *Drosophila melanogaster:* A chromosomal analysis. *Behavior Genetics* 10:291–302.

———. 1998. Genes for normal behavioral variation: Recent clues from flies and worms. *Neuron* 21:463–466.

Spaethe, J., J. Tautz, and L. Chittka. 2001. Visual constraints in foraging bumblebees: Flower size and color affect search time and flight behavior. *Proceedings of the National Academy of Sciences U.S.A.* 98:3898–3903.

Stearns, S. C. 1992. *The evolution of life histories*. New York: Oxford University Press.

Stephens, D. W., and K. C. Clements. 1998. Game theory and learning. In *Game theory and animal behavior*, ed. L. A Dugatkin and H. K. Reeve, 239–260. New York: Oxford University Press.

Stephens, D. W., and J. R. Krebs. 1986. *Foraging theory.* Princeton, NJ: Princeton University Press.

Stephens, D. W., J. F. Lynch, A. E. Sorensen, and C. Gordon. 1986. Preference and profitability: Theory and experiment. *American Naturalist* 127:533–553.

Sullivan, D. J., and W. Völkl. 1998. Hyperparasitism: Multitrophic ecology and behavior. *Annual Review of Entomology* 44:291–315.

Sullivan, H. L., and D. H. Morse. 2004. The movement and activity patterns of similar-sized adult and juvenile crab spiders *Misumena vatia* (Araneae: Thomisidae). *Journal of Arachnology* 32:276–283.

Suter, R. B. 1990. Courtship and the assessment of virginity by male bowl and doily spiders. *Animal Behaviour* 39:307–313.

———. 1999. An aerial lottery: The physics of ballooning in a chaotic atmosphere. *Journal of Arachnology* 27:281–293.

Sutherland, W. J. 2002. Conservation biology—science, sex and the kakapo. *Nature* 419:265–266.

Suttle, K. B. 2003. Pollinator-mediated indirect effects: Do predator effects cascade through pollinators to plants? *Ecology Letters* 6:688–694.

Tallamy, D. W., and W. P. Brown. 1999. Semelparity and the evolution of maternal care in insects. *Animal Behaviour* 57:727–730.

Théry, M., and J. Casas. 2002. Visual systems: Predator and prey views of spider camouflage. *Nature* 415:133.

Théry, M., M. Debut, D. Gomez, and J. Casas. 2004. Specific color sensitivities of prey and predator explain camouflage in different visual systems. *Behavioral Ecology* 16:25–29.

Thornhill, R. 1983. Cryptic female choice and its implications in the scorpionfly *Harpobittacus nigriceps. American Naturalist* 122:765–788.

Thornhill, R., and J. Alcock. 1983. *The evolution of insect mating systems.* Cambridge, MA: Harvard University Press.

Tietjen, W. J. 1977. Dragline-following by male lycosid spiders. *Psyche* 84:165–178.

———. 1979. Tests for olfactory communication in four species of wolf spiders (Araneae, Lycosidae). *Journal of Arachnology* 7:197–206.

Tietjen, W. J., and J. S. Rovner. 1980. Physico-chemical trail-following behaviour in two species of wolf spiders: Sensory and etho-ecological concomitants. *Animal Behaviour* 28:735–741.

———. 1982. Chemical communication in lycosids and other spiders. In *Spider communication: Mechanisms and ecological significance*, ed. P. N. Witt and J. S. Rovner, 249–279. Princeton, NJ: Princeton University Press.

Tolbert, W. W. 1977. Aerial dispersal behavior of two orb-weaving spiders. *Psyche* 84:13–27.

Tovée, M. J. 1995. Ultra-violet photoreceptors in the animal kingdom: Their distribution and function. *Trends in Ecology and Evolution* 10:455–460.

Townes, H., and M. Townes. 1962. Ichneumon-flies of America north of Mexico: 3. Subfamily Gelinae, Tribe Mesosternini. *Bulletin of the United States National Museum* 216(3):1–602.

Trabalon, M., G. Pourié, and N. Hartmann. 1998. Relationships among cannibalism, contact signals, ovarian development and ecdysteroid levels in *Tegenaria atrica* (Araneae, Agelenidae). *Insect Biochemistry and Molecular Biology* 28:751–758.

Tullock, G. 1971. The coal tit as a careful shopper. *American Naturalist* 105:77–80.

Turlings, T. C. J., F. L. Wackers, L. E. M. Vit, W. J. Lewis, and J. M. Tumlinson. 1993. Learning of host-finding cues by hymenopterous parasitoids. In *Insect learning*, ed. D. R. Papaj and A. C. Lewis, 51–78. New York: Chapman and Hall.

Uetz, G. W., J. Bischoff, and J. Raver. 1992. Survivorship of wolf spiders (Lycosidae) reared on different diets. *Journal of Arachnology* 20:207–211.

Uhl, G., S. Schmitt, M. A. Schäfer, and W. Blanckenhorn. 2004. Food and sex-specific growth strategies in a spider. *Evolutionary Ecology Research* 6:523–540.

Vásquez, D. P., and D. Simberloff. 2003. Changes in interaction biodiversity induced by an introduced ungulate. *Ecology Letters* 6:1077–1083.

Vogelei, A., and R. Greissl. 1989. Survival strategies of the crab spider *Thomisus onustus* Walckenaer 1806 (Chelicerata, Arachnida, Thomisidae). *Oecologia* 80:513–515.

Vollrath, F. 1986. Eusociality and extraordinary sex ratios in the spider *Anelosimus eximius* (Araneae: Theridiidae). *Behavioral Ecology and Sociobiology* 18:283–287.

———. 1998. Dwarf males. *Trends in Ecology and Evolution* 13:159–163.

Vollrath, F., and G. A. Parker. 1992. Sexual dimorphism and distorted sex ratios in spiders. *Nature* 360:156–159.

Waage, J. K. 1979. Dual function of the damselfly penis: Sperm removal and transfer. *Science* 203:916–918.

Wade, M. J., and S. Kalisz. 1990. The causes of natural selection. *Evolution* 44:1947–1955.

Wagner, J. D., and D. H. Wise. 1996. Cannibalism regulates densities of young wolf spiders: Evidence from field and laboratory experiments. *Ecology* 77:639–652.

Watson, P. J. 1990. Female-enhanced male competition determines the first mate and principal sire in the spider *Linyphia litigiosa* (Linyphiidae). *Behavioral Ecology and Sociobiology* 26:77–90.

Weatherhead, P. J., and R. J. Robertson. 1979. Offspring quality and the polygyny threshold: The "sexy son" hypothesis. *American Naturalist* 113: 201–208.

Weigel, G. 1941. Färbung und Färbwechsel der Krabbespinne *Misumena vatia*. *Zeitschrift für Vergleichende Physiologie* 29:195–248.

Werner, E. E., and S. D. Peacor. 2003. A review of trait-mediated indirect interactions in ecological communities. *Ecology* 84:1083–1100.

Westneat, D. R., P. W. Sherman, and M. L. Morton. 1990. The ecology and evolution of extra-pair copulations in birds. *Current Ornithology* 7:330–369.

Weyman, G. S., K. D. Sunderland, and P. C. Jepson. 2002. A review of the evolution and mechanisms of ballooning by spiders inhabiting arable farmland. *Ethology, Ecology and Evolution* 14:307–326.

White, M. D. J. 1973. *Animal cytology and evolution*. 3rd ed. Cambridge: Cambridge University Press.

Wigglesworth, V. B. 1934. The physiology of ecdysis in *Rhodnius prolixus*

(Hemiptera). II. Factors controlling molting and "metamorphosis." *Quarterly Journal of Microscopic Science* 77:191–222.

Wiklund, C., and T. Fagerström. 1977. Why do males emerge before females? A hypothesis to explain the incidence of protandry in butterflies. *Oecologia* 31:153–158.

Williams, G. C. 1966. *Adaptation and natural selection.* Princeton, NJ: Princeton University Press.

Williams, N. M., and J. D. Thomson. 1998. Trapline foraging by bumble bees: III. Temporal patterns of visitation and foraging success at single plants. *Behavioral Ecology* 9:612–621.

Willson, M. F., and P. W. Price. 1980. Resource limitation of fruit and seed production in some *Asclepias* species. *Canadian Journal of Botany* 58:2229–2233.

Wise, D. H. 1975. Food limitation of the spider *Linyphia marginata:* Experimental field studies. *Ecology* 56:637–646.

———. 1984. The role of competition in spider communities: Insights from field experiments with a model organism. In *Ecological communities: Conceptual issues and the evidence,* ed. D. R. Strong, D. Simberloff, L. G. Abele, and A. B. Thistle, 42–53. Princeton, NJ: Princeton University Press.

———. 1993. *Spiders in ecological webs.* Cambridge: Cambridge University Press.

Wright, S. 1968. *Evolution and the genetics of populations,* vol. 1. Chicago: University of Chicago Press.

Ydenberg, R. C. 1998. Behavioral decisions about foraging and predator avoidance. In *Cognitive ecology,* ed. R. Dukas, 343–378. Chicago: University of Chicago Press.

Yeargan, K. V. 1975. Factors influencing the aerial dispersal of spiders (Arachnida: Araneida). *Journal of the Kansas Entomological Society* 48:403–408.

Yong, T.-H. 2005. Prey capture by a generalist predator on flowering and nonflowering ambush sites: Are inflorescences higher quality hunting sites? *Environmental Entomology* 34:969–976.

Youngs, M., and T. Stephens. 2004. Flower colour and architecture as visual cues used by female crab spiders *(Misumena vatia)* in hunting site selection. *Ontario Insects* 9:43–47.

Zahavi, A. 1975. Mate selection: Selection for a handicap. *Journal of Theoretical Biology* 53:205–214.

Zhao, J., F. Liu, and W. Chen. 1980. Preliminary studies of the life history of *Misumenops tricuspidatus* and its control of cotton pests. *Acta Zoologica Sinica* 26:255–261. In Chinese, English summary.

Index